Current Topics in Microbiology 152 and Immunology

The Scid Mouse

Characterization and Potential Uses

EMBO Workshop held at the
Basel Institute for Immunology,
Basel, Switzerland, February 20–22, 1989

Organized and Edited by
M. J. Bosma, R. A. Phillips, and
W. Schuler

With 72 Figures

Springer-Verlag
Berlin Heidelberg New York
London Paris Tokyo Hong Kong

MELVIN J. BOSMA
Institute for Cancer Research
Fox Chase Cancer Center
7701 Burholme Avenue
Philadelphia, PA 19111
USA

ROBERT A. PHILLIPS
Division of Hematology/Oncology
Hospital for Sick Children
Departments of Immunology and Genetics
University of Toronto
Toronto, M5G 1X8
Canada

WALTER SCHULER
Basel Institute for Immunology
Grenzacherstraße 487
4005 Basel
Switzerland

ISBN-13: 978-3-642-74976-6 e-ISBN-13: 978-3-642-74974-2
DOI: 10.1007/978-3-642-74974-2

Softcover reprint of the hardcover 1st edition 1989

Library of Congress Catalog Card Number 15-12910

2123/3020-543210

Preface

During routine genetic screening of several immunoglobulin heavy chain congenic mouse strains in 1980, one of us (MB) was surprised to find that several mice in the C.B-17/Icr strain, which was being maintained in a specific-pathogen-free facility of the Fox Chase Cancer Center (Philadelphia, PA), did not express serum immunoglobulin of the appropriate allotype. Fearing an error in the breeding of these mice, the sera of the suspect mice were screened for other allotypes. When these tests revealed a complete absence of serum immunoglobulin, it became apparent that a mutation had probably occurred in the C.B-17/Icr line. Further analysis revealed that a single breeding pair was responsible for all of the immunoglobulin negative mice and that the defect showed recessive inheritance. Thus was the C.B-17/Icr *scid* or severe combined immune deficient (scid) mouse discovered. Although it has taken most animal facilities several years to breed scid mice of high quality for experimental purpose, it was clear by 1987 that many investigators were beginning to exploit the unique qualities of the scid mouse for studies in several areas. Accordingly, the editors of *Current Topics in Microbiology and Immunology* decided to devote an issue to the scid mouse. To enhance the usefulness of the publication the editors of the issue decided to organize a workshop to promote the maximum interaction and to explore a wide range of potential research applications for the mouse. The papers in this book were presented at a European Molecular Biology Organization (EMBO) workshop held at the Basel Institute for Immunology on February 20–22, 1989.

The workshop confirmed the expectations of the organizers that the scid mouse will prove useful in many areas of biomedical research and will create new areas of investigation not possible before the discovery of this mouse mutant. Characterization of the basic genetic defect will undoubtedly provide insight into the mechanisms of gene rearrangement, since the mutation appears to interrupt this process, preventing appropriate expression of antigen receptors on B and T lymphocytes.

The original descriptions of the scid mouse, and several of the early reports, documented that the immune deficiency can be reconstituted by transplantation of normal cells into the mice. It is now clear that even transplantation of human lymphoid cells provides some reconstitution in the mice. Further elaboration of the mechanism of reconstitution promises to allow the development of mice whose entire hematopoietic system is derived from human cells. Such mice will provide powerful new models for the analysis of numerous human diseases, such as AIDS, leukemia, immune deficiency, auto-immunity, and congenital anomalies in the hematopoietic system. For the first time, it may be possible to have an experimental system which mimics most of the cellular interactions known to be important in regulating the development of myeloid and lymphoid cells.

It was encouraging to learn at the meeting that there are productive breeding colonies of scid mice in many research centers in the United States, Canada, Europe, and Japan. In addition, there are hopes for a commercial source of scid mice in the near future. The increased availability of scid mice should allow all interested investigators to begin to explore the interesting properties of this unusual mouse mutant.

The organizers of the meeting wish to express their appreciation to the staff at the Basel Institute for Immunology for their assistance in organizing the meeting and for their logistical help in making the meeting itself a large success. Financial support of the meeting was provided by EMBO and the Basel Institute for Immunology. Finally, we are grateful to Springer-Verlag for their help and advice in editing the papers for this book. It was their dedication and effort which allowed the proceedings of this meeting to be published so soon after the meeting.

M. J. BOSMA
R. A. PHILLIPS
W. SCHULER

Table of Contents

III. Introduction of Functional Ig or TCR Transgenes

IV. Leaky Phenotypes

V. Potential Uses: Engraftment with Syngeneic Mouse Cells

List of Contributors

You will find their addresses at the beginning of their respective chapters

I. General Phenotype/Cellular Characterization

The *scid* Mutation: Occurrence and Effect

Melvin J. Bosma

Institute for Cancer Research, Fox Chase Cancer Center, Philadelphia, PA 19111, USA

INTRODUCTION

The congenital syndrome known as severe combined immune deficiency (SCID) is characterized by a loss of both B and T cell immunity. It was first recognized in human infants (Glanzmann and Riniker 1950; Tobler et al. 1958; Hitzig et al. 1958) and later in Arabian foals (McGuire et al. 1975). More recently, the SCID syndrome was reported in a mutant strain of mice, called scid mice (Bosma et al. 1983). This article briefly reviews the genetics of the murine scid mutation as well as early findings on the effect of scid on the lymphoid system.

RESULTS AND DISCUSSION

The scid mouse genotype

The discovery of the scid mutation resulted from a chance observation in the course of quantitating serum levels of IgG1 and IgG2a allotypes in specific-pathogen-free (SPF) mice of the immunoglobulin heavy chain (Igh) congenic strains, BALB/c and C.B-17 (Bosma et al. 1983). Four of 35 C.B-17 serum samples were unexpectedly found to lack C.B-17 allotype (Igh^b) determinants and were presumed to reflect a genetic contamination of the C.B-17 colony by BALB/c. However, retesting of the 4 serum samples revealed an absence of BALB/c allotype (Igh^a) determinants. Further testing with class-specific antisera showed that all of the major serum Ig classes (IgM, IgG3, IgG2b, IgG2a and IgA) were missing.

It soon became apparent that the Ig deficient (Ig^-) serum samples were from 4 littermates in a litter of 7 (see Figure 1). The male that sired this litter was also found to be Ig^-. As the same male was paired with two females, and the offspring of one female were all normal (Ig^+) while about half of the offspring of the other female were Ig^-, it was clear that the defect was heritable and under the control of a recessive gene. Selective breeding of the Ig^- mice led to the establishment of a colony of mutant mice. Further study showed that these mice lacked both B and T cell immunity and we therefore referred to them as scid mice as their disorder resembled that of SCID infants and foals.

SCID infants that show autosomal recessive inheritance of their disease often have a deficiency of the enzyme, adenosine deaminase (ADA) (reviewed by Thompson and Seegmiller 1980). ADA converts adenosine and deoxyadenosine into inosine and deoxyinosine, respectively. In the absence of ADA, lymphocytes rapidly convert deoxyadenosine into deoxyadenosine triphosphate (dATP) because they contain unusually high phosphorylating kinase activity. The resulting high intracellular concentrations of dATP are toxic and prevent lymphocyte maturation. In view of the above it was of interest early on to know whether the recessively inherited disorder of scid mice was due to a deficiency of ADA. Utilizing a simple method to assay ADA (Ito et al. 1977), it was found that erythrocytes of C.B-17scid/scid and C.B-17+/+ mice contained equivalent ADA activity. This finding, that scid mice were not a model for ADA deficient SCID infants, is consistent with recent genetic mapping studies that show scid is not on chromosome 2, the location of the mouse Ada locus (Siciliano et al. 1984).

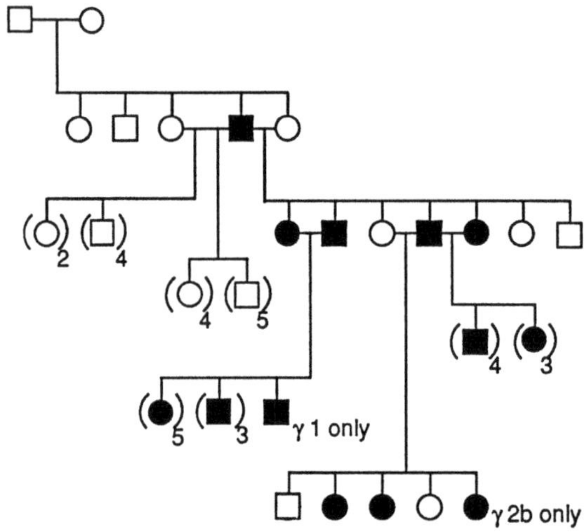

Figure 1. Partial pedigree of C.B-17scid/scid ancestor mice showing recessive inheritance of SCID syndrome. Females (O) and males (□) denote the absence of detectable (<10 μg/ml) serum immunoglobulin (IgM, IgG3, IgG1, IgG2b, IgG2a and IgA). The subscript numbers indicate number of siblings; γ_1 only and γ_2 only refer to two mice that lacked all of the above isotypes except IgG1 and IgG2b, respectively (reprinted by permission from Nature Vol. 301 pp. 527-530 Copyright (c) 1983 Macmillan Magazines Ltd).

To ascertain the chromosomal linkage of scid, scid mice were crossed with different linkage testing mouse stocks (of the Jackson Laboratories) that express one or more markers for genes of known chromosomal location. F_1 offspring were then intercrossed or backcrossed to scid mice to test for possible linkage between the various recessive or dominant test markers and scid. This strategy finally provided evidence that scid is closely linked to mahagnoid (md), a recessive coat color marker on the centromeric end of chromosome 16; the linkage was confirmed in a three-point cross involving scid, md, and the lambda Ig light chain locus (Igl-1) (Bosma et al. 1989).

The location of scid on chromosome 16 is shown in Figure 2. As can be seen, scid maps 7.1 ± 1.9 recombination units from Igl-1; it is placed on the centromeric side of Igl-1 close to md because no recombinants between scid and md were observed in a total of 538 tested progeny. Such recombinants would have been expected if scid were located on the teleomeric side of Igl-1 at a distance of 18 recombination units from md. Also shown is the location of genes for protamine 1 and 2 (Prm-1,-2), which map ~7.2 recombination units from Igl-1 (Reeves et al. 1987) in the vicinity of scid.

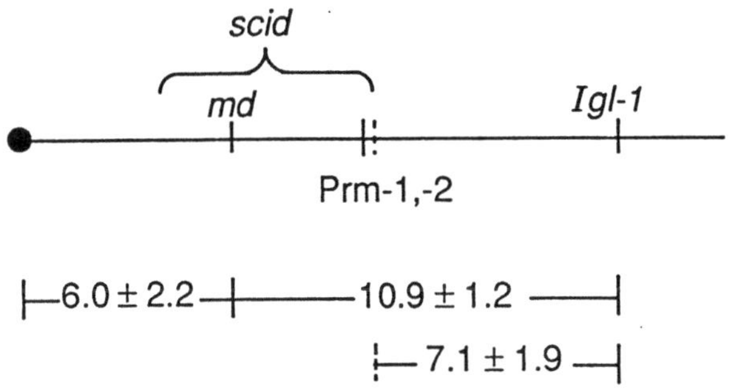

Figure 2. A partial map of the centromeric end of chromosome 16. Recombination distances are based on data from Epstein et al. 1984, 1986; Davisson 1985; Reeves et al. 1987; Bosma et al. 1989. The bracket denotes that the order of and interval between md and scid, and between scid and Prm-1,-2 is uncertain. (Reprinted by permission from Immunogenetics 29: 54-57, 1989, courtesy of Springer-Verlag, Berlin).

The scid mouse phenotype

The serologic and histologic characteristics of scid mice (Bosma et al. 1983; Custer et al. 1985) are as follows: The peripheral blood shows a paucity of lymphocytes and elevated proportions of granulocytes; hematocrits are normal and vary between 40-50%. The serum generally contains less than 0.2 µg/ml of Ig. Lymphoid tissues are one-tenth or less normal size and show the properties illustrated in Figure 3. The thymus lacks a cortex and resembles a rudimentary medulla containing fibroblasts, histiocytes, abundant epithelial cells and relatively few lymphoid cells. Splenic and lymph node follicles are severely deficient in lymphocytes and mainly consist of stromal cell elements, histocytes, granulocytes and occasional plasmacytes. Erythroid hyperplasia is frequently noted in the interfollicular red pulp of the spleen. Also, granulocytic hyperplasia, sometimes associated with marked megakaryocytosis, is commonly seen in spleen and bone marrow of mice with infections, indicating that the myeloid system is not obviously impaired. Peyer's patches and solitary follicles of the intestinal tract are seldom detectable.

The majority of scid thymocytes (>85%) are Thy-1$^+$ and presumably represent early T cells because TCRδ rearrangements and TCRγ/TCRβ transcripts are readily detected in the DNA of scid thymocytes (see articles by Schuler and Carroll in section II of this volume). Although CD4 and CD8 are not detectable on nontransformed scid thymocytes, both of these T cell-specific markers are co-expressed on thymic lymphomas that spontaneously arise in 15-20% of scid mice. The development of T cell lymphomas is further evidence for immature T cells in the thymus of scid mice.

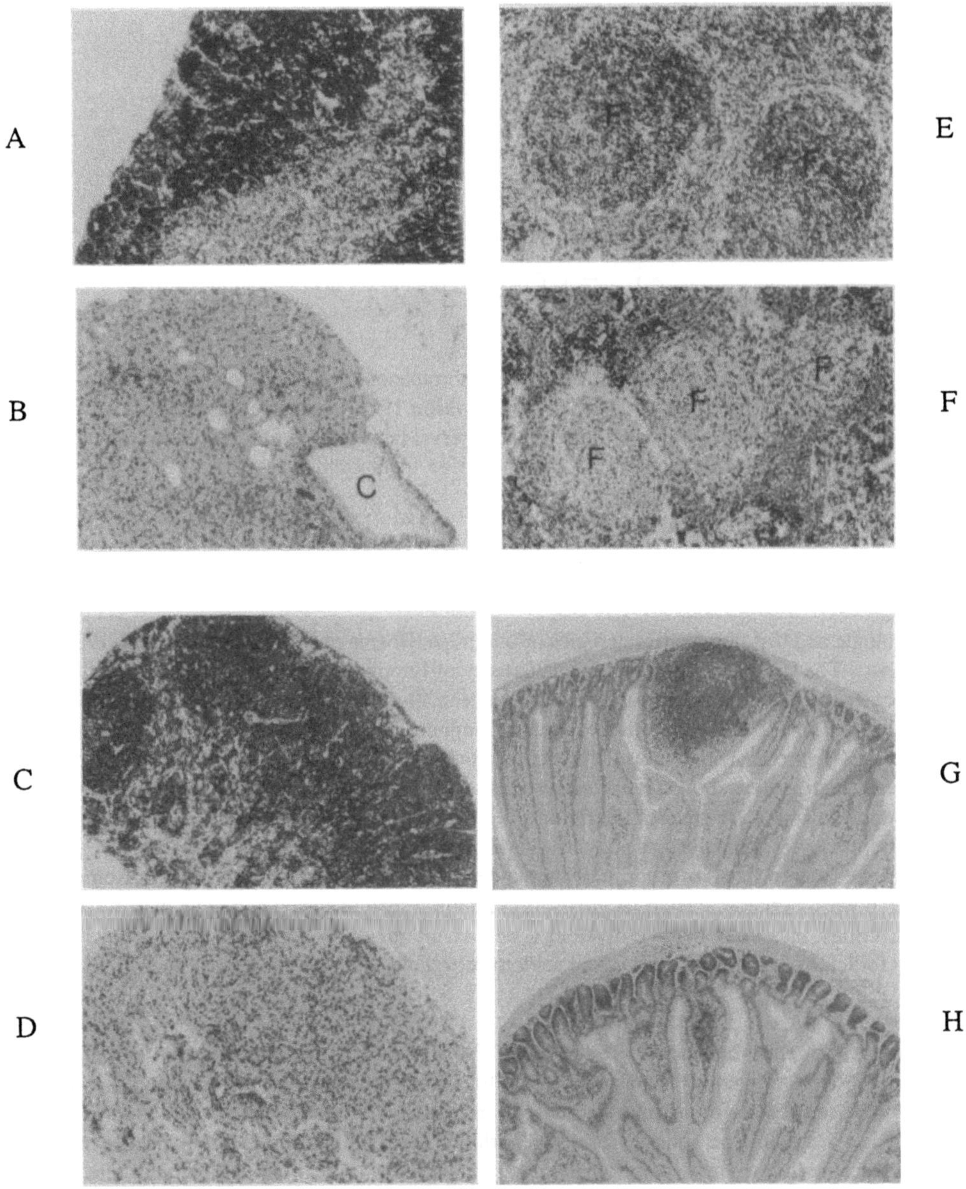

Figure 3. Comparative histology of C.B-17scid/+ and C.B-17scid/scid mice. Tissues were fixed in Carson's buffered neutral formal, embedded in glycol methacrylate (Sorvall), sectioned at 2 µm, and stained with hematoxylin and eosin. **A.** scid/+ thymus appears normal with a dense lymphocytic cortex and sharply defined corticomedullary junction (X80). **B.** scid/scid thymus lacks a cortex, resembling the scid/+ medulla; mucinous cysts (C) are common (X80). **C.** scid/+ lymph node is densely populated with lymphocytes in cortex, paracortical zone and medullary cords (X80). **D.** scid/scid lymph node retains fibrous framework and sinus pattern but is sparsely

populated with lymphocytes (X80). **E.** scid/+ splenic follicles (F) are fully populated with lymphocytes; intervening pulp contains usual hematopoietic tissue in varying proportions (X80). **F.** scid/scid splenic follicles (F) appear relatively empty; the pulp contains active erythroid, granulocytic and megakaryocytic components (X80). **G.** scid/+ intestinal mucosa with well developed lymphoid follicle (X60). **H.** scid/scid intestinal mucosa has poorly populated follicle (X60) (reprinted by permission from American Journal of Pathology 120: 464-477, 1985, courtesy of Harper and Row Publishers, Inc.).

Evidence of early scid B cells first came from the finding that bone marrow cells of scid mice could be transformed with Abelson murine leukemia virus (A-MuLV); the resulting transformants showed a phenotype characteristic of pre-B cells; i.e., rearrangement of their Igh but not Igl loci (Schuler et al. 1986). Interestingly, the frequency of recovered A-MuLV transformants from adult C.B-17scid/scid bone marrow is comparable to that obtained from C.B-17+/+ bone marrow (Fulop et al. 1988). This indicates that adult scid mice contain normal numbers of very early B cells which probably correspond to pro-B cells since $B220^+$ Thy-1^- pre-B cells are undetectable in scid bone marrow (Fried et al., this volume).

The severe deficiency of mature B and T cells can be corrected by engrafting adult scid mice with bone marrow or fetal liver cells from BALB/c normal mice (Bosma et al. 1983; Custer et al. 1985). The scid recipients become reconstituted with donor lymphocytes within 4-6 weeks after cell transfer as evident from the production of donor Ig allotype; host Ig allotype is not detected. Histological examination shows that spleen and lymph node follicles become repopulated with lymphocytes, Peyer's patches and solitary follicles of the intestinal tract become apparent and the thymus displays a prominent cortex and corticomedullary delineation. Clearly, scid recipients can support the differentiation of normal lymphocytes though their reconstitution is often incomplete (Custer et al. 1985). To obtain full reconstitution, it is necessary to sublethally irradiate scid mice prior to cell transfer (Fulop and Phillips 1986) or else use newborn scid recipients (Bosma and Bosma, this volume).

In the converse situation, in which normal recipients [BALB/c or (B6.C-9 X BALB/c)F_1 mice] are lethally irradiated and injected with bone marrow or fetal liver cells of scid mice, the recipients do not apparently become repopulated with donor lymphocytes; i.e., host Ig allotype persists in the absence of donor scid allotype (Bosma et al. 1983). However, recipients do become reconstituted with scid myeloid cells as evidenced by erythrocytes of donor phenotype at 6 and 12 weeks after cell transfer. The engraftment of scid myeloid cells is presumably responsible for rescuing the recipients from radiation death. Thus, the scid defect appears to be an intracellular defect such that scid bone marrow or fetal liver cells cannot differentiate into mature, functional lymphocytes even in the presence of normal lymphocytes and their products.

CONCLUSION

In retrospect, working with an inbred strain of mice under SPF conditions probably favored the chance discovery of the autosomal recessive scid mutation. Mice homozygous for this mutation (scid mice) generally lack mature, functional lymphocytes and are highly susceptible to recurring infections, but otherwise appear normal. Their myeloid cells show normal histologic reactivity to incidental infections and their cellular microenvironment readily supports the generation of lymphocytes by engrafted bone marrow or fetal liver cells of normal histocompatible donor mice. Thus, the defect in scid mice is apparently specific and intrinsic to the lymphoid system. An understanding of the precise nature of this defect will obviously require the cloning of the scid locus which is now known to map on the centromeric end of chromosome 16.

ACKNOWLEDGMENTS

The author's research was supported by NIH grants CA-04946, AI-13323, CA-06927 and RR-05539 and an appropriation from the Commonwealth of Pennsylvania.

REFERENCES

Bosma GC, Custer RP, Bosma MJ (1983) A severe combined immune immunodeficiency mutation in the mouse. Nature 301: 527-530

Bosma GC, Davisson MT, Ruetsch NR, Sweet HO, Shultz LD, Bosma MJ (1989) The mouse mutation severe combined immune deficiency (scid) is on chromosome 16. Immunogenetics 29: 54-57

Custer RP, Bosma GC, Bosma MJ (1985) Severe combined immunodeficiency (SCID) in the mouse: pathology, reconstitution, neoplasms. Amer J Pathol. 120: 464-477

Davisson MT (1985) The Jackson Laboratory. 56th Annual Report, 1984-85, p. 40.

Epstein R, Lehmann K, Cohn M, Buckler C, Rowe W, Davisson M (1984) Linkage of the Igl-1 structural and regulattory genes to Akv-2 on chromosome 16. Immunogenetics 19: 527-537.

Epstein R, Davisson M, Lehmann K, Akeson EC, Cohn M (1986) Position of Igl-1 md, and Bst loci on chromosome 16 of the mouse. Immunogenetics 23: 78-83

Fulop GM, Phillips RA (1986) Full reconstitution of the immune deficiency in scid mice with normal stem cells requires low-dose irradiation of the recipients. J Immunol 136: 4438-4443

Fulop GM, Bosma GC, Bosma MJ, Phillips RA (1988) Early B cell precursors in scid mice - normal numbers of cells transformable with Abelson murine leukemia virus (A-MuLV). Cellular Immunol 113: 193-201

Glanzmann E, Riniker P (1950) Essentielle lymphocytophthise. Ein neues krankheitsbild aus der Säuglings pathologie. Ann Paediat 175: 1-32

Hitzig WH, Biro Z, Bosch H, Huser HJ (1958) Agammaglobinämie und Alyimphocytose met Schwund der lymphatischen Gewebes. Helv Paediat Acta 13: 551-585

Ito K, Sakura N, Usui T, Uchino H (1977) Screening for primary immunodeficiencies associated with purine nucleoside phosphorylase deficiency or adenosine deaminase deficiency. J Lab Clin Med 90: 844-848

McGuire TC, Banks KL, Poppie MJ (1975) Combined immunodeficiency in horses: characterization of the lymphocyte defect. Clin Immunol Immunopathol 3: 555-556

Reeves RH, Gearhert JD, Hecht NB, Yelick P, Johnson P, O'Brien SJ (1987) The gene encoding protamine 1 is located on chromosome 16 in humans and near the proximal end of mouse chromosome 16, where it is tightly linked to protamine 2. Cytogenetics and Cell Genetics 460: 680 (Abstract)

Schuler W, Weiler IJ, Schuler A, Phillips RA, Rosenberg N, Mak T, Kearney JF, Perry RP, Bosma MJ (1986) Rearrangement of antigen receptor genes is defective in mice with severe combined immune deficiency. Cell 46: 963-972

Siciliano MJ, Fournier REK, Stallings RL (1984) Regional assignment of ADA and ITPA to mouse chromosome 2 (Cl→ter). A demonstration of the conserved linkage of enzyme and proto-oncogene loci. J Hered 75: 175-180
Thompson LF, Seegmiller JE (1980) Adenosine deaminase deficiency and severe combined immunodeficiency disease. Adv Enzymol 51: 167-210
Tobler R, Cottier H (1958) Lymphopenie mit Agammaglobulinämie und schwere Moniliasis. Die 'essentiale Lymphocytophthise' als besondere Form der frühkindlichen Agammaglobulinämie. Helv Paediat Acta 13: 313-338

Pleiotropic Effects of the *scid* Mutation: Effects on Lymphoid Differentiation and on Repair of Radiation Damage

R.A. Phillips and G.M. Fulop

Division of Hematology/Oncology, The Hospital for Sick Children and Departments of Medical Genetics and Immunology, University of Toronto, Toronto, Canada, M5G 1X8

INTRODUCTION

The initial description of the scid mutation by Bosma et al. (1983) clearly demonstrated its marked effect on B and T cell differentiation. However, many mutations have pleiotropic effects disrupting the function of numerous systems. For example, the W and Sl mutations affect myelopoiesis, pigment production and gametogenesis (Russell 1979); mutations affecting the lymphoid system also can affect other cell types (Shultz and Sidman 1987). Therefore, in our initial experiments on the scid mouse, we analyzed in detail both myeloid and lymphoid function to identify precisely the stage at which differentiation becomes defective in scid mice and to determine whether the mutation had any effect on cells other than lymphoid cells.

MICE

The initial breeding pairs of C.B-17 scid mice (referred to simply as scid mice) obtained from Bosma were placed in the animal colony at the Ontario Cancer Institute. To protect the mice from potential pathogens in the conventional colony where they were initially housed, the mice were immediately reconstituted with 10^7 bone marrow cells from BALB/c donors; BALB/c mice are congenic partners of C.B-17 mice differing only at the Igh locus. The reconstituted breeding pairs were mated, and a near-term litter was delivered by Caesarian section and foster nursed behind a defined-flora barrier to appropriate foster mothers. The scid mice have been maintained for five years as a defined-flora colony. Routine fecal samples confirm the presence of the eight known bacteria in the defined-flora colony. Periodically, sentinel mice are placed in the same cages to check for unwanted viruses. At the present time, the animals have no known viruses or unwanted bacteria. Under these conditions, the mice have a life span in excess of two years (Fulop and Phillips, in preparation).

FUNCTION OF MYELOID CELLS

Examination of peripheral blood and bone marrow in scid mice revealed normal cell counts and normal morphology with the exception of a complete absence of mature lymphoid cells. Two functional tests of myeloid progenitors were done; Table 1 summarizes the results of these tests. The spleen colony assay (CFU-S) detects a multipotent stem cell committed to myeloid and erythroid differentiation (Till and McCulloch 1961; Till and McCulloch 1980). Scid mice and coisogenic C.B-17 mice contain similar numbers of CFU-S per femur. In addition, an in vitro colony assay

Current Topics in Microbiology and Immunology, Vol. 152

for granulocyte macrophage precursors (CFU-GM) showed no difference between scid and normal mice. In summary, scid mice behave normally in tests for myeloid and erythroid function.

Table 1. Myeloid Progenitors in scid Mice.

Activity	Total Activity C.B-17	Total Activity scid	scid as Per Cent of normal
CFU-S	2,860/femur	3,100/femur	108 %
CFU-GM	16,700/femur	14,000/femur	84 %
	2,970/spleen	3,900/spleen	131 %

FUNCTION OF LYMPHOID CELLS

Our results confirmed the initial data presented by Bosma et al. (1983). We detect no functional B or T lymphocytes in scid mice. Table 2 summarizes the tests used. Two assays of B cell function, response to the mitogen lipopolysaccharide (LPS) and a colony-forming B cell assay (CFU-B) (Kincade et al. 1978), failed to detect activity in scid mice. With the high sensitivity of the CFU-B assay, one can easily detect a frequency of one CFU-B in 10^5 nucleated cells. In tests on many scid mice, we have never detected a functional B cell in a young scid mouse. However, a small proportion of scid mice greater than five months of age are "leaky", having detectable serum immunoglobulin (Bosma et al. 1988) and containing very low levels of CFU-B, less than 0.1% of normal, in the spleen; the bone marrow of "leaky" scid mice contains no functional B cells (Fulop et al. 1988). Pre-B cells, defined as cμ-positive, sμ-negative, are undetectable in the bone marrow of normal or "leaky" scid mice. As shown by Rosenberg and Baltimore (1976), Abelson-murine leukemia virus (A-MuLV) can transform pre-B cells or their immediate precursors. The transformed cells seldom express cytoplasmic immunoglobulin, indicating that the virus transforms a cell at a stage earlier than that detected morphologically by the presence of cμ. Scid mice have normal numbers of cells transformable by A-MuLV (Fulop et al. 1988). Although the A-MuLV-transformed cells from scid mice have severe abnormalities in immunoglobulin gene rearrangement (see following chapters and Schuler et al. 1986; Malynn et al. 1988), the transformed cells have all of the other properties expected of pre-B cells (Malynn et al. 1988).

T lymphocyte function was evaluated in two assays, response to a T cell mitogen, Con-A, and response in a one-way mixed lymphocyte reaction (MLR). Stimulation with ConA gave no significant stimulation above background. The MLR against irradiated C57BL/6J (B6) spleen cells gave no detectable proliferative response, nor was there a detectable cytotoxic response against B6 targets (data not shown).

Despite the absence of functional T cells, scid mice have a small thymus, approximately 1% normal size, and occasionally develop typical T cell thymomas (Bosma et al. 1983). Since NK cells have been suggested to arise as part of the T cell lineage, we examined scid mice for their NK cell activity (Dorshkind et al. 1985). Analysis of NK cell function proved difficult because the defined-flora colony made

NK cell activity undetectable in both scid and normal mice obtained from the animal colony. Injection of poly-IC, a known stimulant of NK cells (Gidlund et al. 1978), increased NK activity to easily detectable levels in both normal and scid mice. Although NK activity in scid mice was higher on a per cell basis than in control mice, the total activity per spleen was equivalent in both scid and normal because the spleen in scid is approximately one-fifth the size of the spleen in C.B-17 mice (Table 2). Thus, there is no defect in NK activity. Other experiments demonstrated that the T cell receptor genes do not rearrange in NK cells isolated from scid mice (Lauzon et al. 1986). These data on NK cells in scid mice provide no additional information on the origin of NK cells or on their relationship to other hematopoietic lineages.

Table 2. Summary of Lymphoid Activities in scid Mice.

Activity	Total Activity C.B-17	Total Activity scid	scid as Per Cent of normal
Spleen size	~1×10^8 cells	~2×10^7 cells	20 %
ConA (spleen)	78,000 cpm	870 cpm	1.1 %
One-way MLR (vs. B6)	21,000 cpm	610 cpm	2.9 %
LPS (spleen)	73,000 cpm	560 cpm	0.7 %
CFU-B	140,000 /spleen	0 /spleen	0 %
	5,600 /femur	0 /femur	0 %
A-MuLV	1100 /femur	790 /femur	72 %

On the basis of the information presented above, the block in scid mice must occur early in the development of B and T cells (Fig. 1). The scid mutation has no detectable effect on the myeloid system. In the lymphoid system, cells appear to develop in normal numbers up to the pro-B or early pre-B stages in the B lineage and up to the double-negative cells in thymus (see Bosma et al. and von Boehmer et al., this volume). However, beyond this stage, there is no detectable differentiation of either lineage in scid mice. Presumably, the absence of functional receptors leads to rapid cell death in the bone marrow and thymus without the accumulation of aberrant lymphoid cells in the peripheral hematopoietic tissues. Interestingly, incubation of scid bone marrow cells in long-term culture leads to the development of large number of defective lymphoid-like cells (Witte et al. 1987; Hirayoshi et al 1987). Apparently, in culture the normal mechanisms for eliminating B cells with non-productively arranged immunoglobulin genes do not function leading to the accumulation of defective B lineage cells (See Lee et al. and Nishikawa et al., this volume).

RADIATION REPAIR DEFECT

In studies on CFU-S, we noticed an abnormality when we attempted to use scid mice as irradiated recipients. Exposure of mice to sublethal doses of irradiation allows the detection of endogenous CFU-S present in the spleens of such mice. Thus, in normal mice, one generally observes endogenous CFU-S between 500 and 900 cGy. With scid mice, however, endogenous colonies first appeared at 200 rads of whole body irradiation and disappeared by 500 rads. This difference between scid and normal could arise by reduced numbers of CFU-S in spleen, or by increased radiation sensitivity of scid stem cells. The experiments shown above (Table 1) indicate that the frequency and function of CFU-S are normal in scid mice. To determine the radiation sensitivity of myeloid cells in scid mice, we exposed normal and scid bone marrow cells to varying doses of ionizing radiation. Following irradiation, the number of surviving CFU-GM were determined. It was clear from these experiments that CFU-GM from scid mice were approximately twice as sensitive to ionizing radiation as the same cells from normal mice (Fulop and Phillips, in preparation).

Because of the sensitivity of CFU-GM, and presumably CFU-S, we also examined the radiation sensitivity of fibroblasts. Fibroblasts were obtained from embryonic kidney of scid and normal mice. Mass cultures of fibroblasts were exposed to varying doses of radiation. Four days after radiation, the cells were examined microscopically for the presence of micronuclei, which have been shown to be a good indicator of radiation sensitivity (Countryman and Heddle 1973). Again, the fibroblasts from scid mice had approximately twice as many micronuclei as control mice at all radiation doses tested. On the basis of these data, we conclude that the genetic defect in scid mice leads to a generalized increase in radiation sensitivity.

Since the increased sensitivity to ionizing radiation is similar to that observed in a human radiation repair defect, ataxia telangiectasia (AT) (for review, see McKinnon 1987), we compared other aspects of the radiation response of scid to evaluate its similarity to AT; it also interesting that patients with AT have varying degrees of immune deficiency (Roifman and Gelfand 1985). A unique feature of cells from patients with AT is the inability of the cells to delay the onset of DNA synthesis following exposure to ionizing radiation (McKinnon 1987). Normal cells generally stop synthesizing DNA following exposure to radiation. However, irradiated scid cells shut down DNA synthesis similar to normal cells. On this basis, we conclude that the defect in scid mice is not analogous to the defect in AT.

DISCUSSION

Fig. 1 outlines our working model for the differentiation of cells in the hematopoietic system and indicates the developmental block resulting from the scid mutation. The genetic defect leading to immune deficiency in scid mice appears to affect the ability of cells to rearrange their immunoglobulin genes and T cell receptor genes (Schuler et al. 1986; Malynn et al. 1988). Thus, it is perhaps not surprising that cells differentiate normally up to the stage when gene rearrangement is required. In the bone marrow, there appear to be normal numbers of pro-B cells and early pre-B cells which are the targets for transformation by A-MuLV. At this stage, these cells begin to rearrange their immunoglobulin heavy chains, but failure to achieve normal rearrangement presumably leads to death of cells without functional receptors, so that no lymphoid cells are observed in the bone marrow of scid mice. Similarly, in the thymus, there are approximately normal numbers of CD4⁻ CD8⁻ cells in the thymus of scid mice. As with B cells, the failure to produce a functional T cell receptor prevents the appearance of $CD4^+$ $CD8^+$ cells and the subsequent normal development of T lymphocytes. The elegant experiments of von Boehmer et al. (this volume) indicate that transgenic scid mice carrying correctly rearranged α and β genes for the T cell receptor can produce normal T cells.

Since gene rearrangement is not involved in the differentiation of myeloid cells or NK cells, these lineages develop normally in scid mice. In terms of function and numbers, no abnormality was observed in any of the myeloid pathways. However, the precise developmental relationship of NK cells remains an enigma. They represent a separate lineage arising directly from S_P. Alternately, they may arise from one of the other restricted stem cells or from more mature progenitors such as thymocytes or monocytes.

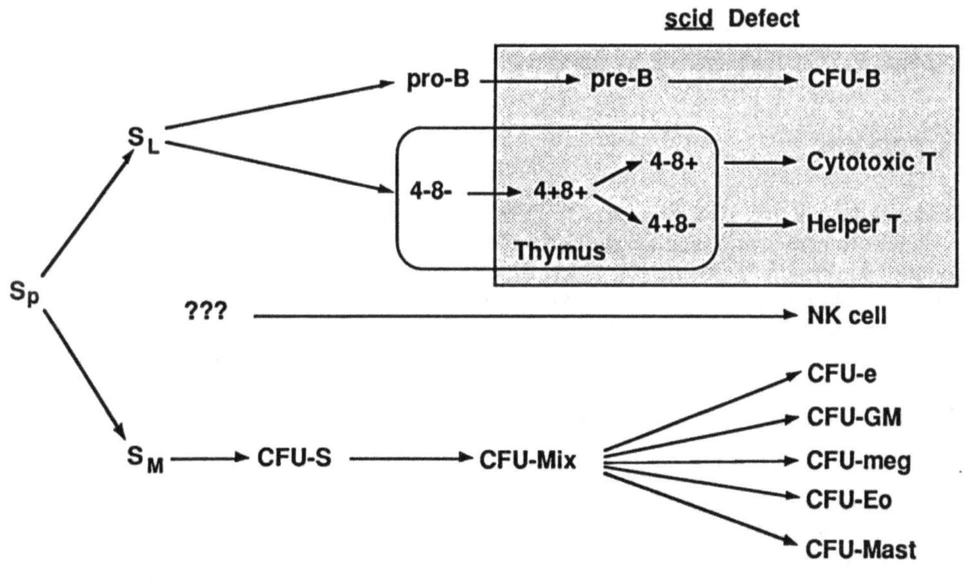

Fig. 1. Model for differentiation of the hematopoietic system and for location of the genetic defect in scid mice. S_P, pluripotent stem cell; S_L, lymphoid-restricted stem cell; S_M, myeloid-restricted stem cell; CFU-S, spleen-colony-forming stem cell; CFU-e, erythrocyte progenitor; CFU-GM, granulocyte-macrophage progenitor; CFU-meg, megakaryocyte progenitor; CFU-Eo, eosinophil progenitor; CFU-Mast, mast cell progenitor.

The increased radiation sensitivity of all cells in scid mice probably reflects an inability to repair certain types of radiation-induced DNA damage. There are numerous examples of heritable defects in radiation repair, but this is the first case where such defects result in both severe combined immune deficiency and a generalized increase in radiation sensitivity. AT patients have a large increase in radiation sensitivity, but the defects in their immune system are relatively minor compared to those observed in scid mice. Other workers have presented data in this symposium that the cells from scid mice can repair double-strand DNA breaks normally, as manifested by the ability to repair double-strand breaks induced by radiation (Weaver et al., this volume) and to join correctly linear pieces of DNA (Gellert et al., this volume). Again, both of these mechanisms are defective in patients with AT. We are presently investigating other types of DNA damage in an attempt to identify the type of repair that is defective in scid mice. The data presented in the following chapters indicate that the defect in immunoglobulin and T cell receptor gene replacement often leads to defective rejoining of the appropriate gene segments and failure to produce a functional receptor. Thus, it is not surprising that the defect associated with immunoglobulin gene production is manifest in other cell types when there is a need to rejoin pieces of DNA. Presumably the defective enzyme system in scid mice is used both in some aspect of rearranging immunoglobulin genes and in repair of radiation damage.

ACKNOWLEDGEMENTS

The research was supported by grants from the Medical Research Council and from Th National Cancer Institute of Canada.

REFERENCES

Bosma GC, Custer RP, Bosma MJ (1983) A severe combined immunodeficiency mutation in the mouse. Nature 301: 527-530

Bosma GC, Fried M, Custer RP, Carroll A, Gibson DM, Bosma MJ (1988) Evidence of functional lymphocytes in some (leaky) scid mice. J Exp Med 167: 1016-1033

Countryman PI, Heddle JA (1976) The production of micronuclei from chromosome aberrations in irradiated cultures of human lymphocytes. Mutation Res 41: 1343-1347

Dorshkind K, Pollack SB, Bosma MJ, Phillips RA (1985) Natural killer (NK) cells are present in mice with severe combined immunodeficiency (scid). J Immunol 134: 3798-3801

Fulop GM, Bosma GC, Bosma MJ, Phillips RA (1988) Early B-cell precursors in scid mice: Normal numbers of cells transformable with Abelson murine leukemia virus (A-MuLV). Cell Immunol 113: 192-201

Gidlund M, Orn A, Wigzell H, Senik A, Gresser I (1978) Enhanced NK cell activity in mice injected with interferon and interferon inducers. Nature 273: 759-761

Hirayoshi K, Nishikawa S, Kina T, Hatanaka M, Habu S, Nomura T, Katsura Y (1987) Immunoglobulin heavy chain gene diversification in the long-term bone marrow culture of normal mice and mice with severe combined immunodeficiency. Eur J Immunol 17: 1051-1057

Kincade PW, Paige CJ, Michael R, Parkhouse EM, Lee G (1978) Characterization of murine colony-forming B cells. I. Distribution, resistance to anti-immunoglobulin antibodies, and expression of Ia antigens. J Immunol 120: 1289-1294

Lauzon RJ, Siminovitch KA, Fulop GM, Phillips RA, Roder JC (1986) An expanded population of natural killer cells in mice with severe combined immunodeficiency (SCID) lack rearrangement and expression of T cell receptor genes. J Exp Med 164: 1797-1802

Malynn BA, Blackwell TK, Fulop GM, Rathbun GA, Furley AJ, Ferrier P, Heinke LB, Phillips RA, Yancopoulos GD, Alt FW (1988) The scid defect affects the final step of the immunoglobulin VDJ recombinase mechanism. Cell 54: 453-460

McKinnon PJ (1987) Ataxia-telangiectasia: an inherited disorder of ionizing-radiation sensitivity in man. Progress in the elucidation of the underlying biochemical defect. Hum Genet 75: 197-208

Roifman CM, Gelfand EW (1985) Heterogeneity of the immunological deficiency in Ataxia-telangiectasia: Absence of a clinical-pathological correlation. In: Gatti RA, Swift M (eds) Ataxia-Telangiectasia. Genetics, Neuropathology, and Immunology of a Degenerative Disease of Childhood, pp 273-285. Alan R. Liss, Inc., New York

Rosenberg N, Baltimore D (1976) A quantitative assay for transformation of bone marrow cells by Abelson murine leukemic virus. J Exp Med 143: 1453-1463

Russell ES (1979) Hereditary anemias of the mouse: A review for geneticists. Adv Genet 20: 357-459

Schuler W, Weiler IJ, Schuler A, Phillips RA, Rosenberg N, Mak TW, Kearney JF, Perry RP, Bosma MJ (1986) Rearrangement of antigen receptor genes is defective in mice with severe combined immune deficiency. Cell 46: 963-972

Shultz LD, Sidman CL (1987) Genetically determined murine models of immunodeficiency. Annu Rev Immunol 5: 367-403

Till JE, McCulloch EA (1961) A direct measurement of the radiation sensitivity of normal mouse bone marrow cells. Radiat Res 14: 213-222

Till JE, McCulloch EA (1980) Hemopoietic stem cell differentiation. Biochim Biophys Acta 605: 431-459

Witte PL, Burrows PD, Kincade PW, Cooper MD (1987) Characterization of B lymphocyte lineage progenitor cells from mice with severe combined immune deficiency disease (SCID) made possible by long term culture. J Immunol 138: 2698-2705

Analysis of Lymphoid Population in Scid Mice; Detection of a Potential B Lymphocyte Progenitor Population Present at Normal Levels in Scid Mice by Three Color Flow Cytometry With B220 and S7

R.R. Hardy§, J.D. Kemp† and K. Hayakawa§

§ Institute for Cancer Research, Fox Chase Cancer Center, Philadelphia, PA 19111

† University of Iowa Hospitals and Clinics, Iowa City, IA 52242

INTRODUCTION

C.B-17*scid* (SCID) mice are characterized by a lack of serum immunoglobulin and the inability to reject skin allografts. The defect has been attributed to an inability of lymphoid cells to make functional immunoglobulin and T cell receptor rearrangements, presumably due to a defect in the recombinase system (Schuler *et al.* 1986). We have employed sensitive multiparameter FACS (fluorescence activated cell sorter) analysis to enumerate the lymphoid populations present in young adult (2-4 month) SCID mice and compared them with those detectable in normal (C.B-17) animals in order to clearly define the baseline levels of lymphoid lineage cells in SCID. Such analysis might also help to pinpoint the stage(s) at which lymphoid lineage cells become defective. We have analyzed tissues expected to include both early (bone marrow and thymus) and later (spleen and peritoneal cell) stages of B and T lineages. In the course of these studies we have found that the recently described S7 antibody (Gulley *et al.* 1988), together with an antibody to B220 (Coffman and Weissman 1981a,b, 1983) defines a previously-unrecognized population of cells in SCID bone marrow that may constitute a set of very early B-lineage (or lymphoid) progenitors.

RESULTS AND DISCUSSION

T lineage cell populations in SCID mice

The left panels of Figure 1 present three color FACS analysis of thymocyte populations derived either from normal or SCID animals. The top panels demonstrate that unlike normal thymus, most of the cells in SCID thymus lack both CD4 and CD8 molecules. As might be expected from this result, the middle panels show that very few cells in SCID thymus bear CD3, a complex of polypeptides that shows obligate co-expression with the T cell antigen receptor, and whose expression marks more mature T lineage cells (Lew *et al.* 1986). Taken together, these data also show that the $CD4^-8^-$ cells in SCID thymus do not include significant numbers of $\gamma\delta$ cells (infrequent T lineage cells that lack both CD4 and CD8) since such cells would be expected to express CD3. Finally, as reported previously (Habu *et al.* 1987), a large fraction of the cells in SCID thymus express the smaller chain of the IL-2 receptor, a characteristic of immature $CD4^-8^-$ thymocytes in normal mice. Thus, phenotypically, the thymocytes in SCID mice resemble the small fraction (0.5-1%) of normal thymocytes ($CD4^-CD8^-CD3^-$) at a very early stage of intrathymic differentiation (Mathieson and Fowlkes 1984). The small size of SCID thymus is also consistent with a lack of expansion and differentiation of such cells.

The right panels of Figure 1 present similar analyses of T cell populations in spleen. Most T cells in spleen can be recognized by co-expression of CD3 together with either CD4 or CD8. We can take advantage of this correlated expression to carefully enumerate the proportion of $\alpha\beta$ T cells in SCID spleen. The results show that even though 0.5% of cells in SCID spleen show staining above background with an anti-CD3 monoclonal antibody, most of these cells are likely not T cells since

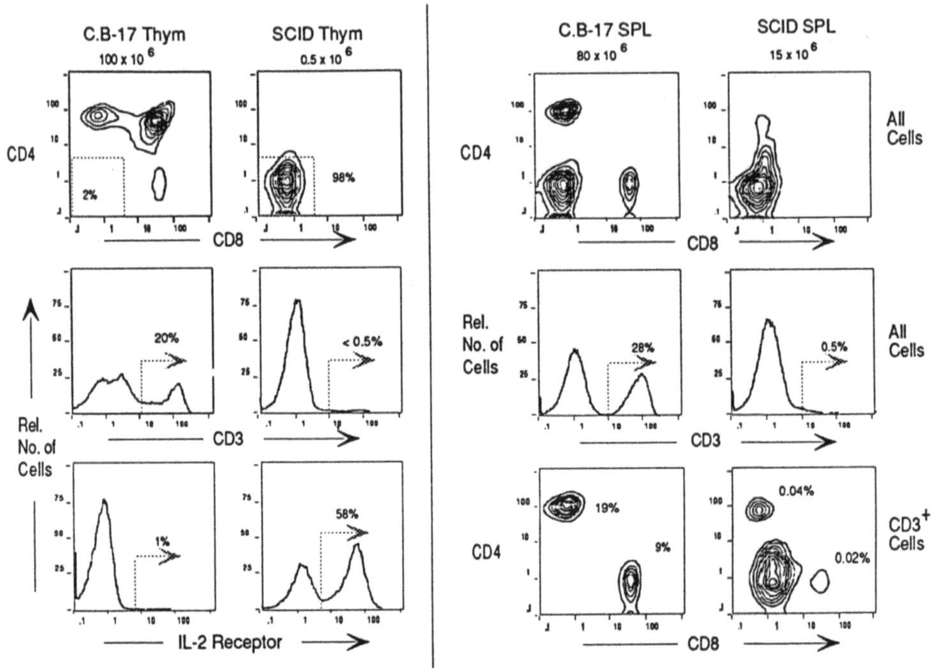

Figure 1. Absence of both immature and mature T lineage cells from thymus and spleen of SCID mice. $5x10^5$ cells from indicated tissues were stained with appropriately labelled monoclonal antibodies and analyzed on a dual-laser, dye-laser FACStar+ equipped with filters for four color immunofluorescence. Reagents employed were GK1.5 (anti-CD4), 53-6 (anti-CD8), PC61 (anti-IL2R) and 500-A2 (anti-CD3). CD3, CD4 and CD8 were used together in a three color combination. Dead cells were excluded by propidium iodide labelling. All cells shown fall within a "lymphoid scatter gate" as defined by both forward and right-angle scatter. Numbers indicate representative cell yields. Data is presented either as histograms (versus relative cell number) or as equal-probability contour plots (with equal percentages of cells enclosed by each contour "level"). All immunofluorescence is displayed with four decade logarithmic amplifiers. 20 to 30 thousand cells were analyzed for two color and 50 to 100 thousand cells were analyzed for three color.

they lack both CD4 and CD8. Although this population could include γδ cells, their low expression of CD3 together with a high background in the CD4 and CD8 channels suggests that most of these cells are simply highly autofluorescent and therefore negligible. At a minimum we can state that the frequency of splenic T cells is decreased by over a hundred-fold. Presumably the rare $CD3^+$ cells occasionally found that *do* express either CD4 or CD8 (a few cells out of 10^4) represent "leaky" T cells that do occur in SCID mice, by an as yet unexplained mechanism (Bosma *et al.* 1988).

B lineage cells in SCID mice

The bone marrow is a site of early B lineage differentiation in the adult animal. Figure 2 presents analysis of expression of cell surface molecules that can be used in defining distinct stages of B lineage differentiation. The determinant recognized by the anti-B220 monoclonal antibody, RA3-6B2 (Coffman and Weissman, 1983), predominantly marks B lineage cells in bone marrow and, together with surface μ heavy chain expression, can be used to define both pre-B and B cells. As is clear from the figure, B cells are severely depleted in SCID bone marrow (over a hundred-fold) and there are also many fewer μ^- immature B lineage cells cells ($B220^+\mu^-$).

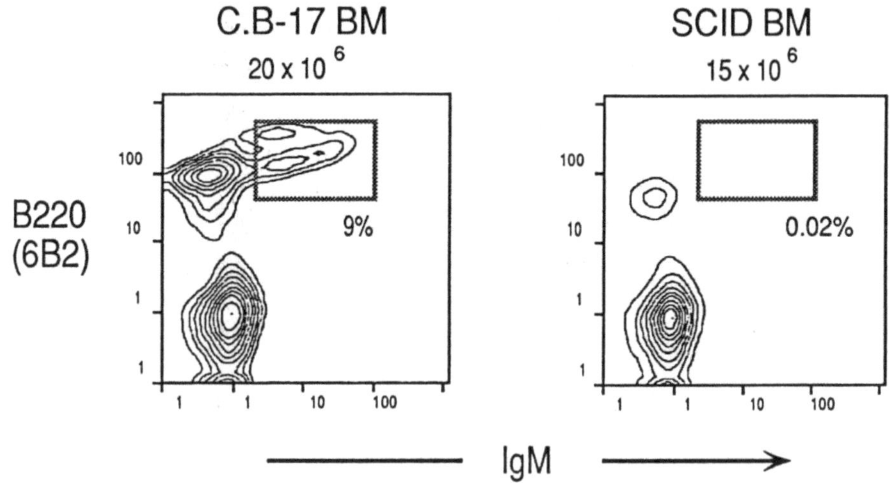

Figure 2. Absence of B cells and 5-fold decrease in putative pre-B cells (B220$^+$/IgM$^-$) in SCID bone marrow. Cells were stained with RA3-6B2 (anti-B220) and 331.12 (anti-IgM), then analyzed as in figure 1. Scatter gate includes all B220$^+$ cells (including somewhat larger cells in bone marrow). Numbers indicate representative cell yields.

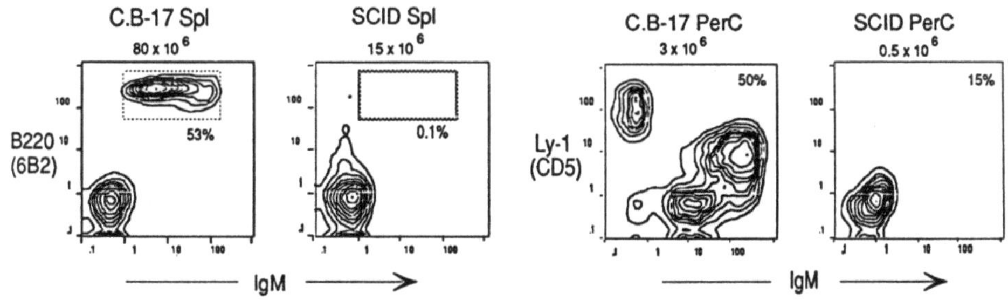

Figure 3. Absence from SCID mice of B cell populations found in spleen and peritoneum. Indicated cell populations were stained either with RA3-6B2 (anti-B220) together with 331.12 (anti-IgM) or with anti-Ly-1 (53-7) together with 331.12 (anti-IgM) and analyzed as in figure 1. Numbers indicate representative cell yields. Percentages in PerC analysis indicate cells falling within a "lymphoid scatter gate".

Next , we have compared normal and SCID peripheral tissues for the presence of more mature B lineage cells. Again, as with T cells, the predominant population of splenic B cells is reduced by over a hundred-fold (Figure 3, left panels). We have also investigated the cell populations present in peritoneum since this tissue is a rich source of Ly-1 B (CD5$^+$ B cells), a B lineage population rather distinct from the majority of splenic B cells (Hardy and Hayakawa, 1986; Hayakawa and Hardy, 1988) and probably independently derived from an early B progenitor source (Hayakawa *et al.* 1985). As is clear from the right panels of Figure 3, both Ly-1$^-$ and Ly-1$^+$ B cells are absent from SCID peritoneum.

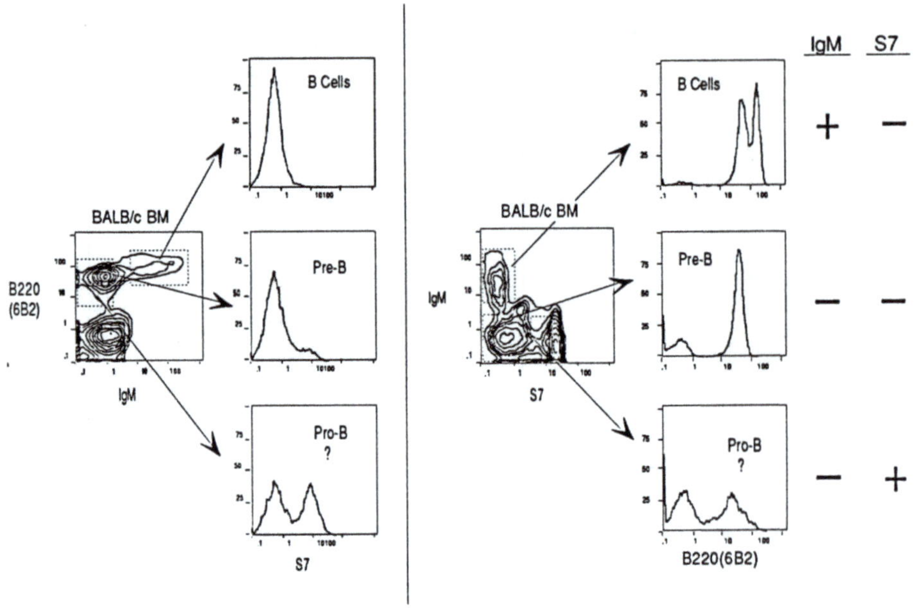

Figure 4. Left panels demonstrate that a significant fraction of $B220^{dull}/IgM^-$ cells in normal bone marrow express a determinant recognized by the S7 monoclonal antibody, a determinant absent from most pre-B and B cells. Right panels demonstrate that whereas many $S7^-$ bone marrow cells bear high or intermediate levels of B220, $S7^+$ cells show only lower expression. $B220^-$ cells in the $S7^+$ population (bottom right panel) are likely not B lineage cells. Bone marrow cells were stained simultaneously with RA3-6B2 (anti-B220), 331.12 (anti-IgM) and S7, then analyzed as in figure 1.

Delineation of a putative lymphoid precursor population by co-expression of B220 and S7

The observation of a small population of $B220^+$ cells in SCID bone marrow was intriguing enough to follow up. There is data suggesting that early B lineage precursors are present in SCID since: 1) Abelson virus transforms SCID bone marrow at frequencies similar to normal (Fulop *et al.* 1988); and 2) B lineage cells can be grown out of SCID bone marrow in long term culture (Dorshkind *et al.* 1984). However both types of cells show aberrant immunoglobulin gene rearrangements and do not progress to the B cell stage (Dorshkind *et al.* 1984; Schuler *et al.* 1986). In light of these data we have focussed on analyzing the $B220^+$ cells in SCID mice with a recently reported antibody that may serve to distinguish very early B lineage progenitors.

The rat monoclonal antibody S7 recognizes a determinant on a single-chain polypeptide of molecular mass 85-95 kDa that is expressed on granulocytes, thymocytes and peripheral T cells, but not most peripheral B cells (Gulley *et al.* 1988). It may be the murine homolog of the molecule recognized by the anti-rat monoclonal, W3/13 that reacts with a 95 kDa protein containing extensive O-linked glycosylation (Brown *et al.* 1981). Since many cells in bone marrow are $S7^+$, one can ask whether the early precursors of B and T cells might express the S7 determinant, it being lost during B (but not T) differentiation. When we measure (by three color FACS) the expression of S7 on different subpopulations of B lineage cells in normal bone marrow, then we see (Figure 4, left panels) that neither immature B cells in bone marrow (IgM^+/IgD^-) nor most pre-B ($B220^+/IgM^-$) stain with S7. Interestingly, we find that a small fraction of $B220^+/IgM^-$ cells (predominantly low for B220) do express S7 (Figure 4, bottom left panel). These cells (3-4% of bone marrow) may represent the

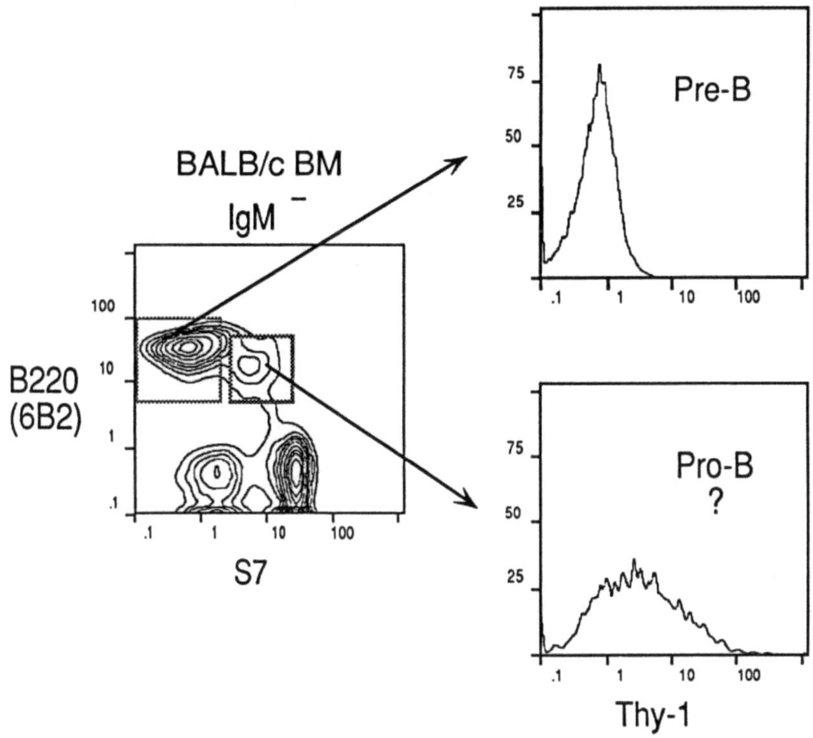

Figure 5. Expression of Thy-1 on a subpopulation of B220^{+} cells in bone marrow delineated by co-expression of the S7 determinant. Cells were simultaneously stained with RA3-6B2 (anti-B220), S7 and 30H12 (anti-Thy-1.2), then analyzed as in figure 1.

earliest B-committed population in bone marrow, potentially "pro-B cells", because: 1) they express very low levels of B220 (Figure 4, bottom right panel) which is known to increase with B cell maturation (Coffman and Weissman 1981b); 2) many cells in this fraction also co-express low levels of Thy-1 (Figure 5), previously reported to be found on hematopoietic stem cells (Muller-Sieburg *et al.* 1986) and very early lymphoid progenitors (Muller-Sieburg *et al.* 1989); and 3) sorted B220^{+}/S7^{+}/IgM^{-} bone marrow cells "progress" to B220^{+}/S7^{-} (including some IgM^{+} cells) in culture on a stromal feeder layer (unpublished observations). Thus it appears that expression of the S7 determinant is lost at an early stage of B cell differentiation, but retained throughout T lymphocyte development.

The B220+/S7+ population is present at normal frequency in SCID bone marrow

Three color analysis comparing the correlated expression of B220 (6B2), IgM and S7 in normal versus SCID bone marrow demonstrates that all of the B220^{+} cells in SCID bone marrow express S7 (Figure 6). Using this type of analysis it is clear that the B220^{+}/IgM^{-}/S7^{-} population is severely depleted in SCID, whereas the B220^{+}/IgM^{-}/S7^{+} is present at normal frequency. Thus it appears that S7 serves to delineate the stage at which the *scid* defect first becomes manifest. If *scid* results in an inability to make functional immunoglobulin and T cell receptor rearrangements, then the defect would presumably become apparent just at the stage where such rearrangements were taking place, the B220^{+}/IgM^{-}/S7^{+} stage. Such a stage has been termed "pro-B" in the B cell differentiation

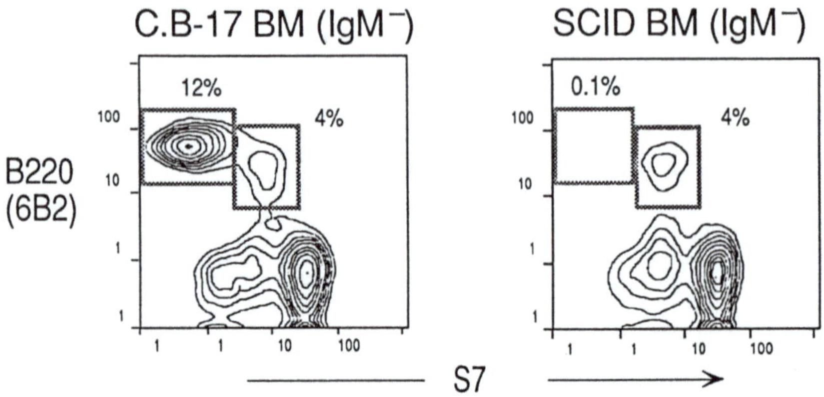

Figure 6. Presence of the S7$^+$/B220$^+$ population in SCID bone marrow at levels comparable with normal strains (such as C.B-17 or BALB/c). Cells were stained with RA3-6B2 (anti-B220), S7 and 331.12 (anti-IgM), then analyzed as in figure 1. Data presented are gated to exclude IgM$^+$ cells.

pathway: committed to B lineage, but not yet rearranged. Alternatively, we cannot rule out that this population may represent a lymphoid-committed fraction just prior to rearrangement of either B or T cell antigen receptor genes. Experiments to determine which of these alternatives is correct are currently underway.

SUMMARY

We have used 3 color FACS analysis (together with dead cell exclusion by propidium iodide) to ascertain the levels of lymphoid lineage cells present in typical young adult (2-5 month) SCID mice. Both mature and early B and T lineage cells are severely decreased (>100X) in thymus, spleen and peritoneum of such mice. Analysis revealed a significant fraction of B220$^+$ cells in bone marrow and, curiously, all such cells co-expressed a determinant recognized by the S7 antibody, likely lost early in B lineage differentiation. Thus, together with B220, expression of S7 may serve to mark the stage at which the SCID defect first becomes manifest, at least in the B lineage. This suggests that B220$^+$/S7$^+$ cells in bone marrow may be pro-B cells or even a lymphoid progenitor population.

ACKNOWLEDGEMENTS

We wish to thank Ms. Susan A. Shinton for excellent technical assistance. This work was supported by grants from the National Institutes of Health (CA-06927, RR-05539), the Pew Charitable Trust (86-5043HE), the Pew Charitable Trust Five Year Award (83-1067HE), and by an appropriation from the Commonwealth of Pennsylvania.

REFERENCES

Brown WRA, Barclay AN, Sunderland CA, Williams AF (1981) Identification of a glycophorin-like molecule at the cell surface of rat thymocytes. Nature 289: 456-460

Bosma GC, Fried M, Custer RP, Carroll AM, Gibson DM, Bosma MJ (1988) Evidence of functional lymphocytes in some (leaky) scid mice. J Exp Med 167: 1016-1033

Coffman RL, Weissman IL (1981) B220: a B cell-specific member of the T200 glycoprotein family. Nature 289:681-683

Coffman RL, Weissman IL (1981) A monoclonal antibody that recognizes B cells and B cell precursors in mice. J Exp Med 153:269-279

Coffman RL, Weissman, IL (1983) Immunoglobulin gene rearrangement during pre-B cell differentiation. J Mol Cell Immunol 1: 31-38

Dorshkind K, Keller GM, Phillips RA, Miller RG, Bosma GC, O'Toole M, Bosma MJ (1984) Functional status of cells from lymphoid and myeloid tissues in mice with severe combined immunodeficiency disease. J Immunol 132:1804-1808

Fulop GM, Bosma GC, Bosma MJ, Phillips RA (1988) Early B cell precursors in SCID mice - Normal numbers of cells transformable with Abelson murine leukemia virus (A-MuLV). Cellular Immunol 113:193-201

Gulley ML, Ogata LC, Thorson JA, Dailey MO, Kemp JD (1988) Identification of a murine pan-T cell antigen which is also expressed during the terminal phases of B cell differentiation. J Immunol 140:3751-3757

Habu S, Kimura M, Katsuki M, Hioki K, Nomura T (1987) Correlation of T cell receptor gene rearrangements to T cell surface antigen expression and to serum immunoglobulin level in scid mice. Eur J Immunol 17: 1467-1471

Hardy RR, Hayakawa K (1986) Development and physiology of Ly-1 B and its human homolog, Leu-1 B. Immunol Rev 93:53-79

Hayakawa K, Hardy RR, Herzenberg LA, Herzenberg LA (1985) Progenitors for Ly-1 B cells are distinct from progenitors for other B cells. J Exp Med 161:1554-1568

Hayakawa K, Hardy RR (1988) Normal, autoimmune and malignant $CD5^+$ B cells: The Ly-1 B lineage? Ann Rev Immunol 6: 197-218

Lew AM, Pardoll DM, Maloy WL, Fowlkes BJ, Kruisbeek A, Cheng S, Germain RN, Bluestone JA, Schwartz RH, Coligan JE (1986) Characterization of T cell receptor gamma chain expression in a subset of murine thymocytes. Science 234: 1401-1405

Mathieson BJ, Fowlkes BJ (1984) Cell surface antigen expression on thymocytes: Development and phenotypic differentiation of intrathymic subsets. Immunol Rev 82: 141-173

Muller-Sieburg CE, Whitlock CA, Weissman IL (1986) Isolation of two early B lymphocyte progenitors from mouse marrow: a committed pre-pre-B cell and a clonogenic Thy-1^{lo} hematopoietic stem cell. Cell 44:653-662

Muller-Sieburg CE, Tidmarsh GF, Weissman IL, Spangrude GJ (1987) Maturation of hematolymphoid cells that express Thy-1. in: Cell Surface Antigen Thy-1: Immunology, Neurology and Therapeutic Applications, Reif AF, Schlesinger M (eds). Marcel Dekker, Inc., New York, p 289

Schuler W, Weiler IJ, Schuler A, Phillips RA, Rosenberg N, Mak TW, Kearney JF, Perry RP, Bosma MJ (1986) Rearrangement of antigen receptor genes is defective in mice with severe combined immune deficiency. Cell 46: 963-972

Phenotype and Differentiation Stage of Scid Mouse Thymocytes

Sonoko Habu, Yoko Norihisa, Takehito Sato, Hideo Yagita and Ko Okumura

Department of Immunology, Tokai University, School of Medicine, Bohseidai, Isehara-shi, Kanagawa, Japan

1. Introduction

Scid mice are severely deficient in T and B cell function whereas other hematopoietic cells develop normally (Bosma et al. 1983; Custer et al. 1985). The effects which the scid mutation manipulates seem to occur after the commitment of lymphoid cells to the T and B cell lineages because early transcription of unrearranged genes for T cell antigen receptor (TCR) and immunoglobulin (Ig) is detected in scid thymus and fetal liver, respectively (Schuler et al. 1988), although these genes are not rearranged in the scid mice (Schuler et al. 1986; Habu et al. 1987a). In addition, experiments involving transfer of bone marrow cells into the scid mice indicate that the thymus and bone marrow microenvironments function normally for T and B cell differentiation (Custer et al. 1985). Thus, identification of the differentiation stage of scid mouse lymphoid cells will provide valuable information for understanding how TCR and Ig gene rearrangements and/or gene products function in early lymphoid differentiation. Here, we will demonstrate the expression of the scid phenotype in thymocytes and discuss the stage in the normal differentiation pathway of T cell lineages at which it occurs.

2. Phenotypes of scid mouse thymocytes

The thymocytes of scid mice are Thy-1 positive but do not express CD4 or CD8 (Bosma et al. 1983; Habu et al. 1987a). The embryonic thymocytes at 14 or 15 days of gestation show the same phenotype as scid thymocytes (Habu et al. 1987a). Since the $CD4^-$ $CD8^-$ (double negative, DN) embryonic thymocytes contain a small proportion of CD3 positive cells which are reported to express T cell antigen receptor (TCR) γ genes (Bluestone et al. 1987), scid thymocytes were examined for CD3 expression. The thymocytes obtained from 4-week old scid mice were stained with fluorescein-conjugated monoclonal antibody against CD3 (anti-CD3, 2C11) and analyzed by flow cytometry, FACStar. As shown in Figure 1, the scid thymocytes were CD3 negative. This finding is consistent with previous observations (Shuler et al. 1986; Habu et al. 1987a) showing that TCR genes are not rearranged in scid mice.

[1] Department of Immunology, Juntendou University, School of Medicine, Bunkyo-ku, Tokyo, Japan

Current Topics in Microbiology and Immunology, Vol. 152

The DN thymocytes may be divided into several subgroups depending on the combination of cell surface molecules they have. Of these, CD2 and interleukin 2 receptor (IL-2R) are useful in defining the differentiation stage, because they are expressed in an ordered manner in mouse DN thymocytes early in gestation (see below). In order to determine the expression of the two molecules, the thymocytes obtained from 1-day-, 7-day- and 12-week-old scid mice were stained with anti-IL-2R and anti-CD2. The fluorescence profiles of the stained thymocytes using FACStar showed that the majority of scid thymocytes express IL-2R (80-90%) and that less than 10% of them are CD2 positive (Fig.1). Two color analysis of scid thymocytes stained with both anti-IL-2R and anti-CD2 showed the presence of a very small proportion of IL-$2R^+CD2^+$ cells.

3. Relationship of phenotype expression and stage in normal differentiation pathway

Thymocytes of normal mice are classified phenotypically into four fractions on the basis of CD4 and CD8 expression; $CD4^-CD8^-$ (double negative, DN), $CD4^+CD8^+$(double positive,DP), $CD4^+CD8^-$(CD4 single positive) and $CD4^-CD8^+$(CD8 single positive) cells. Studies of thymus ontogeny and organ culture experiments using embryonic thymus lobes have shown that DN cells belong to the most immature thymocytes which migrate into the thymus and develop into DP and single positive cells (Habu and Okumura 1983; Ceredig et al. 1983). Although there has been long argument about whether single positive thymocytes,are directly derived from DP thymocytes or DN thymocytes, we recently provided direct evidence that DP adult thymocytes can develop into CD4 or CD8 single-positive cells in the lymph nodes when they are intravenously injected into the athymic nude mouse (Tamauchi et al. 1988). L. Smith has also reported that in vivo injection of monoclonal antibody against Lyt-2 reduced the number of $CD4^+CD8^-$ cells, suggesting that CD4 single positive cells develop from CD8-bearing cells, presumably $CD4^+CD8^+$ cells. Using organ culture of embryonic thymus lobes, we also demonstrated that CD4 and CD8 single-positive cells developed from $CD4^+CD8^+$ thymocytes when DP thymocytes were cultured in 2-deoxyguanosine-treated alymphocytic thymus lobe using Jenkinson's hanging drop method (Jenkinson et al. 1982)(data not shown). In this experiment, the alymphocytic thymus lobes which received DP thymocytes were shown to contain a small number of lymphocytes which was less than the number of cells originally transferred. By contrast, the lymphocyte cell number in the alymphocytic thymus lobes receiving DN cells increased around 50-fold (data not shown). Taken together, it seems that although some DP thymocytes may die in the thymus, a certain proportion of DN thymocytes can develop into CD4 and CD8 single-positive cells via the differentiation stage of DP cells.

In the embryo, Thy-1 is the first cell surface molecule detectable on lymphoid cells which have migrated into the thymus (day 11 or day 12). One or 2 days later, IL-2R (Habu et al. 1985; Ceredig et al. 1985) and subsequently CD2 are found on the thymocytes . The proportion of IL-$2R^+$ thymocytes decreases rapidly after day 16 to 1% and at the same time, CD4 and CD8

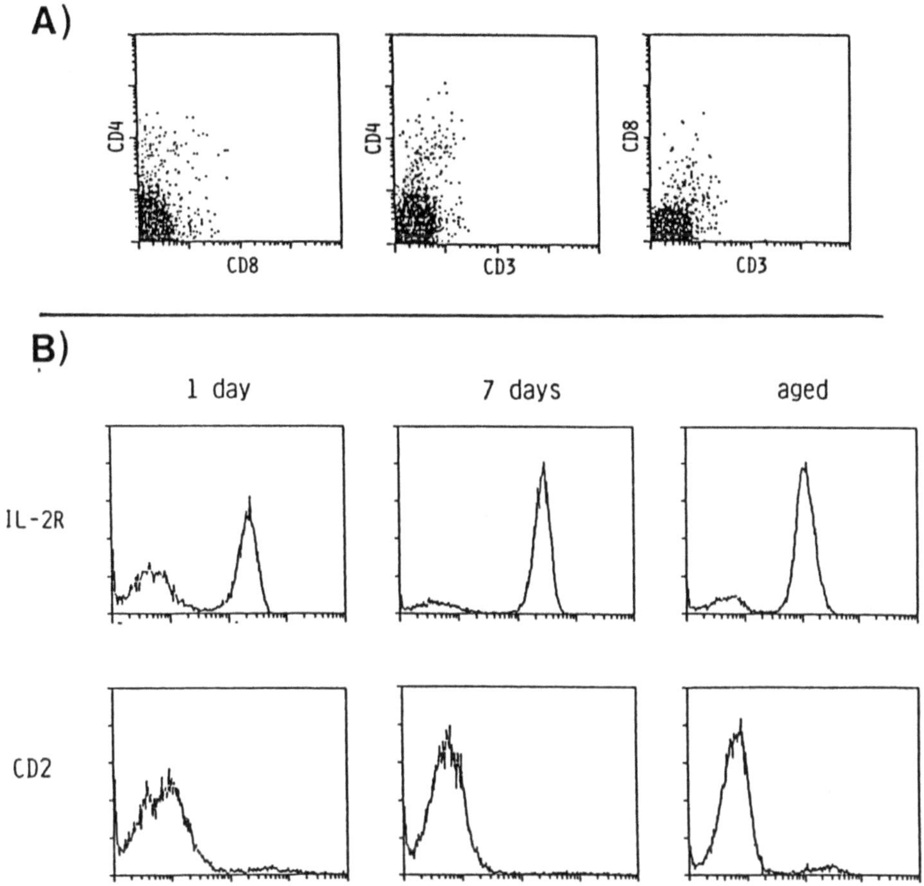

Figure 1. Phenotype expression on thymocytes of scid mice. A) Thymocytes obtained from a 4-week-old scid mouse were simultaneously stained with three antibodies, phycoerythrine-conjugated anti-L3T4 (anti-CD4), fluorescein-conjugated anti-Lyt-2 (anti-CD8) and biotin-conjugated anti-CD3 (2C11), and subsequently reacted with diochrom-conjugated avidin. These stained thymocytes were subjected to three-color analysis by FACStar. B) Thymocytes from 1-day, 7-day and 12-week old scid mice were stained with fluorescein-conjugated anti-IL-2R and anti-CD2. With anti-IL-2R, 70 % to 92 % of the thymocytes were stained but with anti-CD2, less than 10% were stained.

become detectable on IL-2R$^-$ cells. The CD2 molecule which was originally reported to be expressed on immature human thymocytes, was detectable on a few mouse embryo thymocytes on day 13 or 14 and the proportion of the CD2$^+$ thymocytes increased during gestation reaching nearly 100 % at birth (submitted for publication). In addition to the clear inverse relationship between the proportion of IL-2R$^+$ and CD2$^+$ cells during gastation days, a small proportion of the IL-2R$^+$CD2$^+$ cells were found on day 14 or 15. From these observations, it is postulated that IL-2R is expressed transiently on DN thymocytes in advance of CD2 expression during intrathymic differentiation .

4. Dividing and developing ability of IL-2R positive thymocytes in normal and scid mouse

Based on the above hypothesis that IL-2R$^+$ thymocytes are immature cells on the way of differentiation but not subsets which are destined to die in the thymus, they must cease to express the gene during differentiation. To clarify this issue, we examined the fate of IL-2R$^+$ embryonic thymocytes using the hanging drop method. After staining embryonic thymocytes with anti-IL-2R, IL-2R$^+$ and IL-2R$^-$ cells were sorted by FACStar and were cultured in 2-dioxyguanosine-treated alymphocytic thymus for 7 days. The lymphocytes harvested from thymus lobes were stained with anti-CD4, anti-CD8 and anti-IL-2R for flow-cytometry analysis. The lymphocytes obtained from thymus receiving IL-2R$^+$ cells or IL-2R$^-$ cells proved to contain CD4-and CD8-bearingcells (Fig.2). The majority of lymphocytes did express IL-2R. Furthermore, the number of lymphocyte cell in both thymus lobes increased more than 100- fold compared to the number originally transferred. Thus, it is strongly suggested that IL-2R$^+$ DN thymocytes are thymocytes developing into the DP and SP cells, during which IL-2R expression is lost. In fact, our previous immunohistochemical study showed many mitotic figures in embryonic thymocytes stained with anti-IL-2R antibody at various intensities (Habu et al. 1987b).

In the scid mouse, the thymus, which comprises almost exclusively IL-2R$^+$ cells is very tiny and contains few lymphocytes, about 1% of the normal number of thymocytes (Bosma et al. 1983; Habu et al. 1987b). This suggests that scid thymocytes are not able to proliferate in the thymus or die without developing to an advanced stage. In fact, when scid thymocytes were cultured in the 2-dioxyguanosine-treated thymus lobes, they did not proliferate and did not show CD4 or CD8 expression (data not shown). However, when scid thymocytes were cultured with IL-2, they showed high ^{3}H-thymidine incorporation, while CD4 and CD8 expression was not induced. This finding indicates that certain stimulants can induce scid thymocytes to divide but not to differentiate.

Although DN thymocytes of normal embryos and scid mice express some cell surface molecules which are responsible for signaling, such as Thy-1 and CD2, DN thymocytes were not activated by the antibodies against Thy-1 and CD2 (manuscript in preparation). Thus, it seems likely that signals via some molecules other than Thy-1 and CD2 may cause immature embryonic thymocytes to proliferate and differentiate but not scid thymocytes. Although it is as yet unclear what factors these may be, impairment of TCR gene rearrangement in the scid mice may inhibit the IL-2R$^+$ cells from differentiating further by expressing such as CD4 and CD8. If that is correct, TCR gene rearranging events are important for promoting differentiation as well as proliferation of IL-2R$^+$ embryonic thymocytes in early stages of differentiation even before TCR gene products are expressed on the cell surface. Further studies are required to prove this.

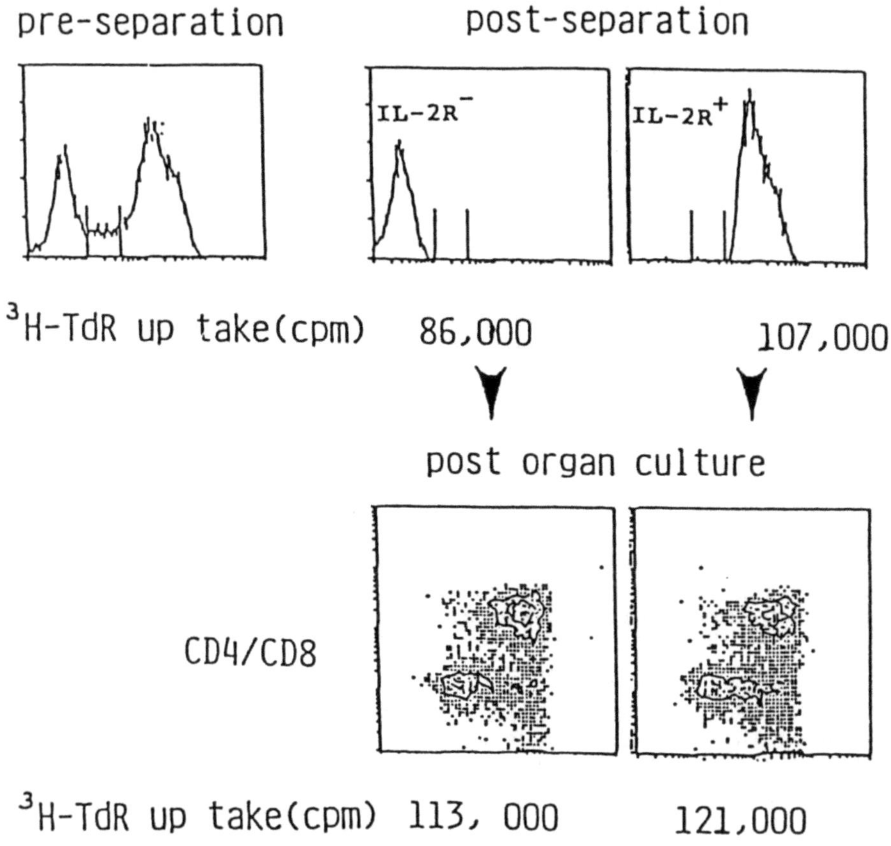

Figure 2. Proliferation and differentiation of IL-2R$^+$ embryonic thymocytes. After staining with fluorescein-conjugated anti-IL-2R, 15 day embryonic thymocytes were sorted into IL-2R$^+$ and IL-2R$^-$ cells by FACStar, and each sorted cell population were cultured in another 2-deoxyguanocyn treated embryonic thymus lobe. The lymphocytes harvested from the cultured thymus lobes were stained with anti-L3T4 and anti-Lyt-2 and were then analyzed by FACStar.

Acknowledgements. The authors are grateful to Mrs.T.Takizawa and Mr.Y.Takeuchi for their excellent technical assistance. This work was supported by a Grant-in-Aid from the Ministry of Education, Culture and Science, Japan.

References

1. Bluestone JA, Pardoll D, Sharrow SO, Fowlkes BJ (1987) Characterization of murine thymocytes with CD3-associated T cell receptor structure. Nature 326:82
2. Bosma GC, Custer RP, Bosma MJ (1983) A severe combined immunodeficiency mutation in the mouse. Nature 301:527
3. Ceredig R, MacDonald HR, Jenkinson EJ (1983) Flow microfluorometric analysis of mouse thymus development in vivo and in vitro. Eur J Immunol 13:185
4. Ceredig R,Lowenthal JW, Nabholz M, MacDonald HR (1985) Expression of interleukin-2 receptors as a differentiation marker on intrathymic stem cells. Nature 314:98
5. Custer RP, Bosma GC, Bosma MJ (1985) Severe combined immunodeficiency (SCID) in the mouse. Am J Pathol 120:464
6. Habu S, Okumura K (1983) Cell surface antigen marking the stages of murine T cell ontogeni and its functional subsets. Immunol Rev 82:117 .
7. Habu S, Okumura K, Diamanstein T, Shevach EM (1985) Expression of interleukin 2 receptor on murine fetal thymocytes. Eur J Immunol 15:456
8. Habu S, Kimura M, Katsuki M, Hioki K, Nomura T (1987a) Correlation of T cell receptor gene rearrangements to T cell surface antigen expression and to serum immunoglobulin level in scid mice. Eur J Immunol 17:1467
9. Habu S, Koji T, Akatsuka A (1987b) Division of IL-2 receptor expressing lymphocytes in murine embryonic thymus. Acta. Cytochem Histochem 20:251
10. Jenkinson EJ, Franch LL, Kingston R, Owen JJT (1982) Effect of deoxyguanosine on lymphopoiesis in the developing thymus rudiment in vitro: application in the production of chimeric thymus rudiments. Eur J Immunol 12:538
11. Schuler W, Weiler IJ, Schuler RA, Phillips PA, Rosenberg N, Mak TW, Kearney JF, Perry RP, Bosma MJ (1986) Rearrangement of antigen receptor genes is defective in mice with severe combined immunodeficiency. Cell 46:963
12. Schuler W, Schuler A, Lennon GG, Bosma GC, Bosma MJ (1988) Transcription of unrearranged antigen receptor genes in scid mice. EMBO J 7:2019
13. Tamauchi H, Tamaoki N, Habu S (1988) $CD4^+$ $CD8^+$ thymocytes develop into CD4 and CD8 single positive cells in the athymic nude mice. Eur J Immunol 18:1859

Growth Requirements of B Lineage Lymphocytes From Scid and Normal Mice

G. Lee, K. Medina, and P.W. Kincade

Oklahoma Medical Research Foundation
825 N.E. 13th Street
Oklahoma City, Oklahoma 73104 USA

Long term culture techniques provide unique approaches to dissecting components of the bone marrow microenvironment. These procedures are particularly effective for studies utilizing mice, where it is possible to selectively propagate either myeloid, or B lineage lymphoid cells for many months (Whitlock et al. 1984). Several laboratories have established and cloned "stromal" cell lines which support myelopoiesis and/or lymphopoiesis, as well as lymphocyte clones which depend on them for growth. Progress is also being made in learning which regulatory cytokines are made in these circumstances, and conditions favoring their production. However, there are indications that these model systems do not fully reflect the cellular and molecular heterogeneity that exists *in vivo* (Kincade et al. 1988). This has become particularly noteworthy from culture studies of genetically determined abnormalities, including SCID. Further comparison of the behavior of such defective cells *in vivo* and *in vitro* should be informative about mechanisms involved in normal lymphopoiesis.

SCID mice must have at least small numbers of lymphocytes committed to both T and B lineages. This follows from the presence of phenotypically defined cells in the thymus and bone marrow, the spontaneous development of thymomas, occasional mice which have a "leaky" phenotype, and a near normal incidence of Abelson virus transformation targets (Bosma et al. 1988). However, it was surprising to find that SCID bone marrow cells readily give rise to long term lymphocyte cultures (Witte et al. 1987a). At a given initial cell density, the fraction of successful cultures is lower than for normal BALB/c bone marrow. However, once growth is established, numbers and characteristics of the cells are unremarkable, with the exception that immunoglobulins are not made. Similar findings were made by Nishikawa and colleagues (Hirayoshi et al. 1987) and this raises the question of why such lymphocytes do not expand significantly *in vivo*. One possibility is that defective SCID lymphocytes are recognized and either eliminated or suppressed through a "quality control" mechanism which is only operational *in vivo*. We and others are learning a great deal about the cells and molecules which are represented in long term cultures and a brief update on those findings will be given here.

Adherent cells which grow in LTBMC under the conditions originally described by Dexter include adipocytes, macrophages, endothelial cells, and fibroblasts in addition to the myeloid cells (mainly neutrophils) whose growth they support (Allen and Dexter, 1984). A majority of the adherent cells in Whitlock-Witte cultures are macrophages, but they are not required for continued lymphocyte growth (Witte et al. 1987b). After macrophage depletion, the lymphocytes can be seen attached to a subpopulation of similar appearing cells which lack macrophage properties. Several laboratories have succeeded in cloning stromal cells on which cultured lymphocytes will grow (reviewed in Kincade et al. 1988). A panel of clones developed in our lab were classified as being support (S) or non-support (NS) when tested with cloned, stromal cell dependent lymphocytes (Pietrangeli et al. 1988). Further studies with additional lymphocyte clones revealed additional heterogeneity in the stromal cells; some lymphocytes grow on particular stromal cell clones, but not others.

Diversity was also apparent by immunoperoxidase staining with monoclonal antibodies. A phenotype that is common and distinctive for support stromal cells was not found and differences were noted between primary cultures and cloned lines. For example, most of the stromal cell clones express Thy-1 antigen, which was not conspicuous on adherent cells in conventional long term cultures (Pietrangeli et al. 1988). The basis for functional and

Current Topics in Microbiology and Immunology, Vol. 152

phenotypic diversity between such stromal cell clones is not understood. It may be that cells which are cloned and extensively propagated assume characteristics unlike their normal counterparts in bone marrow. Alternatively, the available clones may reflect small subpopulations of cells and differentiation stages which contribute to the normal microenvironment.

Close interaction between lymphocytes and stromal cells is necessary for efficient long term growth, and multiple cell adhesion molecules may be involved in that recognition (Kierney and Dorshkind, 1987; Witte et al. 1987b; Kincade, 1987). However, it has been possible to demonstrate at least two factors in stromal cell conditioned medium which can influence lymphocyte replication and/or differentiation (Landreth and Dorshkind, 1988; Hunt et al. 1987; Whitlock et al., 1987; Lemoine et al. 1988). The first of these to be purified and molecularly cloned is termed interleukin 7 (IL-7), and it was derived from an SV40 transformed cell line (Namen et al. 1988). IL-7 has now been shown to be an effective stimulus for normal B lineage precursors (Lee et al. 1989). Large pre-B cells, which could already be in cycle, appear to respond immediately. This was sufficiently prompt and selective for pure pre-B cell colonies to form in semisolid agar cultures within five days and there was little tendency for the cells to mature, i.e. express surface immunoglobulins. Significantly, SCID bone marrow, which lacks pre-B cells, did not form colonies in this assay. Under these conditions, a linear cell dose response curve suggests that the cytokine directly promoted pre-B cell growth. This cloning procedure could be an important new experimental tool for studies of the final events in B cell formation.

A broader spectrum of IL-7 responding cells could be observed with liquid cultures. Virtually all non-adherent cells which arose in IL-7 cultures initiated with enriched B lineage precursors or bone marrow expressed high levels of BP1 antigen. A majority were positive for the CD-45R epitope recognized by the monoclonal 14.8 antibody, and a subset were typical pre-B cells (Lee et al. 1989). Thy-1 and Class II antigens were also found on many of the cells after culture with IL-7. Replication could usually not be maintained for more than three to six weeks despite repeated addition of IL-7. B lineage cells were extensively depleted from marrow cell suspensions with monoclonal 14.8 antibody and the remainder placed in liquid culture with IL-7. A response did occur after a delay of several days and proliferating clusters of 14.8^{+}, $BP1^{+}$ lymphocytes were observed. Similar delayed responses were obtained with mutant SCID marrow cells. IL-7 stimulated SCID lymphocytes also expressed 14.8 and BP1, but in contrast to those from normal cultures, were uniformly $Thy\text{-}1^{+}$ and no pre-B cells were observed. Time may be required for early precursors to mature sufficiently for response to this cytokine, or to accumulate in significant numbers in the cultures.

These findings suggest that IL-7 may be an important replication stimulus for pre-B cells and early B lineage precursors. However, it is unlikely to be the only one and a number of questions arise about its importance and functions in normal lymphopoiesis. Consistent with its effects on fresh bone marrow suspensions, IL-7 mediated stimulation of lymphocyte clones is usually of limited duration (unpublished observations and A. Namen, personal communication). However, the same lymphocytes divide indefinitely when maintained on stromal cells. Northern analyses revealed the presence of IL-7 mRNA in peripheral lymphoid tissues, including the spleen of SCID mice (Namen et al. 1988 and unpublished observations) whereas pre-B cells are normally not present in that tissue. While up to 500 units/ml of IL-7 alone did not cause spleen cell division (Lee et al. 1989), there is evidence that it can be a costimulus for peripheral T cell replication (P. Morrissey et al. manuscript in press and Namen, unpublished observations).

Different production rates of IL-7 could in part account for the effectiveness of stromal cell clones for lymphocyte support. Northern blots of poly A selected RNA from several of our support and nonsupport lines indicated that this might be the case (Gimble et al. submitted). More definitive experiments are being performed on subclones of a single stromal cell line. Of ten such subclones, only one totally failed to support lymphocyte growth and it lacked detectable IL-7 mRNA. Further analysis of these subclones revealed substantial variations in transcript levels for six other genes, but at least some of each was found in the nonsupport subclone(BMS 2.4). BMS2.4 has other unusual features and it actually suppresses lymphopoiesis. That is, IL-7 responding cells do not grow on it in the presence of exogenous IL-7. Cocultures prepared with 25% BMS2.4 cells and 75% parental BMS2 cells have little ability to support lymphocytes, and preliminary studies indicate that BMS2.4 conditioned medium may contain suppressive substances (unpublished observations). These findings give some indication of the phenotypic and functional stability of stromal cells, make it possible to compare support capability with expression of various genes, and suggest that antagonist production may influence lymphocyte growth.

One potential antagonist of lymphopoieis has been defined. Latent precursor forms of TGF-β's are known to be made in many tissues and we have recently found that this includes stromal cells and other elements of long term cultures (Gimble et al. submitted and Hayashi et al. submitted). Active TGF-β's have discrete effects on lymphocyte precursors. They inhibit kappa light chain transcription in pre-B cells, but only when particular inductive stimuli are used, and ongoing sIg expression in mature B cells is unaffected (Lee et al. 1987). These cytokines inhibit IL-7 induced replication of pre-B cells and can also block mitogenesis in mature B and T lymphocytes (Lee et al. 1989). We recently studied the effects of exogenous TGF-β_1 on long term cultures (Hayashi et al. submitted). Myelopoiesis could be prevented in Dexter cultures and this resulted in part from effects on cells of the adherent layer. Adipocyte formation was prevented and treated adherent layers were subsequently unable to support myeloid cell growth. Lymphocyte production was arrested in Whitlock-Witte cultures, but only when moderately high doses were used or when it was added at the initiation of culture. The apparent selectivity of this cytokine for myeloid cells was also indicated in "switch" culture experiments. Precursors in Dexter cultures capable of giving rise to lymphocytes survived exposure to TGF-β_1 and expanded normally when the conditions were made lymphocyte permissive. It is not known to what degree endogenous TGFβ's become activated *in vivo*, but it is clear that they could be important, multifunctional regulators.

In addition to IL-7 and TGF-β, stromal cells can elaborate a number of other cytokines. All clones thus far described express M-CSF transcripts or activity (reviewed in Kincade et al. 1988 and Gimble et al. submitted). This may mean that all lymphocyte supporting stromal cells are potentially multifunctional; i.e. also able to stimulate myeloid cell production (Collins and Dorshkind 1987; Hunt et al. 1987). With appropriate inductive stimulation, many if not all of them produce G-CSF and/or G/M-CSF (for example Rennick et al. 1987). Low levels of mRNA for IL-6 are constitutively made and this is increased in response to a number of stimuli (Gimble et al. submitted). Although no IL-4 message has been detectable in our stromal cell clones, there is indirect evidence that some stromal cells may make it (King et al. 1988). We have preliminary indications that our stromal cells make a chemotactic factor(s) (Miyake, unpublished observations). A pre-B cell inducing activity that is not IL-7 has been identified and possibly cloned (Landreth and Dorshkind, 1988 and K. Landreth, personal communication).

Production of cytokines by stromal cells is probably a carefully regulated function, and they are themselves potentially responsive to their own products. For example, our BMS2 clone makes mRNA for TGF-β, IL-6, and IL-7. However, it also responds to each of these factors when they are added to the culture medium and this suggests that autocrine regulation, or recuitment of neighboring stromal cells into a coordinated response, are possible. Among other factors which induce changes in these cells are IL-1, TNF, EGF, lipopolysaccharide, and gamma interferon (Gimble et al. submitted and unpublished observations). Therefore, two way communication may occur via these mediators between stromal cells and T cells, macrophages and endothelial cells.

Reproduction of some features of the bone marrow microenvironment with cloned cell lines and cytokines represents a major advance. However, the fact that SCID lymphocytes grow well in these situations underscores the incomplete and/or artificial nature of such culture conditions. There is no evidence for the maintenance of natural killer cells in LTBMC (Pam Witte, personal communication), but they are present in SCID mice (Dorshkind et al. 1985). It is possible that they function to limit growth of SCID lymphocytes *in vivo*. While we have no evidence for that possibility thus far, it seems reasonable to assume that some cell type or regulatory network is not represented in the tissue culture models. Mutant SCID, motheaten, and Xid mice, as well as genetically determined polymorphisms (Kincade et al. 1988; Hayashi et al. 1989), may provide sensitive indicators for appreciating such mechanisms.

Acknowledgements:

Our work is supported by grants AI-20069 and AI-19884 from the National Institutes of Health.

References

Allen TD, Dexter TM (1984) The essential cells of the hemopoietic microenvironment. Exp Hematol 12:517-521

Bosma GC, Fried M, Custer RP, Carroll A, Gibson DM, Bosma MJ (1988) Evidence of functional lymphocytes in some (Leaky) scid mice. J Exp Med 167:1016-1033

Collins LS, Dorshkind K (1987) A stromal cell line from myeloid long-term bone marrow cultures can support myelopoiesis and B lymphopoiesis. J Immunol 138:1082-1087

Dorshkind K, Pollock SB, Bosma MJ, Phillips RA (1985) Natural killer (NK) cells are present in mice with severe combined immunodeficiency (SCID). J Immunol 134:3789

Hayashi S-I, Witte PL, Kincade PW (1989) The Xid mutation affects hemopoiesis in long term cultures of murine bone marrow. J Immunol 142:444-451

Hirayoshi K, Nishikawa S, Kina T, Hatanaka M, Habu S, Nomura T, Katsura Y (1987) Immunoglobulin heavy chain gene diversification in the long-term bone marrow culture of normal mice and mice with severe combined immunodeficiency. Eur J Immunol 17:1051-1057

Hunt P, Robertson D, Weiss D, Rennick D, Lee F, Witte ON (1987) A single bone marrow-derived stromal cell type supports the *in vitro* growth of early lymphoid and myeloid cells. Cell 48:997-1007

Kierney PC, Dorshkind K (1987) B lymphocyte precursors and myeloid progenitors survive in diffusion chamber cultures but B cell differentiation requires close association with stromal cells. Blood 70:1418-1424

Kincade PW (1987) Experimental models for understanding B lymphocyte formation. Adv Immunol 41:181-267

Kincade PW, Lee G, Pietrangeli CE, Hayashi S-I,Gimble, JM (1988) Cells and molecules that regulate B lymphopoiesis in bone marrow. Ann Rev Immunol 7:111-143

King AG, Wierda D, Landreth KS (1988) Bone marrow stromal cell regulation of B-lymphopoiesis: 1. The role of macrophages, interleukin-1, and interleukin-4 in pre-B cell maturation. J Immunol 141:2016-2026

Landreth KS, Dorshkind K (1988) Pre-B cell generation potentiated by soluble factors from a bone marrow stromal cell line. J Immunol 140:845-852

Lee G, Ellingsworth LR, Gillis S, Wall R, Kincade PW (1987) B transforming growth factors are potential regulators of B lymphopoiesis.J Exp Med 166:1290-1299

Lee G, Namen AE, Gillis S, Ellingsworth LR, Kincade PW (1989) Normal B cell precursors responsive to recombinant murine interleukin 7 (IL-7) and inhibition of IL-7 activity by TGF-β. J Immunol (in press)

Lemoine FM, Humphries RK, Abraham SDM, Krystal G, Eaves CJ (1988) Partial Characterization of a Novel Stromal Cell-derived Pre-B-cell Growth Factor Active on Normal and Immortialized Pre-B Cells. Exp Hematol 16:718-726

Namen AE, Lupton S, Hjerrild K, Wignall J, Mochizuki DY, Schmierer A, Mosley B, March CJ, Urdal D, Gillis S, Cosman D, Goodwin RG (1988) Stimulation of B-cell progenitors by cloned murine interleukin-7. Nature 333:571-573

Pietrangeli CE, Hayashi S-I, Kincade PW (1988) Stromal cell lines which support lymphocyte growth: Characterization, sensitivity to radiation and responsiveness to growth factors.Eur J Immunol 18:863-872

Rennick D, Yang G, Gemmell L, Lee F (1987) Control of hemopoiesis by a bone marrow stromal cell clone: Lipopolysaccharide-and interleukin-1-inducible production of colony-stimulating factors. Blood 69:682-691

Whitlock CA, Robertson D, Witte ON (1984) Murine B cell lymphopoiesis in long-term culture. J Immunol Methods 67:353-369

Whitlock CA, Tidmarsh GF, Muller-Sieburg C, Weissman IL (1987) Bone marrow stromal cell lines with lymphopoietic activity express high levels of a pre-B neoplasia-associated molecule. Cell 48:1009-10021

Witte PL, Burrows PD, Kincade PW, Cooper MD (1987a) Characterization of B lymphocyte lineage progenitor cells from mice with severe combined immune deficiency disease (SCID) made possible by long term culture. J Immunol 138:2698-2705

Witte PL, Robinson M, Henley A, Low MG, Stiers DL, Perkins S, Fleischman RA, Kincade PW (1987b) Relationships between B-lineage lymphocytes and stromal cells in long term bone marrow cultures. Eur J Immunol 17:1473-1484

Defect of Scid Mouse Revealed in In Vitro Culture Systems

Shin-Ichi Nishikawa, Shin-Ichi Hayashi, Satomi Nishikawa, Minetaro Ogawa, Takahiro Kunisada, Tetsuo Sudo, Hiroaki Kodama and Toshio Suda

Department of Immunopathology, Kumamoto University Medical School, 2-2-1 Honjo, Kumamoto 860, Japan

Since a series of analyses by Bosma's group on the scid mouse demonstrated that there are virtually no lymphocytes in this mutant strain (Bosma et al 1983), it has been attracting those investigating the development of lymphocytes from hemopoietic stem cells. A pioneering study by Whitlock et al. enabled us to follow a considerable part, if not all, of the process during intramarrow B cell development reproducibly in an in vitro culture system (Whitlock and Witte 1982; Whitlock et al. 1984). In this situation, expecting that scid mouse might contribute to the understanding of B cell development, we started to analyze the defect of scid mouse in long-term bone marrow culture. The question which was addressed at the beginning by using the Whitlock-Witte type long-term bone marrow culture (W-LTBC) was whether or not B lineage cells were generated in the scid mouse culture. If the defect of scid mouse affects the commitment process of pluripotent stem cells into lymphoid cells, B lineage cells would not appear in the W-LTBC of scid bone marrow. On the other hand, if the defect affects the later differentiation stage, B lineage cells whose differentiation is arrested would be generated in this culture. In this article, we first briefly describe our previous results on scid mouse which utilized classical long-term bone marrow culture methods, then describe the system which we are currently using for the analysis of B cell differentiation, and finally our analysis on the defect of scid mouse in this culture system.

Early pre-B cells generated in the long term culture

The culture system which we used first for analyzing the scid defect was 'switch culture' in which the bone marrow cells are first cultured under Dexter's type LTBC (D-LTBC: Dexter et al., 1977) conditions, and then the culture is switched to W-LTBC conditions (Dorshkind 1986; Hirayoshi et al. 1987). The advantages of the switch culture over W-LTBC are that: 1) it is possible to check the in vitro myelopoiesis of scid mouse during the primary D-LTBC, 2) all the B lineage cells except early progenitors can be eliminated during the primary D-LTBC, and 3) synchronized B cell differentiation from early B progenitors can be induced by switching the culture conditions, which allows detailed analysis of the early process.

The most important result coming from this analysis was that B220(+) B lineage cells did appear after switching the culture conditions. This result is consistent with the observations of Witte et al. using W-LTBC (Hirayoshi et al. 1987; Witte et al. 1987) and indicates clearly that

---

Other addresses: Biomaterial Research Institute, Taya cho, Yokohama, Dept. Oral Biology, School of Dentistry, Ohu University, Koriyama, and Division of Hematology, Dept. of Medicine, Jichi Medical School,, Tochigi, Japan.
This work was supported in part by Special Grant-in-Aid from Riken, and Grant-in-aid from Ministry of Education, Science and Culture, Japan.

Current Topics in Microbiology and Immunology, Vol. 152

the commitment of hemopoietic stem cells into B lineage cells is normal in scid mouse. Interestingly, although B220(+) B lineage cells were generated in the culture, no mature sIgM(+) B cell appeared. This suggests that maturation arrest did occur in the scid mouse culture. In order to further identify the differentiation stage of B lineage cells in the culture, we next carried out Southern blot analysis of the cells harvested 3 and 5 weeks after the switch by using the JH probe. In the normal mouse culture, rearranged JH bands are detectable 3 weeks after the culture, while in the scid mouse culture they were detected 5 weeks after the switch. In addition to this delay in the process of JH gene rearrangement, it was noted that the distribution of rearranged JH bands of the cells in the scid mouse culture differed completely from those in the normal mouse culture (Hirayoshi et al. 1987). These results suggest that the defect of scid mouse is one that affects the process of immunoglobulin gene rearrangement. This was further confirmed in the DNA sequence analysis of the individual JH rearrangement events in this culture, which demonstrated that almost all the JH gene rearrangement events occurring in LTBC were aberrant, accompanying deletion of the genes around the recombination site, often so large as to delete all JH genes (See Sakano et al. this volume; Okazaki et al. 1988). This result indicates that the stage of B lineage cells in the LTBC of scid mouse are arrested at the pre-pre-B stage, namely the B cells with DJ/DJ Ig gene configuration, and is consistent with the results on Abelson virus transformed pre-B cell lines from scid mouse (Schuller et al. 1986; Hendrickson et al. 1988; Malynn et al. 1988). Thus, the absence of lymphocytes in scid mouse is very likely to be due to the low frequency of productive Ig gene and T cell receptor gene rearrangements. One question which remained was why virtually no B220(+) cells exist in scid bone marrow, although pre-B cells with aberrant Ig gene configurations can be generated in LTBC.

B cell development on the cloned stromal cell lines

Although B cell development occurs in the primary W-LTBC or switch culture, the problem inherent to this system is that there exist so many cellular components within the culture that it is almost impossible to dissect each component individually. To solve this problem, attempts were made to isolate the stromal cells which support B lymphopoiesis in W-LTBC, and several stromal cell clones, including our ST2, have been reported (Collins and Dorshkind 1987; Hunt et al. 1987; Whitlock et al. 1987; Ogawa et al. 1988). At present we are using two stromal cell clones, ST2 and PA6, which differ in their ability to support B lymphopoiesis (Kodama et al. 1984, Nishikawa et al. 1988). When bone marrow cells are cultured on the PA6 layer, active hemopoiesis begins within a week and production of myeloid cells including immature promyelocytes is maintained. B lineage cells, however, have never emerged on the PA6 cell layer. In contrast, B lineage cells are the major cell population generated on the ST2 layer. Because the cells maintained on the PA6 layer can give rise to mature B cells when transferred onto the ST2 layer, PA6 is a stromal cell clone which supports the growth of early B cell progenitors but lacks the signals required for the further maturation step into B220(+) pre-B cells. This means that the process of B cell development is controllable by using these two stromal cell clones with different supporting capacities.

What is the molecular basis for this functional difference between PA6 and ST2 ? During the course of our investigation, Namen et al reported a stromal cell derived pre-B cell growth factor, lymphopoietin 1 / interleukin 7 (IL7) (Namen et al. 1988 a, b, Lee et al. 1989). This report prompted us to test the possibility that IL7 is actually the molecule which determines the stromal cell ability to support B lymphopoiesis. We constructed a cDNA library of ST2 which was stimulated with lipopolysaccharide (LPS) and 4-phorbol-12-myristate-13-acetate (TPA), and screened it with a synthetic oligonucleotide probe prepared according to the sequence of the IL7 coding region. 4 recombinant colonies out of 160,000 were isolated, and the longest clone, CDMmIL7, which had a 2.6Kb insert, was further characterized. This clone is identical with that reported by Namen et al. except for its 1Kb longer 3' non-coding region which corresponds to the dominantly expressed mRNA species in the stromal cell clone. Therefore, our ST2 clone was shown to produce IL7. In order to obtain recombinant IL7, we recloned the IL7 coding region into the expression vector pCDL-SRa (Yokota et al. 1987), and the resulting clone was transfected into COS1 cells. The supernatant of the cells had the ability to induce the proliferation of the ST2 dependent pre-B cell line DW34, and to stimulate normal bone marrow cells to form colonies consisting of B220(+) B lineage cells. These two tools, IL7 gene and recombinant form of IL7, finally enabled us to investigate the possibility that IL7 is the factor determining the ability of stromal cells to support B lymphopoiesis. If IL7 is the one and only necessary factor, it is predicted that PA6 would be a nonproducer of IL7, and addition of recombinant IL7 into the culture of PA6 would render this in vitro microenvironment B lymphopoietic. Northern blot analysis with the IL7 probe demonstrated that ST2 produced IL7 upon stimulation with LPS/TPA or IL1, while the same stimuli could not induce IL7 in PA6. Thus, stromal cell ability to support B lymphopoiesis correlates very well with the production of IL7. Next, we prepared the cell population which was cultured on the PA6 layer for two weeks in order to remove mature B lineage cells. This cell population was then cultured in various in vitro environments: 1) IL7 alone, 2) PA6 alone, 3) PA6 + IL7, or 4) ST2. 3 weeks after the secondary culture, the cells were harvested, counted and analyzed for the surface antigens. Although previous studies demonstrated that IL7 per se can induce pre-B cell proliferation, the bone marrow cells precultured on the PA6 layer did not proliferate in response to IL7 alone. In contrast, PA6 + IL7 was effective in inducing active cell proliferation, and the cells generated were B220(+) B lineage cells including mature sIgM(+) B cells. The proportion of these cells is very similar to those generated on the ST2 layer. This result demonstrates that IL7 is an essential molecule for the stromal cell dependent B cell development without which the process of B cell differentiation does not proceed. In addition, the following two important implications came from this experiment. First, there should be a stage of B cell differentiation which requires both IL7 and unknown stromal cell molecule(s) for proliferation. Secondly, once B cell differentiation is induced by IL7, the maturation into sIgM(+) B cells occurs spontaneously through proliferation. Previous studies (Lee et al. 1989) and our unpublished observations indicate that some B220(+)sIgM(-) pre-B cells proliferate in response to IL7 without the aid of other molecules and differentiate into mature sIgM(+) cells.

Taking all these observations together, we propose a model for the stromal cell dependent process of B cell development (Fig. 1). In this model, the

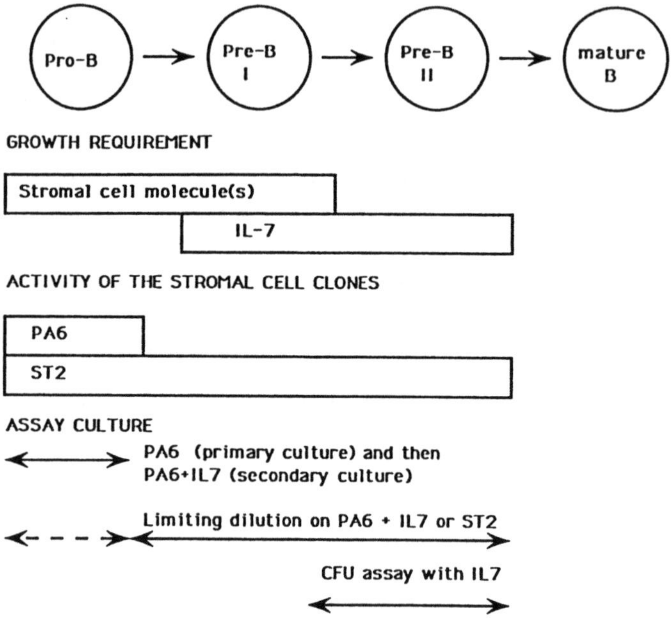

Figure 1 A model for stromal dependent B cell development

process of intramarrow B cell development is subdivided into three stages, the first pro–B cell stage which is independent of IL7, the second immature pre–B cell stage which requires both IL7 and other stromal cell molecule(s) present in PA6, and the final mature pre–B stage which proliferates in response to IL7 alone. From this model, it is predicted that cultures with PA6 alone, IL7 alone, and PA6+IL7 should support pro–B cell, mature pre–B cell, and all stages from pro–B to mature pre–B cells respectively. When bone marrow cells precultured on the PA6 layer are used for the starting cell population, subsequent transfer to PA6+IL7 is able to pick up only pro–B cells, because the cells at the later stages are eliminated during the preculture.

The frequency of the cells detected by the culture with PA6 alone, PA6+IL7 and IL7 alone in normal bone marrow cells were approximately 1/600–1500, 1/250–800, and 1/500–1000 respectively. The frequency of pro–B cells in the cells precultured on the PA6 layer is about 1/500–1000 and virtually no CFU–IL7 is present there. The problem involved in these frequency figures is that our assay might have covered only a minor population of bone marrow B220(+) cells. The previous reports demonstrated that the proportion of B220(+)sIgM(–) cells in the normal bone marrow is 10–20%. This means that the cells detectable in our system are at most 1–2% of this cell population. The rest of the B220(+) cells may belong to different cell lineages requiring other as yet unknown signals for their growth. Alternatively, it is also possible that majority of the cells generated on the stromal cell lose the ability to proliferate after the exposure to IL7. To test this possibility, bone marrow cells precultured on the PA6 layer were transferred to secondary culture with PA6 +IL7. Two weeks later, the majority of cells in the culture became B220(+). This cell population is regarded as B lineage cells generated *de novo* in response to the stromal cells and IL7. The proliferation capacity of these B220(+) cells was measured by limiting dilution analysis by using PA6+IL7. The frequency of proliferative cells was 1/20 – 40 rather than 1/1, suggesting that the majority of cells lose the proliferating ability very rapidly. If such loss of proliferative activity is also occurring at the same rate in the bone marrow, this result indicates that our assay may cover 20–80% of intramarrow pre–B cells.

Our model also predicts that the culture with PA6+IL7 contains B lineage cells of heterogeneous differentiation stages which may appear as a difference in the number of progeny from individual clonable B lineage cells. To test this the expansion size of the clonable B cells in culture with PA6+IL7 was measured on the 10th day of assay and compared with that of pro–B cells in culture with PA6 alone. Every single clonable B cell in the culture with PA6 alone could give rise to 50,000–200,000 B220(+) cells in 10 days. On the other hand, the expansion size of the clones already exposed to IL7 varied greatly from 300 to 50,000. This result indicates that the B lineage cells in culture with PA6 + IL7 are at heterogeneous differentiation stages, while the B lineage cells in culture with PA6 alone are homogeneous cells with high expansion potential. In conclusion, we think that the stromal cell dependent culture system represents the major pathway of B lineage cells in normal bone marrow, and the assay system described above is a reasonable system for estimating each stage of B lineage cells separately.

Defect of scid mouse affects early stage of B cell differentiation

The previous studies with W–LTBC and switch culture showed that pre–B cells are generated in LTBC. This result does not necessarily mean, however, that stromal cell dependent differentiation from pro–B to pre–B cells is normal in scid mouse because these culture systems are not quantitative. Moreover, it has been claimed that the cells in LTBC are under strong selective pressure preferential to the cells with high proliferating capacity, particularly when the culture proceeds for more than 3 weeks (Whitlock et al. 1984; Hirayoshi et al. 1987). Therefore, the number of cells generated does not properly represent the frequency of clonogenic B cells in the culture. On the other hand, our limiting dilution assay described above is a quantitative analysis of clonogenic B cells at various differentiation stages, and duration of the assay is short as 10–14 days.

According to the previous analysis on JH gene rearrangement in scid mouse, the disorder of scid mouse appears as soon as Ig gene rearrangement starts. In our assay system, this stage corresponds to the point when differentiation of pro–B cells maintained on PA6 is induced by the addition of IL7. Therefore, as the pro–B cell source, we used bone marrow cells precultured on PA6 for two weeks, and measured the frequency of cells clonable with PA6+IL7. An important advantage of this cell source over fresh bone marrow cells is that it allows us to exclude the possible presence of leaky mature B cells (Bosma et al. 1983,1988) as they can be removed during the preculture. As shown in Table 1, the frequency of PA6+IL7 clonable B cells in scid mouse culture was 1/5 of that in normal mouse culture. In contrast, when the primary culture was with PA6+IL7, almost no difference was observed between scid and normal mouse. Interestingly, the expansion size of the clonogenic pro–B cells of scid

Table 1. Frequency of clonogenic B cells in the culture of scid mouse

1° culture	Mouse	Limiting dilution PA6+IL7 (Ave. expansion size)	CFU-IL7(/10^5)
PA6	normal	1/630 (149000)	0
	scid	1/3500 (24000)	0
PA6+IL7	normal	1/35 (5300)	21.5
	scid	1/22 (7400)	0

mouse was lower than that of normal mouse although the expansion size of clonogenic pre-B cells (the cells harvested from the culture with PA6+IL7) did not differ between the two mice. Because wells containing more than 300 B220(+) cells were regarded as positive in our assay, our result indicates that 4/5 of pro-B cells in scid mouse culture lose proliferating activity before giving rise to more than 300 progeny (approximately 8 divisions). On the other hand, once the cells have passed through this phase, they seem to become more stable. These cells may be those selected in the classical LTBC. The figures obtained in the this experiment seem to be able to explain the absence of B lineage cells in scid bone marrow. Namely, only 1/5 of pro-B cells pass through the first phase and the expansion capacity is less than 1/5 of normal. Therefore, B220(+) cells in scid bone marrow should be no more than 1/25 of those in normal bone marrow. Assuming that the proportion of B220(+)sIgM(-) cells in normal bone marrow is 20%, those in the bone marrow of scid mouse are calculated to be 0.8%. We notice that this value needs further correction by several parameters. First, previous studies showed that almost all pre-B cells in the LTBC of scid mouse which might have survived through the first phase are those which have undergone non-productive Ig gene rearrangement. Sooner or later these useless B cells are eliminated from the bone marrow, although the mechanisms underlying such a quality control for useless B cells remain to be investigated. Secondly, it is still difficult to exclude the argument that the artificial in vitro microenvironment stimulates the cells to proliferate more than normal. In fact, it is reasonable to assume that in contrast to the culture system, the number of stromal cell is limited in normal bone marrow, and it is very likely that this limitation plays an important role in determining the total amount of daily hemopoiesis. At present too little is understood to make the proper correction of the figures obtained in our culture system. To this end, we have to start to look at the actual bone marrow environment in the light of our understanding obtained in an in vitro system.

Concluding remarks

The long term B cell culture system first developed by Whitlock and Witte has now developed into a quantitative and reliable method for studying the intramarrow process of B cell development in vitro. This progress is due to the establishment of various stromal cell clones and cloning of IL7 genes.

Other molecules playing a role in this process are expected to be revealed in the near future, which will certainly bring the in vitro system closer to the actual process, and allow us to achieve fine control of the differentiation of B cells. On the other hand, it is now certain that the contribution of scid mouse as an ideal host for cells and genes will increase. When scid mice carrying various transgenes are developed (as they are being,see other contributions in this volume), we expect that such a culture system as described here will be the best choice for undertaking the in vitro analysis.

References

Bosma,GC, Custer,RP, Bosma,MJ (1983) A severe combined immunodeficiency in the mouse. Nature 301: 527–530

Bosma,GC, Fried M, Carroll A, Gibson DM, Bosma MJ (1988) Evidence of functional lymphocytes in some(leaky) scid mice. J Exp Med 167: 1016–1033

Collins LS, Dorshkind K (1987) A stromal cell line from myeloid long-term bone marrow cultures can support myelopoiesis and B lymphopoiesis. J Immunol 138: 1092–1087

Dexter TM, Allen TD, Lajtha LG (1977) Conditions controlling the proliferation of haematopoietic stem cells in vitro. J Cell Physiol 91: 335–344

Dorshkind K (1986) In vitro differentiation of B lymphocytes from primitive hematopoietic precursors present in long-term bone marrow cultures. J Immunol 136: 422–429

Hendrickson EA, Schatz DG, Weaver DT (1988) The scid gene encodes a transacting factor that mediates the rejoining event if Ig gene rearrangement. Genes and Development 2: 813–829

Hirayoshi K, Nishikawa SI, Kina T, Hatanaka, M, Habu S, Nomura T, Katsura Y (1987) Immunoglobulin heavy chain gene diversification in the long-term bone marrow culture of normal mice and mice with severe combined immunodeficiency. Eur J Immunol 17: 1051–1057

Hunt P, Robertson D, Weiss D, Rennick D, Lee F, Witte ON (1987) A single bone marrow-derived stromal cell type supports the in vitro growth of early lymphoid and myeloid cells. Cell 48: 997–1007

Kodama H, Sudo H, Koyama H, Kasai S, Yamamoto S (1984) In vitro hemopoiesis within a microenvironment created by MC3T3-G2/PA6. J Cell Physiol 118: 233–240

Lee G, Namen AE, Gillis S, Kincade PW (1989) Normal B cell precursors responsive to recombinant murine interleukin 7 (IL-7) and factors inhibitory to IL-7 activity. J Immunol (in press)

Malynn BA, Blackwell TK, Fulop GM Rathbun GA, Furley AJW, Ferrier P, Heinke LB, Phillips RA, Yancopoulos GD, Alt FW (1988) The scid defect affects the final step of the immunoglobulin VDJ recombinase mechanism. Cell 54: 453–460

Namen AE, Schmierer AE, March CJ, Overell RW, Park LS, Urdal DL, Mochizuki DY (1988a) B cell precursor growth promoting activity. Purification and characterization of a growth factor active on lymphocyte precursors. J Exp Med 167: 988–1002

Namen AE, Lupton S, Hjerrild K, Wangnall J, Mochizuki DY, Schmiere A, Mosley B, March CJ, Urdal D, Gillis S, Cosman D, Goodwin RG (1988b) Stimulation of B–cell progenitors by cloned muorine interleukin–7. Nature 333: 571–573

Nishikawa SI, Ogawa M, Nishikawa S, Kunisada T, Kodama H (1988) B lymphopoiesis on stromal cell clone: stromal cell clones acting on different stage of B cell differentiation. Eur J Immunol 18: 1767–1771

Ogawa M, Nishikawa S, Ikuta K, Yamamura F, Naito M, Takahashi K, Nishikawa SI (1988) B cell ontogeny in murine embryo studied by a culture system with the monolayer of a stromal cell clone, ST2: B cell progenitor develops first in the embryonal body rather than in the yolk sac. EMBO J 7: 1337–1343

Okazaki K, Nishikawa SI, Sakano H (1988) Abberant immunoglobulin gene rearrangement in scid mouse bone marrow cells. J Immunol 141: 1348–1352

Schuller W, Weiler IJ, Schuller A, Phillips RA, Rosenberg N, Mak TW, Kearney JF, Perry RP, Bosma MJ (1986) Rearrangement of antigen receptor genes is defective in mice with severe combined immune deficiency. Cell 46: 963–972

Whitlock CA, Witte ON (1982) Long–term culture of B lymphocytes and their precursors from murine bone marrow. Proc Natl Acad Sci USA 79: 3608–3612

Whitlock CA, Robertson D, Witte ON (1984) Murine B cell lymphopoiesis in long–term culture. J Immunol Methods 16: 353–369

Whitlock CA, Tidmarsh GF, Muller–Sieburg C, Weissman IL (1987) Bone marrow stromal cell lines with lymphopoietic activity express high levels of a pre–B neoplasia–associated molecule. Cell 48: 1009–1021

Witte PL, Burrows PD, Kincade PW, Cooper MD (1987) Characterizaion of B lymphocyte lineage progenitor cells from mice with severe combined immunedeficiency disease (SCID) made passible by long term culture. J Immunol 138: 2698–2705

Yokota T, Coffman RL, Hagiwara H, Rennick DM, Takebe Y, Yokota K, Gemmell L, Shrader B, Yang G, Meyerson P, Luh J, Hoy P, Pene J, Briere F, Spits H, Banchereau J, de Vries J, Lee FD, Arai N, Arai KI (1987) Isolation and characterization of lymphokine cDNA clones encoding mouse and human IgA–enhancing factor and eosinophil colony–stimulating factor activities: Relationship to interleukin 5. Proc Natl Acad Sci USA 84: 7388–7392

Natural Killer Cells and Their Precursors in Mice With Severe Combined Immunodeficiency

Vinay Kumar, J. Hackett, Jr., M. M. Tutt, B. A. Garni-Wagner, W. A. Kuziel, P. W. Tucker, and M. Bennett[1].

Department of Pathology, University of Texas Southwestern Medical Center, 5323 Harry Hines Blvd., Dallas, Texas 75235, USA.

Natural Killer (NK) cells are capable of lysing tumor cells and virally infected cells and of regulating immune responses in the absence of apparent sensitization (Trinchieri, 1989). The precise relationship of NK cells to other lympho-hemopoietic cells is uncertain. Previous studies have failed to provide definitive evidence of NK cell lineage, since NK activity has been ascribed to a variety of cell types, including pre-T cells, basophils, and promonocytes. Indeed it could be argued that NK cells do not represent a homogeneous population with a distinct lineage. However, several recent studies have documented that NK killing by freshly isolated murine spleen cells is mediated by a morphologically homogeneous population (Hackett, et al. 1986a) that has a characteristic phenotypic profile. Murine NK cells are NK-1.1$^+$/NK2.1$^+$, Qa-5$^+$, Thy-1±, AsGM-1$^+$, CD4$^-$, CD8$^-$, CD3$^-$, J11d$^-$, and IL-2R alpha (p55)$^-$.

To address the issue of NK cell lineage we examined the characteristics of NK cells and their progenitors in mice with severe combined immunodeficiency (scid). As described in detail elsewhere in this volume, mice with the scid mutation are severely deficient in mature B or T cells since the scid mutation impairs immunoglobulin (Ig) and T cell receptor (TCR) gene rearrangements (Schuler, et al. 1986). Three approaches were utilized in our studies: (i) bone marrow cells of scid and +/+ mice were examined for the content of transplantable NK-progenitor cells; (ii) NK cells were purified from spleens of scid mice and their functional, phenotypic, and molecular characteristics were investigated; (iii) the thymus of scid mice was analyzed for the presence of NK cells or their progenitors. These studies are summarized below.

TRANSPLANTABLE PROGENITORS OF NK CELLS IN BMC OF SCID MICE

The NK progenitor cells contained in the bone marrow of C.B-17 scid or +/+ mice were detected by their ability to generate mature NK cells, following transfer of bmc into NK-depleted and lethally irradiated mice, as previously described (Hackett, et al. 1985). Regeneration of NK activity in the recipient mice was monitored by two different assays: the ability to rapidly clear infused YAC-1 tumor cells *in vivo* and the ability of spleen cells to lyse YAC-1 cells *in vitro*. One such experiment is depicted in Table 1. In this experiment, B6D2F1 mice were depleted of host NK cells by injection with rabbit anti-Asialo GM1 serum (day -2), irradiated (900R) on day 0 and infused with two doses of C.B-17 scid or C.B-17 +/+ bmc. Thirteen days after cell transfer, the extent of NK activity generated in various

[1]Supported by grants AI-20451, CA-36921, CA-36922, and CA-09082 from the National Institutes of Health, USA, and by grant 3128 from Tobacco Council for Research USA, Inc.

Table 1

TRANSPLANTABLE NK PROGENITORS IN BONE MARROW CELLS OF C.B-17 SCID MICE*

Bone marrow cells transferred	% ^{125}IUdR retained in lungs Geom. mean (95% C.L.)
None	21.8 (17.5 - 27.1)
1 x 10^6 C.B-17 scid/scid	10.0 (6.5 - 15.5)
5 x 10^6 C.B-17 scid/scid	1.9 (1.0 - 3.4)
1 x 10^6 C.B-17 +/+	8.3 (5.8 - 12.0)
5 x 10^6 C.B-17 +/+	1.4 (0.8 - 2.4)

*(C57BL/6 x DBA/2)F1 mice were pretreated with anti-Asialo GM1 serum, lethally irradiated, and injected with no bmc, 1 or 5 x 10^6 scid/scid or +/+ bmc. Thirteen days after cell transfer the mice were injected with ^{125}IUdR labelled YAC-1 cells; their lungs were removed 4 hours after tumor cell challenge.

groups was assessed by injecting the recipient mice with 1 x 10^6 ^{125}I-labelled YAC-1 cells. Four hours after challenge with tumor cells, the lungs were removed and the ^{125}I radioactivity retained in the lungs was measured. The amount of radioactivity remaining in the lungs is an index of YAC-1 cell survival, and therefore NK activity *in vivo*. Higher retention of ^{125}IUdR labelled YAC-1 cells indicates greater survival of tumor cells and hence low NK cell function. As noted in Table 1, recipients of C.B-17 scid/scid or C.B-17 +/+ bmc generated equivalent and dose-dependent NK activity. In parallel experiments other recipient mice were monitored for thymus repopulation, and generation of mitogen-responsive splenic T and B cells. As expected neither thymocytes nor mature T and B cells were generated from scid bmc. To examine whether the NK cells generated from scid bmc were similar to NK cells in normal mice, we examined cell surface markers on NK cells generated in the spleens of recipient mice by antibody plus complement depletion assays. Splenic NK cells derived from scid bmc were phenotypically indistinguishable from those derived from +/+ bmc, i.e. they were NK-2.1$^+$, Asialo GM1$^+$ (Hackett, et al 1986b). Together these results prove that the transplantable NK progenitor cells are not defective or deficient in C.B-17 *scid* mice and argue in favor of a non-T, non-B lineage of NK cells.

ANALYSIS OF NK CELLS IN THE SPLEENS OF C.B-17 SCID MICE

Although the experiments described above strongly suggested that NK cells are quite distinct from T cells, other reports seemed to indicate that NK cells belong to the T lineage. The latter is based on the observations that certain cloned cell lines with NK activity rearrange and express T cell receptor (TCR) genes (Yanagi, et al. 1985; Ritz, et al. 1985). To resolve this discrepancy we examined whether TCR genes are rearranged and expressed in short term cultures of highly purified NK cells. NK cells were obtained from the spleens of scid mice to avoid possible contamination with T cells. Purification of NK cells was accomplished by sorting NK-2.1$^+$ cells. Approximately 40-60% of lymphoid cells in the spleens of scid mice expressed the NK specific marker NK-2.1 and all the NK activity of the scid spleen cells was contained in the sorted NK-2.1$^+$ population. These cells proliferated in response to human recombinant IL-2 (500U/ml) and were maintained in culture for 2-3 weeks. Cultured NK-2.1$^+$ cells were highly lytic, but the pattern of lysis was similar to that of fresh NK cells. In addition they displayed a phenotype typical of freshly isolated NK cells (Asialo $GM1^+$, $CD3^-$, $CD4^-$, $CD8^-$). Analysis of the DNA and RNA extracted from NK cells failed to show any rearrangment of the TCR-gamma or TCR-beta loci (Tutt, et al. 1987), and there were no transcripts of the CD3 epsilon, gamma, or delta chain genes. These data lead to the conclusion that the expression of TCR is not necessary for functional NK activity, and NK cells are distinct from both mature cytotoxic T cells and the earliest identifiable cells of the T lineage.

Table 2

PROLIFERATION AND ACTIVATION OF SCID AND +/+ NK CELLS*

Factor	Proliferation	Lytic Activation
ConA	-	-
IL-2	+++	+++
IL-1 alpha	-	-
IFN-alpha/beta	-	++
IFN-gamma	-	-
TNF-alpha	-	-
IL-4	-	-
IL-4 + IL-2	Suppressed	Suppressed
GM-CSF	-	-

*$NK\text{-}2.1^+$ cells isolated from spleens of scid mice or $NK\text{-}1.1^+$ cells sorted from spleens of B6D2F1 mice were placed in culture with several concentrations of the indicated factors. To assess proliferation, cells were incubated for 3-5 days prior to labelling with ^{3}H-thymidine. Activation was assessed by testing lytic activity against YAC-1 cells after a 2-4 hour preincubation. (-) no effect; (+) activation of lysis or induction of proliferation.

It could be argued, however, that in normal mice NK activity is mediated by both T cells and non-T cells, and scid mice are lacking the former (Lauzon, et al. 1986). This seems unlikely for several reasons. First, in normal mice, virtually all of the NK activity of freshly obtained spleen cells is mediated by $NK\text{-}1.1^+/NK\text{-}2.1^+$, $CD4^-$, $CD8^-$, $CD3^-$ cells. Molecular analysis of such cells also reveals TCR in germline configuration (Tutt, et al. 1986) and absence of CD3 transcripts (Biron, et al. 1987). Thus even in mice with unimpaired T cell differentiation, little, if any NK activity of freshly isolated spleen cells can be attributed to T cells. Second, the response of NK cells isolated from scid mice and +/+ mice to a variety of growth and activation factors is indistinguishable (Table 2). In particular, it should be noted that as in normal mice 30-50% of NK cells of scid mice also express Thy-1.2 (Hackett, et al. 1986b; Tutt, et al. 1987), contrary to an earlier report that scid mice lack $Thy\text{-}1.2^+$ NK cells (Dorshkind, et al. 1985). In unpublished experiments we have noted that $Thy\text{-}1.2^-$ NK cells sorted from spleens of scid mice rapidly acquire this marker upon activation with rIL-2. Thus Thy-1.2 appears to be an activation, rather than a lineage marker on NK cells.

In view of these findings and similar observations in humans a clear distinction should be made between NK cells and T cells which display an NK-like or non-MHC restricted cytotoxicity (Lanier, et al. 1986). It is now clear that there exists a small population of $CD3^+$ (TCR alpha/beta or TCR gamma/delta) cells that exhibits non-MHC-restricted cytotoxicity especially after growth and activation in IL-2 containing media. These cells account for only a small proportion of human peripheral blood NK activity, and are undetectable in freshly isolated murine spleens.

NK CELLS IN THE THYMUS OF SCID MICE

The data obtained by analyzing the spleen and bone marrow of scid mice argues against a relationship between T cells and NK cells. However, it cannot be ruled out that NK cells and T cells share a common (? bipotential) progenitor that diverges prior to TCR gene rearrangement.

We chose to look for the common NK/T cell progenitor in the $CD3^-$, $CD4^-$, $CD8^-$ thymocyte population known to contain T cell progenitors. Instead of selecting $CD3^-$, $CD4^-$, $CD8^-$ cells from normal mice we have used thymocytes from scid mice, which are an enriched source of $CD3^-$ double negative cells.

Table 3

PHENOTYPE OF THE NON-MHC-RESTRICTED CYTOTOXIC CELLS IN THE SCID THYMUS*

Source of scid cells	% Lysis of YAC-1 cells at E:T of 25:1	
	Experiment I	Experiment II
Whole Thymus	16.6	7.1
Whole Spleen	21.0	ND
$IL\text{-}2R^+$ Thymocytes	0.5	ND
$IL\text{-}2R^-$ Thymocytes	58.8	ND
$J11d^+$ Thymocytes	ND	0.5
$J11d^-$ Thymocytes	ND	18.9

*Scid thymus cells were stained with either J11d or 7D4 mAB (anti-IL-2R alpha) and sorted by flow cytometry. Lysis of YAC-1 was determined on both sorted and whole unsorted populations after 1 hour incubation in 1000U/ml rIL-2 and 4000U/ml alpha/beta IFN.

Thymus in scid mice contains approximately $1\text{-}2\times10^6$ cells; these were analyzed by flow cytometry. As expected scid thymocytes were $CD3^-$, $CD4^-$, and largely $CD8^-$. However, 70-80% were IL-2R $alpha^+$, $J11d^+$, and $PgP\text{-}1^+$. These three markers are characteristic of fetal thymocytes and T cell progenitors within adult double negative thymocytes (Crispe, et al. 1987). We next determined whether freshly isolated scid thymus cells were cytotoxic against NK-sensitive tumor YAC-1(Table 3). Freshly isolated thymocytes from scid mice lysed YAC-1 cells. Furthermore, by cell sorting all the cytotoxic activity was found within IL-2R $alpha^-$, $J11d^-$ subpopulation that forms 20-25% of the thymocytes. Thus, the scid thymus seems to contain at least two cell populations: 1) $CD3^-$, $CD4^-$, $CD8^-$, IL-2R $alpha^+$, $J11d^+$, $PgP\text{-}1^+$ that is non-lytic and resembles fetal thymus cells, 2) $CD3^-$, $CD4^-$, $CD8^-$, IL-2R $alpha^-$, $J11d^-$ that lyses YAC-1 and resembles mature NK cells. This distinction is further supported by the response of $J11d^+$ and $J11d^-$ scid thymocytes to various growth stimuli. $J11d^+$ cells, like fetal thymocytes, failed to respond to rIL-2 alone despite the fact that they are IL-2R $alpha^+$. They do respond to rIL-2 in the presence of PMA and Ionomycin. In contrast, J11d$^-$ cells behaved like mature NK cells by responding to high amounts (500 U/ml) of rIL-2 alone.

NK cells found in the scid thymus could have been derived from their precursors in the bone marrow or from a common NK/T progenitor within the J11d$^+$ thymocyte population. To investigate this possibility J11d$^+$ thymocytes were cultured in 40U/ml rIL-2 with PMA and Ionomycin. J11d$^-$ cells containing mature NK cells were cultured in 500U/ml of rIL-2 only. After 8-10 days in culture both groups were tested for cytotoxic activity and cell surface antigens. As expected J11d$^-$ cells cultured in IL-2 were highly cytotoxic; in addition, J11d$^+$ cells also acquired cytotoxic activity, although to a lower extent. Following culture J11d$^+$ cells became largely J11d$^-$ and IL-2R alpha$^-$, but they remained CD3$^-$, CD4$^-$, and mostly CD8$^-$. Thus these cells acquired some functional and phenotypic similarities to NK cells. The cultured J11d$^-$ cells showed the typical antigenic profile of mature NK cells.

Although cultured J11d$^+$ cells acquired non-MHC-restricted cytotoxic activity and some phenotypic attributes of NK cells, we wished to determine whether they resembled NK cells or non-MHC-restricted CTL. mRNA extracted from cultured J11d$^+$, J11d$^-$ thymocytes, and mature splenic NK cells was probed for expression of CD3 and TCR genes. The J11d$^+$ cells transcribed abundant message for the gamma, delta, and epsilon chains of CD3, a feature that clearly distinguishes them from J11d$^-$ cells and mature splenic NK cells (Biron, et al. 1987). It appears therefore that most J11d$^+$ cells belong to the T cell lineage. However, they do not express functional transcripts of TCR genes since the scid mutation prevents normal rearrangement of TCR genes. While the J11d$^+$ cells do transcribe CD3 genes, the products of these genes are not expressed on the surface in the absence of functional TCR gene transcripts.

These data suggest that the majority of J11d$^+$ cells in the scid thymus are committed to the T lineage and acquire non-MHC-restricted cytotoxicity even though they cannot successfully rearrange their TCR genes. However, we cannot rule out the possibility that a minority of J11d$^+$ cells may indeed be bipotential NK/T cell progenitors. If such cells exist it will require a study of cloned J11d$^+$ cells.

SUMMARY AND CONCLUSIONS

Our studies with scid mice have clarified the relationship between T cells and NK cells. C.B-17 scid mice have normal frequency of transplantable NK progenitors in their bone marrow which develop into fully functional NK cells. Spleens of scid mice contain mature NK cells which are phenotypically and functionally indistinguishable from NK cells found in normal mice. These cells retain their TCR genes in germline configuration and do not transcribe the CD3 genes. Thus, NK cells are distinct from the earliest identifiable cells committed to the T-lineage. In addition to the spleen, the thymus of scid mice also contains mature NK cells. These cells constitute a small proportion of the thymus cell population and can be clearly distinguished from the majority of cells, which have the phenotype and molecular characteristics of very early T-lineage cells. There is no evidence that NK cells within the thymus are derived *in situ* from a common NK/T precursor. Together these data support the hypothesis that NK cells form an independent lineage.

REFERENCES

Biron CA, Van Den Elsen P, Tutt MM, Medveczky P, Kumar V, Terhorst C (1987) T cell receptor gene expression by murine natural killer cells and T cells stimulated *in vivo*. J Immunol 139:1704-1710

Crispe IN, Moore MW, Husmann LA, Smith L, Bevan MJ, Shimonkevitz RP (1987) Differentiation potential of subsets of $CD4^-8^-$ thymocytes. Nature 329:336-339

Dorshkind K, Pollack SB, Bosma MJ, Phillips RJ (1985) Natural killer (NK) cells are present in mice with severe combined immunodeficiency (*scid*). J Immunol 134:3798-3901

Hackett J, Bennett M, Kumar V (1985) Origin and differentiation of natural killer cells I. Characteristics of a transplantable NK-Precursor. J Immunol 134:3731-3738

Hackett J, Tutt MM, Lipscomb M, Bennett M, Koo G, Kumar V (1986a) Origin and differentiation of natural killer cells II. Functional and morphologic studies of purified $NK\text{-}1.1^+$ cells. J Immunol 136:3124-3131

Hackett J, Bosma GC, Bosma MJ, Bennett M, Kumar V (1986b) Transplantable progenitors of natural killer cells are distinct from those of T and B lymphocytes. Proc Natl Acad Sci 83:3427-3431

Lanier LL, Phillips JH, Hackett J, Tutt MM, Kumar V (1986) Opinion. Natural Killer cells: definition of a cell type rather than a function. J Immunol 137:2735-2739

Lauzon RJ, Siminovitch KA, Fulop GM, Phillips RA, Roder JC (1986) An expanded population of natural killer cells in mice with severe combined immunodeficiency (*scid*) lack rearrangement and expression of T cell receptor genes. J Exp Med 164:1797-1802

Ritz J, Campen TJ, Schmidt RE, Royer HD, Hercend T, Hussey E, Reinherz EL (1985) Analysis of T-cell receptor gene rearrangement and expression in human natural killer clones. Science 228:1540-1544

Schuler W, Weiler IJ, Schuler A, Phillips RA, Rosenberg N, Mak TW, Kearney JF, Perry RP, Bosma MJ (1986) Rearrangement of antigen receptor genes is defective in mice with severe combined immune deficiency. Cell 46:963-972

Trinchieri G (1989) Biology of Natural Killer cells. Adv Immunol 47:(in press)

Tutt MM, Kuziel WA, Hackett J, Bennett M, Tucker PW, Kumar V (1986) Murine natural killer cells do not express functional transcripts of the alpha, beta, or gamma-chain genes of the T cell receptor. J Immunol 137:2998-3001

Tutt MM, Schuler W, Kuziel WA, Tucker PW, Bennett M, Bosma MJ, Kumar V (1987) T cell receptor genes do not rearrange or express functional transcripts in natural killer cells of SCID mice. J Immunol 138:2338-2344

Yanagi Y, Caccia N, Kronenberg M, Chin B, Roder J, Rohel D, Kiyohara T, Lauzon R, Toyonaya B, Rosenthal K, Dennert G, Acha-Orbea H, Hengartner H, Hood L, Mak TW (1985) Gene rearrangement in cells with natural killer activity and expression of the beta-chain of the T-cell antigen receptor. Nature 314:631-633

II. General Phenotype/Molecular Characterization

Nature of the *scid* Defect: A Defective VDJ Recombinase System

W. Schuler[1,2] and M.J. Bosma[2]

[1]Basel Institute for Immunology, Grenzacherstrasse 487
4058 Basel, Switzerland

1. Introduction

Lymphocyte antigen receptors include immunoglobulins (Ig) which consist of heavy (Igh) and light (Igl) chains, and T cell receptors (Tcr) which are composed of either α- and β-chains or of γ- and δ-chains. Early in lymphocyte development separate germline DNA segments, i.e., V, D and J or V and J, are assembled in recombinatorial fashion to form the genes that encode the variable, antigen-binding region of both Ig and Tcr proteins. (For reviews on this subject see: Tonegawa 1983; Kronenberg et al. 1986; Blackwell and Alt 1988; Davis 1988). It is thought that rearrangement of antigen-receptor genes proceeds in an ordered fashion although exceptions have been found. In general, D-to-J recombination occurs first followed by joining of a V gene segment to DJ. In B cell development, rearrangement of Igh genes preceeds that of Igl genes. Likewise, in developing T cells the Tcrδ genes are rearranged as early as or earlier than Tcrγ and Tcrβ genes and followed by Tcrα gene rearrangement.

Highly conserved heptamer-nonamer DNA sequence motifs flank V, D and J coding sequences and serve as recognition signals (signal sequences) for a site-specific recombination system (the so-called VDJ-recombinase system) which is utilized by both T and B cells (Yancopoulos et al. 1986). This recombinase system has not been characterized and the underlying mechanism of VDJ assembly is only partly understood. A multi-step mechanism has been proposed by Alt and Baltimore (1982) in which the first step is an endonucleolytic cut at the signal and coding sequence junctions of the two gene segments involved in recombination, e.g., of a given D and J. The signal sequences of each coding segment are subsequently fused back-to-back (signal joints) resulting in circularization and deletion of the intervening DNA. However, rearrangement by inversion has also been observed for Iglκ, Tcrδ and Tcrβ genes where the signal joints are retained on the chromosome. While signal joint formation is generally very precise, the ends of the coding regions can incur minor modifications by loss and/or addition of a few nucleotides before they are also ligated (coding joints). If the respective V(D)J-joining events at two different loci (i.e., Igh, and Igl, Tcrα and Tcrβ, Tcrγ and Tcrδ) are functional such that each gene segment is in the same translational reading frame and there are no in-phase nonsense codons, then synthesis, assembly, and expression of heterochain receptors follows. If a functional joining does not occur on either allele of a critical locus (e.g., Igh, Tcr), the affected lymphocyte will be non-functional for lack of an antigen-specific receptor.

[2]Institute for Cancer Research, Fox Chase Cancer Center
Philadelphia, PA 1911, USA

Current Topics in Microbiology and Immunology, Vol. 152

Inability to undergo *functional* antigen receptor gene rearrangement appears to be the explanation for the lack of T and B lymphocytes in scid mice. This article briefly reviews our observations that led, and give further support, to this idea (Schuler et al. 1986).

2. Status of antigen receptor genes in scid mice

2.1 Transcription of antigen-receptor genes

Before antigen-receptor genes undergo rearrangement they become transcriptionally active early in lymphocyte development and are transcribed in a lineage-specific fashion from germline configuration. For example, it has been shown that germline *Igl*-Cκ and *Igh*-V_H558 transcripts are exclusively found in B-lineage cells (Van Ness et al. 1981; Yancopoulos and Alt 1985) while germline *Tcrβ* transcription is restricted to T-lineage cells (Yanagi et al. 1984; Cook and Balaton 1987). These germline transcripts therefore appear to reflect early commitment of developing lymphocytes to the T or B cell lineage. They also might be a prerequisite for recombination or simply reflect an "opening" of the respective antigen-receptor genes which now become accessible to the VDJ-recombinase system (Van Ness et al. 1981; Yancopoulos and Alt 1985).

Since early attempts to demonstrate T and B lineage-committed cells in scid mice by either serological or functional assays were unsuccessful (Bosma et al. 1983; Dorshkind et al. 1984; Habu et al. 1987), we looked by means of Northern blot analysis for germline transcription of antigen receptor genes as evidence for early T- and B-lineage cells. We found in 17 and 18 day fetal liver, and in adult bone marrow, transcripts of the *Igh*-Cμ constant region gene, the *Igh*-V558 variable region gene family, and the *Igl*-Cκ constant region gene (Schuler et al. 1988). All of the transcripts found corresponded to germline genes; transcripts of rearranged *Ig* genes could not be detected. Likewise, in thymocytes of young adult scid mice, we were able to demonstrate a high level of *Tcrγ* and a low level of *Tcrβ* gene transcripts while *Tcrα* transcripts were undetectable. This transcription pattern is reminiscent of that of day 15 normal fetal thymocytes. Again, transcripts derived from fully assembled *Tcr* genes could not be detected.

The above results indicated that T and B lineage cells do indeed develop in scid hematopoietic tissues. They further suggested that developing scid lymphocytes reach a stage at which antigen receptor genes become accessible to the VDJ-recombinase system and normally undergo rearrangement.

2.2 Apparent absence of antigen-receptor gene rearrangement in DNA from lymphopoietic tissues of scid mice

Although developing scid lymphocytes seem to reach a stage where antigen-receptor gene rearrangement normally occurs, we could not directly demonstrate *Tcrγ* or *Tcrβ* gene rearrangement in

DNA from scid thymocytes (Schuler et al. 1986). In DNA from normal thymocytes we could easily detect, by means of Southern blot analysis, novel, non-germline restriction fragments hybridizing to Tcrγ or Tcrβ specific DNA probes. In contrast, the hybridization pattern obtained with DNA from scid thymocytes was indistinguishable from that obtained with liver (germline) DNA; i.e., both Tcrγ and Tcrβ loci appeared in germline configuration.

To address the question of rearrangement at the immunoglobulin heavy chain locus we generated hybridomas from fetal liver or adult bone marrow from both normal and scid mice. DNA of individual hybridomas was analyzed using Igh specific DNA probes. While we detected Igh gene rearrangement in 4 out of 15 hybridomas derived from normal mice, we could not detect Igh gene rearrangement in any of 40 hybridoma lines generated from scid fetal liver or bone marrow cells (Schuler et al. 1986). The above results were consistent with our inability to detect transcripts derived from fully assembled Tcr and Ig genes.

2.3 Abnormal antigen-receptor gene rearrangements in transformed lymphoid cells of scid mice

A quite different situation with regard to antigen-receptor gene rearrangement was found when we analyzed transformed lymphoid cells derived from scid mice. The material studied included spontaneous thymic lymphomas (see M. Bosma, Part I, this volume) as well as bone marrow cell lines transformed by Abelson-murine leukemia virus [A-MuLV] (Fulop et al. 1988).

Southern blot analysis of DNA from scid thymic lymphomas (Schuler et al. 1986) revealed rearrangement of the Tcrβ genes at both alleles in all twelve lymphomas studied. However, about 60% of the rearranged alleles had deleted the entire cluster of Jβ2 coding segments, including 3'-flanking regions, while retaining the Cβ2 constant region gene. Employing various restriction enzymes and Tcrβ specific probes we could demonstrate that Tcrβ alleles which had lost the Jβ2 elements contained DNA sequences upstream from Dβ1 or Dβ2 elements that were juxtaposed to the Cβ2 constant region gene. This suggested that the observed Jβ2 associated deletions in scid thymic lymphomas resulted from attempted site-specific Dβ-to-Jβ recombinations and not from recombination to DNA sequences from different chromosomes, or from viral integration. Deletion of Jβ2 is normally not found in T lymphocytes (Kronenberg et al. 1985). Therefore, the observed Jβ2 associated deletions were regarded as abnormal.

We recently confirmed and extended the above results by cloning and DNA sequence analysis of the recombination sites of rearranged Tcrβ alleles from scid thymic lymphomas (Schuler, Amsler, Bosma; unpublished results). We found instances in which attempted D-to-J recombination resulted in the deletion of either both or only one coding segment. In the latter case, recombination appeared normal with respect to one partner in that one coding segment (either Dβ or Jβ2) was properly cut and modified in normal fashion at the coding region/signal sequence junction and was ligated to non-coding DNA sequences in the flanking region of the other coding segment. This type of defective rearrangement could not be detected by Southern blot analysis. Therefore, the frequency of defective D-to-J recombinations at the Tcrβ locus is even higher than previously indicated (Schuler et al. 1986). Though our analysis did not reveal a common DNA sequence at the abnormal recombination sites, we noted in several instances isolated heptameric sequences which matched the consensus heptamer sequence of the classical heptamer-nonamer recognition signal.

Defective D-to-J recombination was also observed in A-MuLV transformed B lineage cells from scid mice (Schuler et al. 1986). Southern blot analysis of DNA from nine scid derived A-MuLV transformed cell lines showed rearrangement of both Igh alleles in all cell lines. But, similar to the situation found for transformed scid T lineage cells, about 80% of the Igh alleles had deleted the entire J_H gene cluster. Again we could show by Southern blotting (Schuler et al. 1986) and confirm by DNA cloning and sequencing (Kim et al. 1988) that these J_H associated deletions resulted from attempted D_H-to-J_H recombinations causing the deletion of the respective coding segments including large stretches of flanking DNA. J_H-associated deletions accompanying D_H-to-J_H recombinations were not observed in A-MuLV transformed cell lines derived from normal mice. Such deletions are clearly abnormal since they render the Igh genes non-functional. Our results were recently confirmed and extended in several laboratories (Hendrickson et al. 1988; Malynn et al. 1988; Okazaki et al. 1988). These authors also demonstrated that, as discussed above for Tcrβ gene rearrangement, attempted D_H-to-J_H rearrangement can be normal with respect to one or the other coding segment. Rearrangements of Igh variable region genes or of Igl genes have not been detected in most transformed scid B lineage cells studied (Schuler et al. 1986; Kim et al. 1988; Hendrickson et al. 1988; Okazaki et al. 1988).

Not only is D-to-J recombination defective in transformed scid lymphoid cells but also V-to-J rearrangement. This is evident from studies of the Igl locus in A-MuLV transformed cell lines (Blackwell et al., Part II, this volume) and from our recent studies of Tcrγ genes in scid thymic lymphomas. Tcrγ genes lack D elements. Rearrangement of these genes is very restricted; normally, a limited number of novel EcoRI fragments ranging in size from 14 to 18kb are generated which hybridize to both variable and constant region gene probes. In contrast, when EcoRI digested DNA from scid thymic lymphomas was probed with a probe specific for Tcrγ constant region genes (Cγ) the majority of novel Cγ-positive EcoRI fragments were considerably shorter than 14 kb. Reprobing the same Southern blot filters showed that most of the rearranged Cγ-positive restriction fragments failed to hybridize to probes specific for Tcrγ variable (Vγ) and/or J region (Jγ) gene segments. These results suggest that in the scid thymic lymphomas Vγ and/or Jγ gene segments are deleted upon attempted V-to-J recombination at the Tcrγ loci, explaining the unusually short Cγ-positive restriction fragments. This interpretation has been confirmed by cloning and sequencing rearranged Tcrγ genes from the scid thymic lymphomas (Schuler W, Schuler A, Bosma; manuscript in preparation).

3. Hypothesis

To account for the preceeding results we proposed that a component of the VDJ-recombinase system, common to both T and B lineage cells, might be altered or missing in scid mice (Schuler et al. 1986). As a result, developing scid lymphocytes are generally unable to make functional VDJ rearrangements and therefore fail to express an antigen receptor. We assume that early lymphocytes with abnormal Ig, Tcrβ or Tcrγ rearrangements are prematurely and rapidly eliminated from scid lymphopoietic tissues. This would explain why the detection of such cells has depended on their transformation and clonal expansion in vivo or in vitro (Schuler et al. 1986) or on their selective outgrowth in long term bone marrow cultures (Hirayoshi et al. 1987; Witte et al. 1987).

4. Is the scid mutation leaky?

By genomic cloning or gene amplification (polymerase chain reaction) and subsequent DNA-sequence analysis we have recently identified five scid thymic lymphoma lines among the twelve originally described (Schuler et al. 1986) that in addition to the typical defective "scid type" rearrangements have undergone normal rearrangements (Schuler, Amsler, Bosma; manuscript in preparation). In one particular line, for instance, in which all rearranged Tcrβ and Tcrγ alleles have been cloned and sequenced, we found two perfectly normal V-to-J recombinations; the remaining six recombinations showed the typical "scid type" deletions of one or both coding elements. One explanation for these observations is that the scid mutation itself is leaky, i.e., due to residual activity of the affected protein, normal rearrangements are still possible, although at a low frequency. A residual enzymatic activity could be best explained by a point mutation that alters the wildtype scid protein. Whether this low frequency of normal rearrangements can explain the occurrence of functional lymphocytes in some scid mice ("leaky" mice, see Part IV, this volume) is an open question since it takes three independent functional rearrangements to generate the genes encoding a functional receptor protein (D-to-J followed by V-to-DJ on one chromosome and V-to-J on the other chromosome). The occurrence of functional scid lymphocytes might therefore be explained by normalization of the scid recombinase activity in early lymphocyte precursor cells (Carroll et al., Part IV, this volume). Thus, an alternative explanation for the occurrence of both normal and defective rearrangements in a given lymphocyte would be normalization of the scid recombinase activity during the development of this particular cell, after several unsuccessful attempts to undergo functional rearrangement. In an attempt to distinguish between these alternatives, pre-T cell lines harboring both conventional and defective "scid-type" rearrangements will be transfected with extrachromosomal recombination substrates (Lieber et al. 1988). This should enable us to ascertain whether the VDJ recombinase activity in these lines is normal or abnormal.

Acknowledgements

Our work reviewed here was supported by NIH grants AT-13323, CA-04946 and CA-06927, by an appropriation from the Commonwealth of Pennsylvania, and by a research fellowship from the Deutsche Forschungsgemeinschaft to W.S. The Basel Institute for Immunology was founded and is supported by F. Hoffmann-La Roche Ltd, Basel Switzerland.

5. References

Alt FW, Baltimore D (1982) Joining of immunoglobulin heavy chain gene segments, implications from a chromosome with evidence of three D-J_H fusions. Proc Natl Acad Sci USA, 79:4118-4122

Blackwell TK, Alt FW (1988) Immunoglobulin genes. In: Hames BD, Glover DM (eds) Molecular Immunology, Chap 1. IRL Press, Oxford

Bosma, GC, Custer RP, Bosma MJ (1983) A severe combined immunodeficiency mutation in the mouse. Nature 301:527-530

Bosma GC, Fried M, Custer RP, Carroll A, Gibson DM, Bosma MJ (1988) Evidence of functional lymphocytes in some (leaky) scid mice. J Exp Med 167:1016-1033

Carroll A, Bosma MJ (1988) Detection and characterization of functional T cells in mice with severe combined immune deficiency. Eur J Immunol 18:1965-1971

Cook WD, Balaton AM (1987) T-cell receptor and immunoglobulin genes are rearranged together in Abelson virus-transformed pre-B and pre-T cell. Mol Cell Biol 7:266-272

Davis MM, (1988) T cell antigen receptor genes. In: Hames BD, Glover DM (eds) Molecular Immunology, Chap [illegible] IRL Press, Oxford

Dorshkind K, Keller GM, Phillips RA, Miller RG, Bosma GC, O'Toole M, Bosma MJ (1984) Functional status of cells from lymphoid and myeloid tissues in mice with severe combined immunodeficiency disease. J Immunol 132:1804-1808

Fulop GM, Bosma GC, Bosma MJ, Phillips RA (1988) Early B-cell precursors in scid mice: Normal numbers of cells transformable with Abelson murine leukemia virus (A-MuLV). Cell Immunol 113:192-201

Habu S, Kimura M, Katsuki M, Hioki K, Tatsuji N (1987) Correlation of T cell receptor gene rearrangements to T cell surface antigen expression and to serum immunoglobulin level in scid mice. Eur J Immunol 17:1467-1471

Hendrickson EA, Schatz DG, Weaver DT (1988) The scid gene encodes a trans-acting factor that mediates the rejoining event of Ig gene rearrangement. Genes & Development 2:817-829

Hirayoshi K, Nishikawa S-I, Kina T, Hatanaka M, Habu S, Nomura T, Katsura Y (1987) Immunoglobulin heavy chain gene diversification in the long-term bone marrow culture of normal mice and mice with severe combined immunodeficiency. Eur J Immunol 17:1051-1057

Kim MG, Schuler W, Bosma MJ, Marcu, KB (1988) Abnormal recombination of Igh D and J gene segments in transformed pre-B cells of scid mice. J Immunol 141:1341-1347

Kronenberg M, Goverman J, Haars R, Malissen M, Kraig E, Phillips L, Delovitch T, Suciu-Foca N, Hood L (1985) Rearrangement and transcription of β-chain genes of the T cell antigen receptor in different types of murine lymphocytes. Nature 313:647-653

Kronenberg M, Sui G, Hood L, Shastri N (1986) The molecular genetics of the T cell antigen receptor and T cell antigen recognition. Ann Rev Immunol 4:529-591

Lieber MR, Hesse JE, Lewis S, Bosma GC, Rosenberg N, Mizuuchi K, Bosma MJ, Gellert M (1988) The defect in murine severe combined immune deficiency: Joining of signal sequences but not coding segments in V(D)J recombination. Cell 55:7-16

Malynn BA, Blackwell TK, Fulop GM, Rathbun GA, Furley AJW, Ferrier P, Heinke LB, Phillips RA, Yancopoulos GD, Alt FW (1988) The scid defect affects the final step of the immunoglobulin VDJ recombinase mechanism. Cell 54:453-460

Okazaki K, Nishikawa SI, Sakano H (1988) Aberrant immunoglobulin gene rearrangement in scid mouse bone marrow cells. J Immunol 141:1348-1352

Schuler W, Weiler IJ, Schuler A, Phillips RA, Rosenberg N, Mak TW, Kearney JF, Perry RP, Bosma MJ (1986) Rearrangement of antigen receptor genes is defective in mice with severe combined immune deficiency. Cell 46:963-972

Schuler W, Schuler A, Lennon GG, Bosma GC, Bosma MJ (1988) Transcription of unrearranged antigen receptor genes in scid mice. EMBO J 7:2019-2024

Tonegawa S (1983) Somatic generation of antibody diversity. Nature 302:575-581

Van Ness BG, Weigert M, Coleclough C, Mather EL, Kelly DE, Perry RP (1981) Transcription of the unrearranged mouse Cκ locus: Sequence of the initiation region and comparison of activity with a rearranged Vκ-Cκ gene. Cell 27:593-602

Witte PL, Burrows PD, Kincade PW, Cooper MD (1987) Characterization of B lymphocyte lineage progenitor cells from mice with severe combined immune deficiency disease (SCID) made possible by long term culture. J Immunol 138:2698-2705

Yanagi Y, Yoshikai Y, Leggett K, Clark SP, Aleksander I, Mak TW (1984) A human T cell-specific cDNA clone encodes a protein having extensive homology to immunoglobulin chains. Nature 308:145-149

Yancopoulos GD, Alt FW (1985) Developmentally controlled and tissue-specific expression of unrearranged V_H gene segments. Cell 40:271-281

Yancopoulos GD, Blackwell TK, Suh H, Hood L, Alt FW (1986) Introduced T cell receptor variable region gene segments in pre-B cells: evidence that B and T cells use a common recombinase. Cell 44:251-259

Rearrangement of T Cell Receptor Delta Genes in Thymus of Scid Mice

Ann M. Carroll and Melvin J. Bosma

Institute for Cancer Research, Fox Chase Cancer Center, Philadelphia, PA 19111, USA

INTRODUCTION

Previous studies have indicated that developing lymphocytes in C.B-17/Icr scid/scid mice (hereafter referred to as scid mice) reach the stage at which antigen receptor genes become transcriptionally active and normally undergo rearrangement (Schuler et al. 1988). However, the demonstration of scid lymphocytes with antigen receptor gene rearrangements has required that such cells be transformed and clonally expanded (Schuler et al. 1986) or selectively outgrown in culture (Witte et al. 1987; Hirayoski et al. 1987). To explain the apparent paucity of these cells and their abnormal antigen receptor gene rearrangements, it was proposed that developing scid lymphocytes die prematurely as a result of a defective recombinase system (Schuler et al. 1986). This would, for example, account for the absence of detectable rearrangement of immunoglobulin genes and T cell receptor (TCR) β and γ genes (TCRβ and TCRγ) in scid lymphopoietic tissues (Schuler et al. 1988).

The interest of this report is that we provide direct evidence that developing scid T cells reach the stage at which TCRδ genes undergo rearrangement. We show that TCRδ rearrangements can be readily detected in the apparent absence of TCRβ and TCRγ rearrangements in DNA of whole scid thymus and in DNA of scid thymocyte hybridomas. This represents the first in vivo evidence of ongoing antigen receptor gene rearrangement in non-transformed early lymphocytes of scid mice. We anticipate that these results and those of studies in progress will offer new insights into the nature of the scid defect, and also provide new information on the recombination of distinct TCR loci and their ordered expression in normal T cell development.

RESULTS AND DISCUSSION

The TCRδ locus consists of two D and two J elements (Chien et al. 1987), and at least six V genes (Elliott et al. 1988). Vδ to Dδ recombination may occur prior to Dδ to Jδ recombination; also, Dδ1 to Dδ2 recombination occurs frequently (Chien et al. 1987; Elliott et al. 1988). The TCRδ locus rearranges as early or earlier than TCRβ and TCRγ genes in developing T cells of normal fetal thymus and rearrangements at this time (~ day 14 of gestation) frequently involve Dδ2 and Jδ1 (Chien et al. 1987). Rearrangement of TCRα genes occurs several days later. As TCRδ genes are located within the TCRα locus between Vα and Jα elements, rearrangement of these elements results in the deletion of the TCRδ locus.

As shown in Figure 1, a probe specific for a region downstream of Jδ1 (p3′Jδ1) was used to detect Dδ2/Jδ1-associated rearrangements in Eco R1 restricted DNA from thymus of normal mice (C.B-17+/+) and scid mice. Recombination involving the Dδ2 and/or Jδ1 elements results in the loss of the Eco R1 site 5′ of Dδ2 (see map of Figure 1) and the generation of novel size fragments.

As can be seen, almost identical sets of multiple non-germline hybridization fragments are present in newborn and adult thymus of normal mice. One major difference is a predominant 6.6 kb fragment present only in newborn mice. In contrast, DNA from newborn and adult scid thymus shows a set of four hybridization fragments which were common to all mice examined (>20 mice). In addition to a prominent germline fragment of 7.5 kb, three smaller fragments of 6.6, 5.4 and 4.3 kb are seen. The 6.6 kb fragment may correspond to a Dδ2-Jδ1 recombination commonly seen in thymocyte DNA of normal fetal mice (Chien <u>et al</u>. 1987). The 5.4 and 4.3 kb fragments appear unique to scid thymus. Further analysis of these non-germline fragments is in progress.

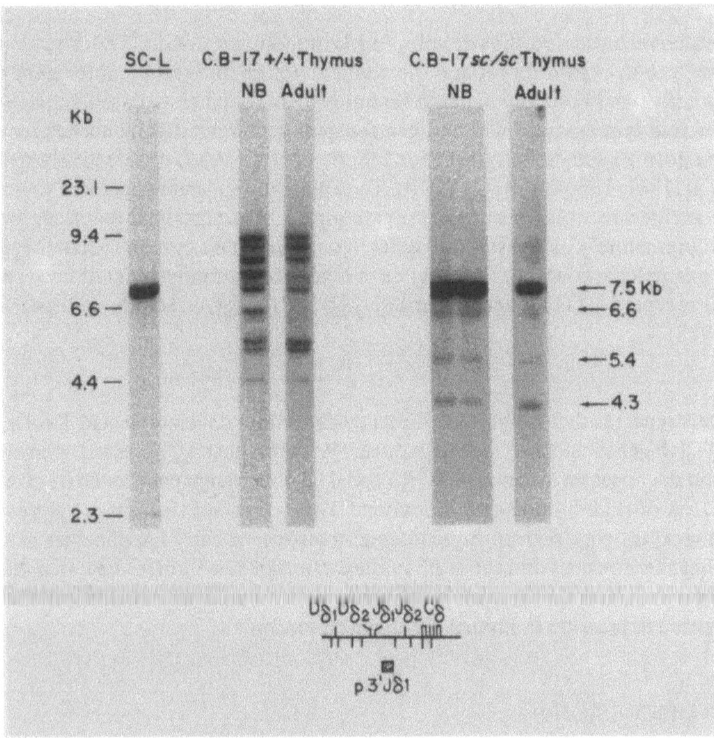

Figure 1. Southern blot analysis of <u>Eco</u> R1 restricted DNA from scid liver (SC-L), and from individual preparations of normal (C.B-17+/+) and scid (C.B-17<u>sc/sc</u>) newborn (NB) and young adult thymus. The DNA was hybridized with p3'Jδ1, a 2 kb <u>Sac</u> I fragment obtained from Y. Chien. A map of the <u>TCR</u> δ locus and p3'Jδ1 (derived from Chien <u>et al</u>. 1987) is shown below the Figure.

Consistent with an earlier study (Schuler <u>et al</u>. 1986), <u>TCRβ</u> and <u>TCRγ</u> loci appeared in germline configuration in scid thymocyte DNA (data not shown). If recombination of these loci were infrequent and resulted in fragments of random and abnormal sizes, such fragments might not be detected by Southern blot analysis of whole tissue DNA. Therefore, to address this possibility, young adult scid thymocytes (stimulated or unstimulated with phorbol ester and ionomycin) were fused to the BW5147 thymic lymphoma in which both TCRδ alleles have been deleted. The particular BW5147 line used was a non-expressing TCRα/β variant kindly supplied by W. Born (National Jewish Hospital, Denver, CO). Control hybridomas were obtained by fusing BW5147 to thymocytes of normal age-matched donors. Hybridization of Eco R1 restricted DNA from these hybridomas to the appropriate probes showed that most control hybridomas contained TCRβ and TCRγ rearrangements; further, most hybridomas deleted both TCRδ alleles, presumably as a result of TCRα rearrangement (data not shown). The scid hybridomas were strikingly different: the majority (17/23 analyzed) had TCRδ rearrangements, but none of those examined contained detectable rearrangements of TCRβ (18/18 analyzed) or TCRγ (10/10 analyzed).

The results for five scid thymocyte hybridomas are illustrated in Figure 2. As indicated in Figure 2A, hybridization to p3'Jδ1 shows three hybridomas (SA1, SA2 and SA8) contained one or two rearranged TCRδ alleles; SA6 contained only a germline fragment and SA3 lost both TCRδ alleles either by gene rearrangement or chromosome loss. Various size TCRδ fragments were seen in addition to the 6.6, 5.4 and 4.3 kb fragments characteristic of whole scid thymus. It is noteworthy that a minority of the recombination fragments (4/21), such as the smaller fragment in SA1, were too small to contain an upstream Jδ1 coding sequence, indicating that at least some scid TCRδ rearrangements were aberrant.

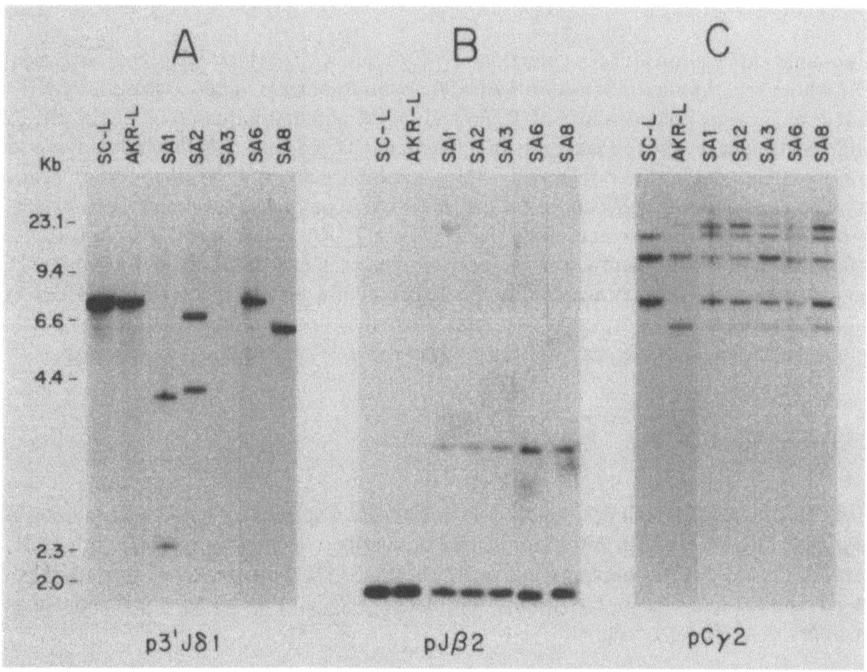

Figure 2. Southern blot analysis of Eco R1 restricted DNA of scid liver (SC-L), AKR liver (AKR-L) and several scid thymocyte hybridomas (designated SA). The blot was sequentially hybridized with p3'Jδ1 (Chien et al. 1987) pJβ2 (Malissen et al. 1984) and pCγ2 (Iwamoto et al. 1986). The AKR-derived BW5147 fusion partner contained one TCRβ fragment of 3.9 kb and three TCRγ fragments of 6.6, 16 and 22 kb.

The results of sequential hybridization of the blot in Figure 2A with probes specific for TCRβ (pJβ2) and TCRγ (pCγ2) are shown in Figure 2B and 2C, respectively. No TCRβ or TCRγ scid rearrangements were detected. It should be noted in Figure 2C that pCγ2 cross-hybridizes to Cγ1 and Cγ3 and that most hybridomas analyzed contained three germline scid fragments (7.5, 10.5 and 13.4 kb) and three fragments contributed by BW5147 (6.6, 16 and 22 kb). The BW5147 fragments of 16 and 22 kb correspond in size to conventionally rearranged TCRγ alleles of C.B-17+/+ mice. Thus, it is possible that a conventional rearrangement involving JCγ1 or JCγ2 occurred at one TCRγ allele and was masked by the 16 or 22 kb fragment of BW5147.

In conclusion, the detection of the same three TCRδ recombination fragments (6.6, 5.4 and 4.3 kb) in the DNA of whole scid thymus is puzzling as we would have not expected a common set of fragments based on the apparently random and abnormal sizes of TCRβ and TCRγ recombination fragments in scid thymic lymphomas (Schuler et al. 1986; Schuler and Bosma, this volume). The common TCRδ fragments may correspond to the most frequent recombinations at the scid TCRδ locus. Alternatively, the cells containing these fragments may have a selective advantage over other developing scid thymocytes with TCRδ recombination fragments of a different size; evidence of the latter cells is clear from our results with scid thymocyte hybridomas. Whether the 6.6, 5.4 and 4.3 kb fragments (or any other TCRδ recombination fragments) contain normal recombination joints remains to be ascertained.

Three possible explanations for the detection of TCRδ but not TCRβ or TCRγ rearrangements in DNA of whole scid thymus and in scid thymocyte hybridomas are under consideration: (i) that TCRδ rearrangements precede those of TCRγ and TCRβ and that initiation of TCRγ and TCRβ rearrangements requires the full assembly of a functional TCRδ gene, which in developing scid T cells rarely occurs due to their defective VDJ recombinase activity; (ii) that certain spatial or structural features of the TCRδ locus allow the scid VDJ recombinase system to recombine TCRδ genes more frequently and/or successfully than TCRγ or TCRβ genes; or (iii) that the effect of the *scid* mutation is not fully manifest in the early stages of T cell development at which TCRδ rearrangements occur; this may account for the detection of a common, non-random set of TCRδ recombination fragments in different scid mice. Whether one or more of these explanations is likely to be valid clearly awaits the results of further analyses and additional experiments.

ACKNOWLEDGEMENTS

We thank N. Ghobadi for technical help, M. Piatek and A. Capriotti for help in preparation of the manuscript and M. Fried and J. Petrini for helpful discussions. We thank Y. Chien for p3'Jδ1, and W. Born for the TCR α/β non-expressing variant of BW5147. This work was supported by NIH grants AI-13323, CA-04946, CA-06927 and RR-05539, and by an appropriation from the Commonwealth of Pennsylvania.

REFERENCES

Chien Y, Iwashima M, Wettstein DA, Kaplan KB, Elliott JF, Born W, Davis MM (1987) T-cell receptor δ gene rearrangements in early thymocytes. Nature 330: 722-727

Elliott JF, Rock EP, Patten PA, Davis MM, Chien Y (1988) The adult T-cell receptor δ-chain is diverse and distinct from that of fetal thymocytes. Nature 331: 627-631

Hendrickson EA, Schatz DG, Weaver DT (1988) The scid gene encodes a transacting factor that mediates the rejoining event of Ig gene rearrangement. Genes Dev 2: 817-829

Hirayoshi K, Nishikawa S-I, Kina T, Hatanaka M, Habu S, Nomura T, Katsura Y (1987) Immunoglobulin heavy chain gene diversification in the long-term bone marrow culture of normal mice and mice with severe combined immunodeficiency. Eur J Immunol 17: 1051-1057

Iwamoto A, Rupp F, Ohashi PS, Walker CL, Pircher H, Joho R, Hengartner H, Mak TW (1986) T cell specific γ genes in C57BL/10 mice. J Exp Med 163: 1203-1212

Lieber MR, Hesse JE, Lewis S, Bosma GC, Rosenberg N, Mizuuchi K, Bosma MJ, Gellert M (1988) The defect in murine severe combined immune deficiency: joining of signal sequences but not coding segments in V(D)J recombination. Cell 55: 7-16

Malissen M, Minard K, Mjolness S, Kronenberg M, Goverman J, Hunkapillar T, Prystowsky MB, Yoshikai Y, Fitch F, Mak TW, Hood L (1984) Mouse T cell antigen receptor: structure and organization of constant and joining gene segments encoding the β polypeptide. Cell 37: 1101-1110

Malynn BA, Blackwell TK, Fulop GM, Rathbun GA, Furley AJW, Ferrier P, Heinke LB, Phillips RA, Yancopoulis GD, Alt FW (1988) The scid defect effects the final step of the immunoglobulin VDJ recombinase mechanism. Cell 54: 453-460

Schuler W, Weiler IJ, Phillips RA, Rosenberg N, Mak TW, Kearney JF, Perry RP, Bosma MJ (1986) Rearrangement of antigen receptor genes is defective in mice with severe combined immune deficiency. Cell 46: 963-972

Schuler W, Schuler A, Lennon GC, Bosma GC, Bosma MJ (1988) Transcription of unrearranged antigen receptor genes in scid mice. EMBO J 7: 2019-2024

Witte PL, Burrows PD, Kincade PW, Cooper MD (1987) Characterization of B lymphocyte lineage progenitor cells from mice with severe combined immune deficiency disease (SCID) made possible by long term culture. J Immunol 138: 2698-2705

Abnormal V(D)J Recombination in Murine Severe Combined Immune Deficiency: Absence of Coding Joints and Formation of Alternative Products

ABNORMAL V(D)J RECOMBINATION IN MURINE SEVERE COMBINED IMMUNE DEFICIENCY: ABSENCE OF CODING JOINTS AND FORMATION OF ALTERNATIVE PRODUCTS

Michael R. Lieber, Joanne E. Hesse, Susanna Lewis, Gayle C. Bosma[1]
Naomi Rosenberg[2], Kiyoshi Mizuuchi, Melvin J. Bosma[1] and Martin Gellert

Laboratory of Molecular Biology
National Institute of Diabetes and Digestive and Kidney Diseases
National Institutes of Health, Bethesda, MD 20892

INTRODUCTION

We have studied the VDJ recombination reaction in early B and T cells from mice homozygous for the severe combined immune deficiency (scid) mutation. Schuler et al. (1986) previously found abnormal deletions by Southern blot analysis at rearranged IgH and TCR β alleles in transformed B and T lymphocytes from scid mice. We have recently completed a study (Lieber et al. 1988a) in which we transfected scid lymphoid cell lines with VDJ recombination substrates that remain extrachromosomal. These substrates allow large numbers of recombinant reaction products to be collected (Hesse et al. 1987; Lieber et al. 1987), thereby permitting detailed analysis of the VDJ recombination reaction in scid lymphoid cells.

SUMMARY OF VDJ RECOMBINATION REACTIONS IN NORMAL AND SCID LYMPHOID CELLS

The extrachromosomal VDJ recombination substrates that we transfected into pre-B and pre-T cell lines allow all the reactions depicted in Fig. 1 to be studied. The arrangement of coding and signal sequences at most Ig and TCR loci is depicted in line 2 of this figure. In a rearrangement event, the two coding ends are joined to form a coding joint. Two signal sequences and the intervening DNA between them are deleted from the genome on a piece of DNA circularized by the joining together of the two signal sequences to form a signal joint. The two halves of the reaction, signal joint and coding joint formation, occur at approximately equal frequencies in normal cells. Inversional VDJ recombination occurs at some Ig and TCR loci (line 3). In this arrangement, signal and coding sequences remain on the same DNA molecule.

In normal cells signal joints lose nucleotides very rarely (well below 1%) (Lieber et al. 1988b). Nucleotide addition occurs in up to 25% of signal joints formed in normal lymphoid cells and is dependent in frequency on terminal transferase activity levels. In lymphoid cells from scid mice, signal joint formation is abnormal in that about 50% of the signal joints show nucleotide loss from either the 12-signal, the 23-signal, or both (Fig. 2). The endpoints of the deletions in signal joints from scid lymphoid cells are interesting. Many end at or within one base pair of the end of the heptamer that borders the spacer region. Some end at the spacer-nonamer border in one or both signals. Taken together, these results indicate that, though the VDJ recombination is present in scid cells, it is abnormal. Nevertheless, the altered activity can complete the signal joint even when portions of one or both signals have been deleted.

---

[1] Institute for Cancer Research, Fox Chase Cancer Center, Philadelphia, PA 19111

[2] Department of Pathology, Tufts University School of Medicine, Boston, MA 02111

Current Topics in Microbiology and Immunology, Vol. 152

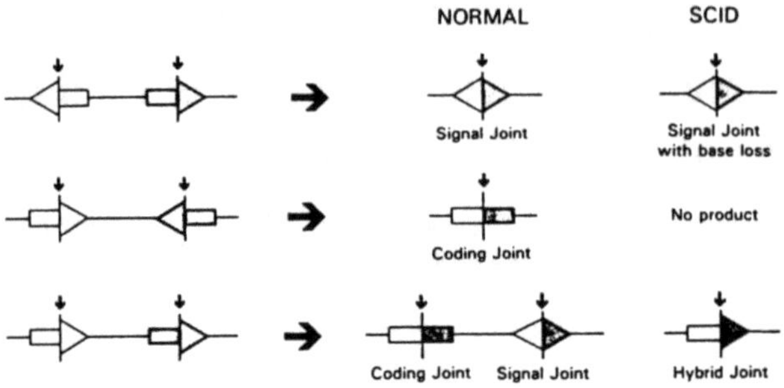

Figure 1. Summary of VDJ Recombination Reactions in Normal and Scid Lymphoid Cells

The VDJ recombination substrates are depicted here with the following symbols: open triangle, joining signal with a 12-base spacer between the heptamer and nonamer (hereafter abbreviated "12-signal"); closed triangle, joining signal with a 23-base spacer between the heptamer and nonamer ("23-signal"); open rectangle, coding end sequences originally attached to the 12-signal; and closed rectangle, coding end sequences originally attached to the 23-signal. Vertical arrows indicate the sites of recombination.

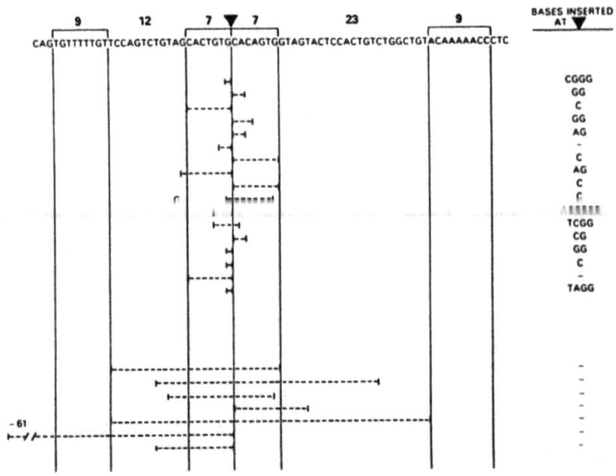

Figure 2. Deletions in Signal Joints from Scid Cell Lines

The structure of those signal joints found to have deletions of the signal sequences are presented. The sequence shown is that of two signal ends joined in the normal manner without base loss. The solid triangle indicates the position of the junction. The bracketed dashes show the extent of the deletions, with each dash corresponding to one base. Bases inserted at the joint are listed at the right of the figure.

The abnormal signal joints formed in scid cells can be further subdivided into those with deletions of 8 bp or less from either signal end and those greater than 8 bp. The group with smaller deletions has an increased incidence of non-germline nucleotide addition at the signal joint, whereas the group with larger deletions has no base additions.

We have detected no coding joint formation in scid cells using substrates of the design in line 2 of Fig. 1. Based on our level of screening, coding joint formation is depressed at least 1000-fold below signal joint formation.

Because of the larger deletions found at endogenous loci (Schuler et al. 1986), we also screened for coding joints with several hundred base pairs deleted, but found none (Lieber et al. 1988a). If their frequency had been comparable to that of signal joint formation, several examples should have been seen. This suggests that the large deletions found at endogenous loci may represent a small fraction of the total number of attempted coding joint formation events.

We also have detected no coding joint formation in inversion substrates. However, inversion substrates can undergo two kinds of alternative reactions in which coding ends participate. One of these reactions is a VDJ recombination event that we call hybrid joint formation (Fig. 3). It occurs when a signal end is joined not to the other signal but rather to the coding end attached to the other signal. This alternative reaction is a minor product in the recombination of inversion substrates in normal cells, occurring at about 20% of the frequency of inversion (Lewis et al. 1988). The fact that coding ends can form hybrid joints indicates that these ends are not systematically destroyed in scid cells, but fail to join to one another.

There is a second way in which coding ends made by the VDJ rearrangement system are used in scid cells. If the coding ends share sequence homology, it is possible for them to be joined by homologous recombination. In normal lymphoid cells, when the two coding ends of the substrate share blocks of homology, as shown in Fig. 4, the signal and coding joints are formed as usual, retaining the two regions of homology. However, in scid cells, though the signal joint is formed by VDJ recombination, the coding ends are resolved by homologous recombination, which leaves only a single copy of the region. Again, this indicates that the coding ends are able to participate in reactions that are alternative to coding joint formation, even when they are not able to join to one another. This particular reaction also indicates that signal joint and coding joint formation can be dissociated from one another, the signal joint being formed by the VDJ recombination reaction but the coding joint being formed by homologous end joining (Roth et al. 1985; Roth and Wilson 1986).

We have excluded the possibility that the defect is a general failure to join DNA ends. Linear DNA transfected to scid lymphoid cells was circularized as well as DNA transfected into lymphoid cells from normal mice.

We find identical VDJ recombination defects in both pre-B and pre-T cells from scid mice, firmly establishing that both lymphoid lineages use a common recombination activity which can be affected by a single mutation.

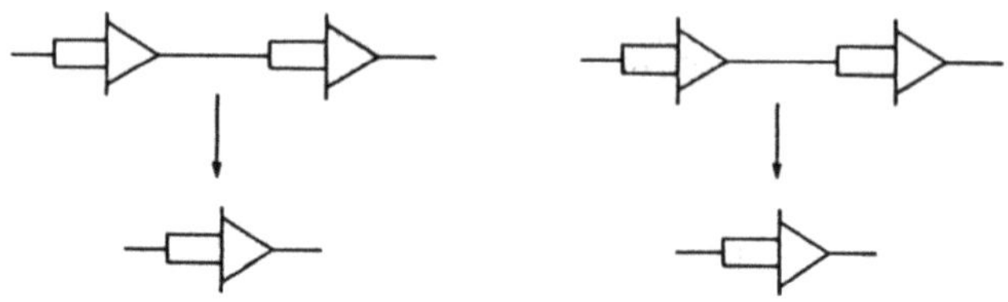

Figure 3. Occurrence of Hybrid Joints in Scid Cells

Inversion substrates with the 12-signal on the left (left substrate) or right (right substrate) can give rise to hybrid joint deletion events in scid cells (as they can in normal cells).

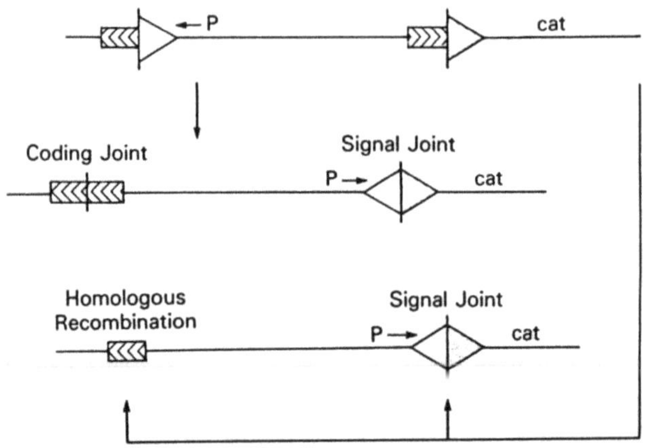

Figure 4. Recombination of an Inversion Substrate with Homologous Coding Ends

In an inversion substrate where the coding regions flanking the signals have initially inverted homology (top drawing), two kinds of inversional recombination are possible. If the complete reaction proceeds by V(D)J recombination, joining of the signals is accompanied by joining of the coding ends as a tandem duplication (middle drawing). When coding joint formation is blocked by the scid defect, the recombination process initiated in the V(D)J pathway may be completed by homologous recombination between the coding ends, leaving a single copy of the coding flank in the product (bottom drawing).

SIGNIFICANCE OF EXTRACHROMOSOMAL SUBSTRATE STUDIES FOR ENDOGENOUS LOCI IN SCID LYMPHOID CELLS

Our results demonstrating the failure of the half of the VDJ recombination reaction responsible for coding joint formation explains several features of the Ig and TCR loci found in scid lymphoid cells. A number of end resolutions in scid (about one-fifth of them) appear to be alternative VDJ recombination reactions, instead of coding joint formation. As we said above, scid lymphoid cells can form hybrid joints and signal joints. Several recombinants previously reported by others at Ig loci in scid lymphoid cells have features apparent on re-examination of the sequences that suggest the occurrence of one of these two alternative reactions and use cryptic signal sequences at one or both sites of recombination. These possible instances of hybrid joints include one recombination at the kappa locus in scid early B cells, SC24.1.13 of Fig. 5B of Blackwell et al. (1989), and several events at the heavy chain locus: SBM4 of Figure 3d of Okazaki et al. (1988); 1PR(a) of Figure 3B of Kim et al. (1988); and SC7-2.6 of Fig. 3b of Okazaki et al. (1988). Possible signal joining events mediated by the scid VDJ recombination activity at the heavy chain locus, but in which cryptic signal sequences near one or both sites of recombination may have been used, are SC7-2.6 of Fig. 3B of Okazaki et al. (1988) and Fig. 3C of Kim et al. (1988).

Based on the results with the extrachromosomal substrates, most coding ends would be expected to remain unjoined, resulting in chromosomal breaks (Fig. 5). Chromosomal breaks would reduce cell viability unless some secondary process restored a functional chromosome structure. In the absence of alternative V(D)J reactions, such restoration would have to occur by illegitimate recombination, which is typically associated with large deletions from one or both free DNA ends (Roth et al., 1985; Roth and Wilson, 1986). Schuler et al. (1986) showed that large deletions covering the J regions at IgH and TCR occurred in scid cell lines. Many such deletions now have been sequenced (Kim et al. 1988; Okazaki et al. 1988; Malynn et al. 1988; Hendrickson et al. 1988). There are few rules or hallmarks of illegitimate recombination to firmly establish that this is the mechanism of resolution in these cases. It does appear that in some cases of illegitimate recombination, duplication of a few nucleotides nearby sequences are generated in the resolution process (Roth and Wilson 1986). Such duplications are seen in some end resolutions in scid cells (Okazaki et al. 1988; Malynn et al. 1988).

The observation of signal joint but not coding joint formation explains various aspects of murine severe combined immune deficiency. In the absence of coding joint formation Ig and TCR genes can not successfully be rearranged. Therefore, mature B and T cells bearing surface Ig or TCR can not be generated (Bosma et al. 1983; Custer et al. 1985).

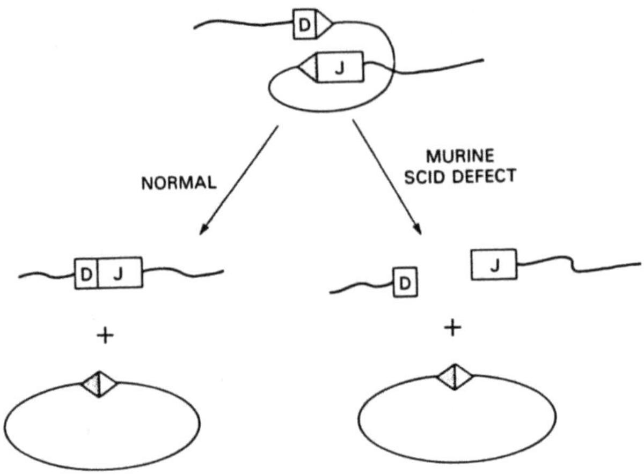

Figure 5. Failure of Coding Joint Formation at Endogenous
Ig or TCR Loci.

The formation of signal joints but failure to form coding joints suggests that the two coding ends remain unjoined in the large majority of cases. This would result in chromosomal breakage. In some instances, illegitimate recombination appears to circumvent chromosome breakage; but this would be expected to result in large deletions from one or both coding ends, as has been observed (see text).

REFERENCES

Blackwell, TK, Malynn, BA, Pollack, RR, Ferrier, P, Covey, LR, Fulop, GM, Phillips, RA, Yancopoulos, GD, and Alt, FW (1989) Isolation of scid pre B cells that rearrange kappa light chain genes: formation of normal signal and abnormal coding joins. EMBO J 8: 735-742

Bosma, GC, Custer, RP and Bosma, MJ (1983) Severe Combined immunodeficiency mutation in the mouse. Nature 301: 527-530

Custer, RP, Bosma, GC and Bosma, MJ (1985) Severe combined immunodeficiency (SCID) in the mouse: pathology, reconstitution, neoplasms. Am J Path 120: 464-477

Hendrickson, EA, Schatz, DG, and Weaver, DT (1988) The scid gene encodes a trans-acting factor that mediates the rejoining event of Ig gene rearrangement. Genes and Devel 2: 817-829

Hesse, JE, Lieber, MR, Gellert, M and Mizuuchi, K (1987) Extrachromosomal DNA substrates in pre-B cells undergo inversion or deletion at immunoglobulin V-(D)-J joining signals. Cell 49: 775-883

Kim, MG, Schuler, W, Bosma, MJ and Marcu, KB (1988) Aberrant Igh locus rearrangement in A-MuLV pre-B lines of SCID mice: evidence for a deregulated D-J recombinase. Curr Top Micro Immunol 137: 203-210

Lewis, S, Hesse, JH, Mizuuchi, K and Gellert, M (1988) Novel strand exchanges in V(D)J recombination. Cell 55: 1099-1107

Lieber, MR, Hesse JE, Mizuuchi, K and Gellert, M (1987) Developmental stage specificity of the lymphoid V(D)J recombination activity. Genes Dev 1: 751-761

Lieber, MR, Hesse, JE, Lewis, S, Bosma, GC, Rosenberg, N, Mizuuchi, K, Bosma, MJ and Gellert M (1988a) The defect in murine severe combined immune deficiency: joining of signal sequences but not coding segments in V(D)J recombination. Cell 55: 7-16

Lieber, MR, Hesse, JE, Mizuuchi, K and Gellert, M (1988b) Lymphoid V(D)J recombination: nucleotide insertion at signal joints as well as coding joints. Proc Natl Acad Sci USA 85: 8588-8592

Maylnn, BA, Blackwell, TK, Fulop, GM, Rathburn, GA, Furley, AJW, Ferrier, P, Heinke, LB, Phillips, RA, Yancopoulos, GD, and Alt, FW (1988). The scid defect affects the final step of the immunoglobulin VDJ recombinase mechanism. Cell 54: 453-460

Okazaki, K, Nishikawa, S and Sakano, H (1988) Aberrant immunoglobulin gene rearrangement in scid mouse bone marrow cell. J Immunol 141: 1348-1352

Roth, DB and Wilson, JH (1986) Nonhomologous recombination in mammalian cells: role for short sequence homologies in the joining reaction. Mol Cell Biol 6: 4295-4304

Roth, DB, Porter, TN and Wilson, JH (1985) Mechanisms of non-homologous recombination in mammalian cells. Mol Cell Biol 5: 2599-2607

Schuler, W, Weiler, IJ, Schuler, A, Phillips, RA, Rosenberg, N, Mak, TW, Kearney, JF, Perry, RP and Bosma, MJ (1986) Rearrangement of antigen receptor genes is defective in mice with severe combined immune deficiency. Cell 46: 963-972

The *scid* Mutation Disrupts Gene Rearrangement at the Rejoining of Coding Strands

David Weaver and Eric Hendrickson

Dana-Farber Cancer Institute, and Department of Microbiology and Molecular Genetics, Harvard Medical School, Boston, MA. USA.

Introduction

The gene rearrangement mechanism in lymphocytes is a highly ordered and regulated process, containing distinct stages and tissue specificity for most of the V(D)J rearrangement events. Several lines of evidence indicate that a common "recombinase" function mediates the rearrangement process for all the immunoglobulin (Ig) and T cell receptor (TCR) gene elements. Whereas significant progress has been acheived in the molecular mapping of the organization of the Ig and TCR gene families and the structure of rearrangement products, very little progress has been made in identifying the proteins important to the recombination pathway.

The *scid* mutation in the mouse blocks the progression of lymphoid differentiation for both B and T cells, although it is apparent that the precursors for these cells are present in fetal liver, fetal thymus and bone marrow. Inappropriate deletions of Ig μ gene and TCR β gene loci are found associated with the precursor stage cells isolated from Abelson murine leukemia virus transformation of pre-B cells in bone marrow and from transformed pre-T or immature T cells isolated from spontaneous thymic lymphomas in *scid* mice (Schuler et al., 1986; Hendrickson et al., 1988). Determination of the molecular mode of action of the *scid* gene product will therefore be valuable in the understanding of the gene rearrangement pathway of lymphoid cells. We have previously shown that *scid* pre-B cells have attempted Ig gene joining events on both endogenous Ig μ heavy chain chromosomal alleles (Hendrickson et al., 1988). These rearrangements had the following properties: 1) rearrangement was restricted to D_H to J_H joining only, the first stage of Ig gene recombination; 2) large deletions occurred in the chromosomal regions containing and flanking the J_H and D_H elements; 3) sequence analysis at the D_H-J_H breakpoints indicated that an alternate recognition element for the recombination events was not utilized; and 4) all of the D_H to J_H joins were defective such that no functional μ protein could be synthesized. In total, these data suggested that the *scid* mutation may have interrupted the normal gene rearrangement mechanism, although a primary recombination defect could not be distinguished from a primary differentiation defect.

Scid pre-B cells aberrantly rearrange recombination templates ***in trans***

In order to specifically examine the role of the *scid* mutation in gene rearrangement and to determine the stage within the recombination pathway at which the *scid* mutation is likely to act, a group of recombinant retrovirus vectors that contain recombination templates were utilized. Wildtype pre-B cell lines integrate and rearrange the DGR viral vector in a characteristic pattern yielding inversional rearrangements (Landau et al., 1987). *Scid* pre-B cell lines integrate and aberrantly rearrange the DGR vector at high frequency, as evidenced by Southern blots generated from mass infections (Hendrickson et al., 1988). The *scid* mutation therefore stops lymphoid differentiation as a trans-acting factor by a direct recombination defect. The *scid* cell lines were subcloned to examine independently the variety of rearrangement

Current Topics in Microbiology and Immunology, Vol. 152

events. Since DGR contains a pBR322 origin of replication and the *neo* gene that can be used for kanamycin-selection in E. coli, size-selected populations of restriction digests containing the proviral DNA were directly cloned. These proviral DNAs were then analyzed by Southern blot analysis, restriction enzyme mapping, and DNA sequencing.

Three categories of gene rearrangements were observed in *scid* cell lines infected with DGR. The predominant class of events were deletions (12/15), where a large percentage of the DGR cassette was absent (Fig. 1A)

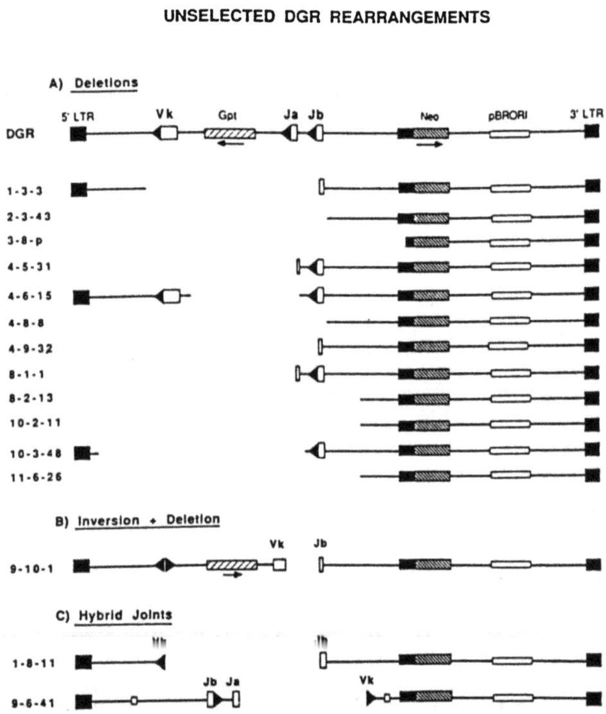

Figure 1. DGR vectors rescued from *scid* pre-B cells predominately exhibit deletional rearrangements. Genomic DNA from the various *scid* subclones was cleaved with XbaI, size-selected by gel electrophoresis, circularized, transfected into bacterial cells and plated under kanamycin selection. Kanamycin-resistant clones were analyzed by extensive restriction mapping and in every case the breakpoint of the deletion was determined by DNA sequencing. DGR contains two LTRs (shaded box), the nonamer-heptamer signal sequence (black triangle) of V_K and two J_K elements (open box), the *gpt* gene (hatched box), the SV40 promoter-enhancer (solid box), the *neo* gene (hatched rectangle), and the pBR322 origin of replication (pBRORI, open rectangle). Various *scid* DGR clones have undergone either: (A)deletions, (B)inversion and deletion or (C)hybrid joint formation. The blank space indicates the extent of the deletion. If the 5'-LTR is missing the deletion extended into genomic DNA 5' to the DGR construct.

In most of these instances, the deletions removed the 5'-side of the construct and extended into the flanking mouse DNA sequences, whereas the 3'-side, which contained the *neo* selection marker, was conserved. These events are formally identical to the large deletions that accompany gene rearrangement at endogenous Ig or TCR alleles in *scid* lymphoid cells. A second vector, JVD, that undergoes deletional rather than inversional rearrangements, was also found to promote aberrant rearrangements in the *scid* pre-B cells (data not shown). The recombination breakpoint DNA sequences were compiled and no consensus signal sequence at the breakpoint borders among this group was found. It is interesting that two *scid* rearrangements, one in DGR (4-6-15, Fig. 1A) and one in JVD (not shown), have deletions that are restricted to the region including one heptamer-nonamer signal only. Cleavage and rejoining at a single heptamer-coding element junction can occur without overall DNA rearrangement, a process called "open/shut" (Lewis et al., 1988). Thus, these two rearrangements may have arisen by cleavage at one heptamer with the *scid*-mediated inability to repair this cleavage resulting in deletion over that region. A second class of *scid* rearrangement, found in only 1/15 of the subclones, resembled the normal inversional DGR rearrangements in wildtype cells (Fig. 1B). Unlike the wildtype inversions, however, clone 9-10-1 was an aberrant rearrangement since the "coding joint" resulting from the inversion was deleted by about 400 bases. A third class of events were hybrid joints, defined as the joining of a heptamer element of a V with another coding element (e.g. J) instead of to another heptamer element (Lewis et al., 1988). Hybrid joints occur at low frequencies in wildtype gene rearrangements. In DGR, which is constructed for inversions, hybrid joints between V and Ja or Jb elements will form large deletions. We found hybrid joints in 2/15 *scid* DGR rearrangements (1-8-11, $V_{heptamer}$ + Jb; 9-6-41, $V_{heptamer}$ + Ja, Fig. 2C), a level that is not significantly different from wildtype levels in similar retroviral constructs. It is interesting that hybrid joints and inversions occurred at a similar frequency in *scid* rearrangements.

Point mutations in recombination heptamer elements disrupt scid rearrangements

Two lines of evidence suggest that *scid* rearrangements are dependent on the recombination heptamer-nonamer elements. First, *scid* endogenous Ig gene rearrangements are centered over the chromosomal regions where the D_H and J_H coding elements and heptamer-nonamer elements reside (Hendrickson et al., 1988). Secondly, X-rays were used to generate random double-stranded chromosomal breaks in viable cells, and no significant differences in the ability of wildtype or *scid* pre-B cells to repair these breaks were detected (not shown). Point mutations in either position 1 or 2 of the heptamer elements of JVD were constructed, yielding three derivative JVD vectors (Figs. 2 and 3). JVD-J_{hep1} has a single point mutation in the first position of the J heptamer, JVD-V_{hep2} has a single point mutation in the 2nd position of the V heptamer, and JVD-VJ_{hep} has a point mutation in both the J heptamer and the V heptamer at these positions. The vectors were introduced into wildtype and *scid* pre-B cells, selected using the G418-resistance of the *neo* gene, and DNA rearrangement assessed after 6 weeks in culture by Southern blot analysis (Fig. 2). Either the JVD-J_{hep1} or the JVD-V_{hep2} single-point mutant vectors showed essentially no rearrangement of the construct in wildtype pre-B cells (0/20 independent infections for each). In contrast, *scid* pre-B cells continued to aberrantly rearrange these vectors, but at reduced levels (5/20 for JVD-J_{hep1}, and 4/20 for JVD-V_{hep2}). In the double point mutant vector, JVD-VJ_{hep}, none of the infected *scid* cell lines appeared to rearrange the vector, just like the wildtype pre-B cell controls (Fig. 3). This experiment demonstrated that the *scid* gene product does require recognition of heptamer elements for its rearrangement mechanism, albeit it is less sensitive to the exact structure of the heptamers than its wildtype counterpart.

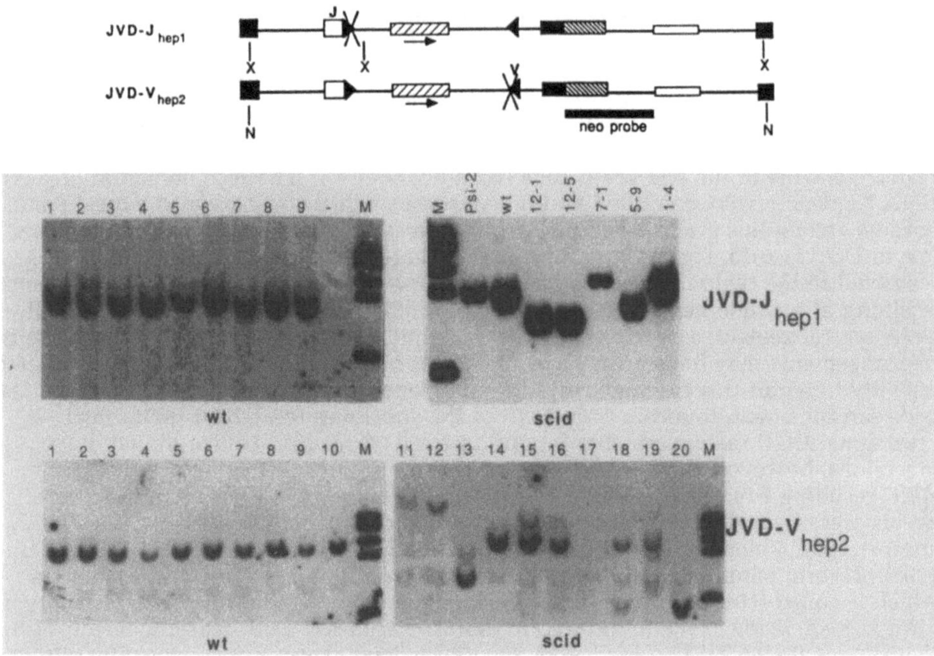

Figure 2. *Scid* pre-B cells are less sensitive than wildtype cells to heptamer single-point mutations. JVD constructs containing point mutations (large cross) in the J-heptamer (JVD-J_{hep}1) or the V-heptamer (JVD-V_{hep}2) are shown. The symbols are the same as for DGR, except for the thymidine kinase gene (hatched box). Restriction sites: (X) XbaI, (N) NheI. Genomic DNA Southern blots were prepared from wildtype (wt) or *scid* cells infected with either JVD-J_{hep}1 (top) or JVD-V_{hep}2 (bottom). JVD-J_{hep}1 DNAs were digested with XbaI and JVD-V_{hep}2 were digested with NheI prior to *neo* probe hybridization. M, markers; Psi-2, Psi-2 JVD-Jhep1 producer cell line.

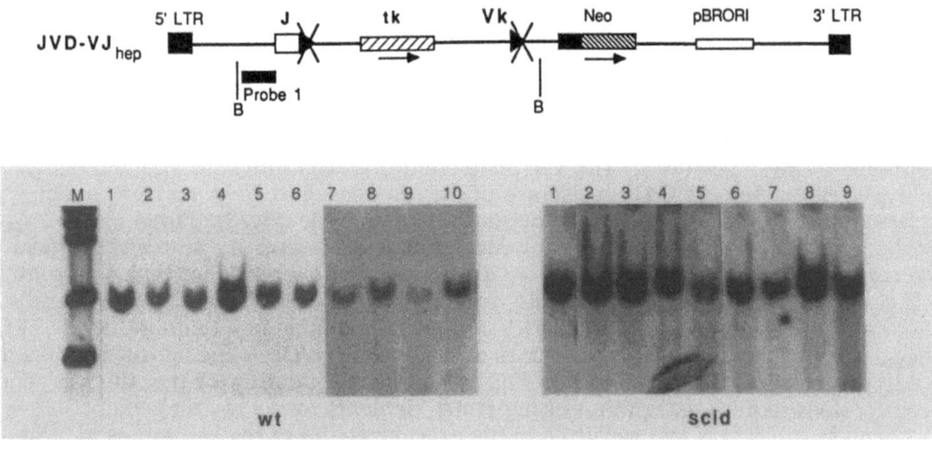

Figure 3. Double heptamer point mutations abolish gene rearrangement in both *scid* and wildtype pre-B cells. JVD-VJ_{hep} contains point mutations in both heptamers.

The coding joint is aberrantly formed in scid rearrangements

The DGR construct is sufficiently destroyed during unselected *scid* rearrangements such that it is difficult to determine both the step of the mechanism and the DNA junction or junctions that are primarily affected. We took advantage of the normal inversional rearrangement properties of DGR, where a positive selection is generated by inversion of the *gpt* gene (Fig. 1; Landau et al., 1987). The *gpt* gene in its new orientation is activated for transcription and is selected for by growth in mycophenolic acid (MPA)(Lewis et al., 1984; 1985). Neither the exact formation of the reciprocal joint nor the coding joint is required for the *gpt* selection. We therefore reasoned that we could select for inversions of DGR in *scid* pre-B cells if the extent of the *scid* deletions was limited. *Scid* pre-B cells containing integrated and unrearranged DGR were placed in MPA selection. Seven MPA-resistant cell lines grew out of the cultures at a frequency that was several orders of magnitude lower than the appearance of unselected rearrangement in DGR for *scid* cell lines, and far below the level of inversions generated in wildtype pre-B cell lines. Southern blot analysis revealed that inversions, but no large deletions, had occurred (data not shown). To determine the structure of the two recombination junctions, the proviral DGR DNA was cloned, and the DNA sequence of the rearrangement junctions was obtained (Fig.4).

SCID INVERSIONAL REARRANGEMENTS

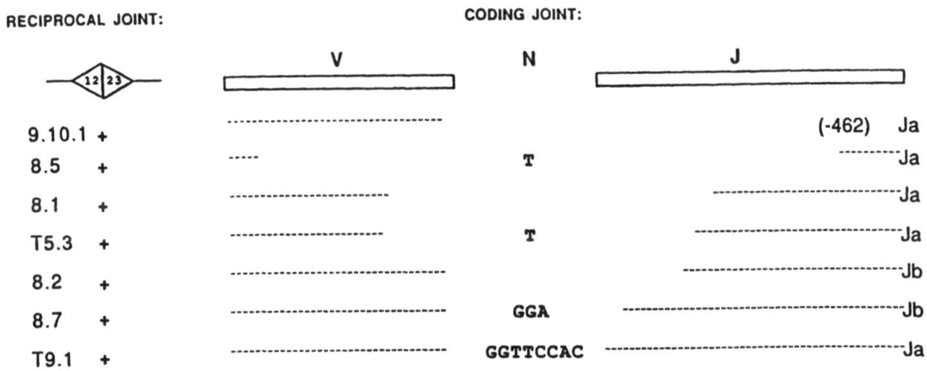

Figure 4. The *scid* mutation acts primarily at coding joints. *Scid* pre-B cells were infected with DGR and MPA-resistant colonies were selected. The rearranged DGR templates were cloned for DNA sequencing. Both junctions formed from a single rearrangement event. Reciprocal Fragment: (+) indicates a wildtype junction without the deletion or addition of nucleotides. Coding Joint: The nucleotide sequence of the unrearranged V and J gene elements is shown boxed. Dashed lines indicate nucleotides retained in the *scid* inversions, and blank space indicates the nucleotides deleted. N-region addition (N) in the coding joint junction is shown. Ja or Jb refers to whether the inversion occurred at the DGR Ja or Jb gene element.

In wildtype DNA rearrangements, the two "product" junctions are the reciprocal joint, which is an exact fusion of the two heptamer elements, and a coding joint (Fig. 5). The coding joints are characterized by small deletions (<5 nucleotides) or insertions (N regions, 1-10 nucleotides) in the joint. Generally, reciprocal joints do not have either deletions or additions of nucleotides. There are several striking features of the selected *scid* inversional rearrangements. Like wildtype rearrangements, the reciprocal joints form perfectly with no addition or deletion of nucleotides in 7/7 of the *scid* rearrangement events (Fig. 4). Interestingly, the coding joints of these same recombination events are frequently aberrant with deletions in the coding joints. In 5/7 cases examined, the *scid* coding joint deletions (ranging from -462 to -12 bases) are larger than normal for wildtype pre-B cell lines. We found no examples where the reciprocal joint was aberrantly rearranged in the presence of a normal coding joint These data demonstrate that the *scid* mutation primarily affects the formation of coding joints. *Scid* inversional rearrangements were indistinguishable from wildtype DNA rearrangments in 2/7 cases. Both had a perfect reciprocal joint, small deletions in the coding joint, and the addition of N region bases in the coding joints (Fig. 4). Thus, homozygous *scid* pre-B cell lines are entirely capable of generating wildtype rearrangements at very low frequencies. A low level of normal rearrangements are one possible explanation for the observation that the *scid* mutation has a leaky phenotype (Bosma et al., 1983). An infrequent functional rearrangement of both an Ig μ and κ gene would lead to the completion of clonal B cell differentiation and production of serum Ig as proposed earlier (Schuler et al., 1986; Hendrickson et al., 1988).

Discussion

The recombination pathway of V(D)J gene assembly may be described as a multi-step process (Fig. 5). We have compared each step for wildtype and *scid* rearrangement to identify the critical points for the *scid* mutation. The position of action of the *scid* mutation in this pathway can be examined by analyzing the variety of rearrangement events generated using retroviral recombination vectors. The *scid* defect consistently produces aberrant deletions in DNA, and can be explained as acting at any stage of rearrangement where strands are cleaved by heptamer recognition. In fact, *scid* DGR deletions centered over one heptamer element only were observed. These rearrangements may be the *scid* equivalent of wildtype "open/shut" events; *scid* rearrangement is initiated by cleavage of one heptamer, but the subsequent rejoining of the same strands is defective. Most of the *scid* DNA rearrangements of DGR would appear to involve cleavage at two heptamer elements, although we cannot distinguish between attempted inversional rearrangements resulting in deletions and those rearrangements where there may have been cleavage at only one heptamer followed by extensive deletion. Two experiments show that attempted rearrangement does occur. First, when the size of the deletions were small enough, inversions in DGR were recovered (Fig. 4). Secondly, hybrid joints were observed in *scid* cells, and the formation of hybrid joints are products of rearrangement rather than cleavage and deletion (Lewis et al., 1988).

Recognition of the heptamer-spacer-nonamer signal sequences is a prerequisite for the *scid* phenotype (Figs. 2 and 3). We never observed *scid* deletions in DGR or JVD or endogenous chromosomes that did not include a heptamer element in the deletion. The double heptamer point mutation is sufficient to inactivate JVD-VJ_{hep} for any rearrangements, including aberrant *scid* deletions, suggesting that the *scid* gene product is part of the lymphoid-specific recombination mechanism. The double point mutant has a more severe effect than single point mutants, suggesting that there is an altered recombination complex formed with the *scid* gene product. Alternatively, the recombinase may be much more active in *scid* pre-B cells than wildtype cells, and thus accentuate apparent differences in sensitivity to heptamer mutants.

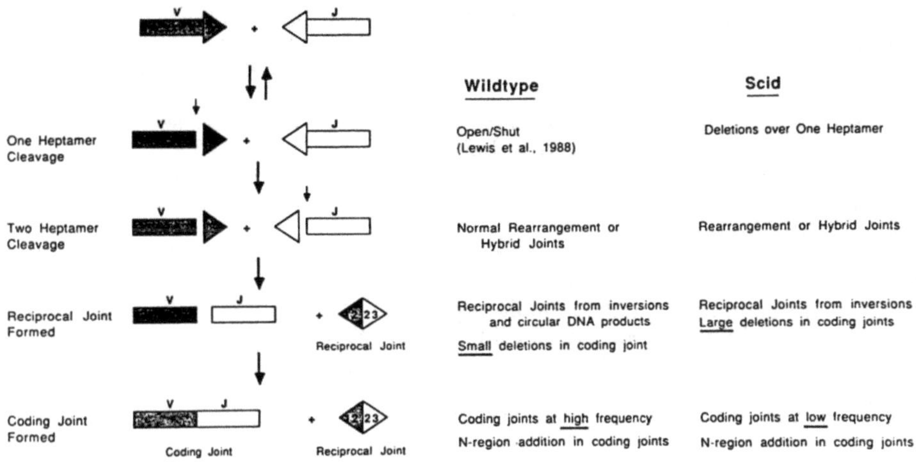

Figure 5. Pathway for Ig gene rearrangement and the intermediates observed during wildtype and *scid* rearrangements. See the text for discussion.

The *scid* defect is uniquely observed on the coding joints, under conditions where both the reciprocal and coding joints are recovered on the same molecule. Therefore, *scid* has a comparable asymmetry to wildtype junctions in which the deletions and N-regions are ordinarily restricted to the coding junction. Possibly, all of the DGR deletions we have observed are unrestricted coding joint deletions where the reciprocal junction had initially formed Therefore, two joining steps in the recombination mechanism probably exist, or at least are differentially recognized by the recombinase. In contrast to our results, in a transient transfection assay for rearrangement,*scid* deletions and N-region insertions in the reciprocal joints were observed (Lieber et al., 1988). Thus, there may be topological or structural differences for a recombination template that is integrated in the chromosome versus an extrachromosomal plasmid element.

In summary, our data strongly suggest that the *scid* gene product is a component of the rearrangement pathway. The *scid* phenotype is most consistent with it being a recombinase defect in which the generation of a coding joint is hampered. We plan to investigate whether the *scid* gene encodes a recombinase by a direct test using genomic DNA transfection and rearrangement of Ig genes in fibroblasts (Schatz and Baltimore, 1988).

References

Bosma G, Custer R, and Bosma M (1983) A severe combined immunodeficiency mutation in mice. Nature 301: 527-530

Hendrickson E, Schatz D, and Weaver D (1988) The *scid* gene encodes a trans-acting factor that mediates the rejoining event of Ig gene rearrangement. Genes and Development 2: 817-829

Landau N, Schatz D, Rosa M, and Baltimore D (1987) Increased frequency of N-region insertion in a murine pre-B cell line infected with a Terminal Deoxynucleotidyl Transferase retroviral expression vector. Mol Cell Biol 7: 3237-3243

Lieber M, Hesse J, Lewis S, Bosma G, Rosenberg N, Mizuuchi K, Bosma M, and Gellert M (1988) The defect in murine severe combined immune deficiency: joining of signal sequences but not coding segments in V(D)J recombination. Cell 55: 7-16

Lewis S, Gifford A, and Baltimore D (1984) Joining of Vk to Jk gene segments in a retroviral vector introduced into lymphoid cells. Nature 308: 425-428

Lewis S, Gifford A, and Baltimore D (1985) DNA elements are asymmetrically joined during the site-specific recombination of kappa immunoglobulin genes. Science 228: 677-685

Lewis S, Hesse J, Mizuuchi K, and Gellert M (1988) Novel strand exchanges in V(D)J recombination. Cell 55: 1099-1107

Schatz D and Baltimore D (1988) Stable expression of immunoglobulin gene V(D)J recombinase activity by gene transfer into 3T3 fibroblasts. Cell 53: 107-115

Schuler W, Weiler I, Schuler A, Phillips R, Rosenberg N, Mak T, Kearney J, Perry R, and Bosma M (1986) Rearrangement of antigen receptor genes is defective in mice with severe combined immune deficiency. Cell 46: 963-972

The Effect of the *scid* Mutation on Mechanism and Control of Immunoglobulin Heavy and Light Chain Gene Rearrangement

T. Keith Blackwell, Pierre Ferrier, Barbara A. Malynn, Roberta R. Pollock,* Lori R. Covey, Heikyung Suh, L. Bruce Heinke, Gabrielle M. Fulop,* Robert A. Phillips,* George D. Yancopoulos, and Frederick W. Alt

Howard Hughes Medical Institute and Departments of Biochemistry and Microbiology, Columbia University College of Physicians and Surgeons, 630 W. 168 Street, New York, NY 10032

*Department of Medical Genetics, University of Toronto and Division of Haematology/Oncology, The Hospital for Sick Children, Toronto, Canada

SUMMARY

Most Abelson murine leukemia virus (A-MuLV)-transformed cell lines derived from scid (severe combined immune deficient) mice actively rearrange their endogenous immunoglobulin (Ig) heavy (H), but not light (L) chain variable region genes. Such cell lines express germline V_H segments and other RNA transcripts that are characteristically produced by early precursor (pre)-B lymphocytes, but do not express high levels of transcripts from the germline kappa (κ) constant region (C_κ) locus. However, we have derived scid A-MuLV transformants that express germline C_κ transcripts and attempt κ gene assembly. In one case κ gene expression and rearrangement occurred in the absence of μ H chain expression, and in another was not induced efficiently by introduction of a μ-expression vector. Although the vast majority of scid H and L chain coding sequence joins are grossly aberrant, scid A-MuLV transformants can form normal coding joins at a very low frequency. In contrast, these cells form generally normal signal sequence joins at an approximately normal efficiency. Thus, these findings mechanistically distinguish coding and signal join formation. Subcloning analyses suggest that scid A-MuLV transformants that do not attempt chromosomal coding sequence joining may have a relative survival advantage, and therefore that these events may often result in unrepaired chromosomal breakage and cell death.

INTRODUCTION

During pre-B cell differentiation, Ig H chain variable region genes are assembled from variable (V), diversity (D), and joining (J) segments, and L chain variable region genes from analogous V_L and J_L segments (Tonegawa, 1983). These DNA rearrangement events are thought to be catalyzed by a site-specific recombinase machinery (V(D)J recombinase) that recognizes consensus signal sequences that flank variable region gene segments (reviewed in Alt, et al., 1987). V(D)J recombination appears to involve a non-reciprocal mechanism (Alt and Baltimore, 1982, Lieber, et al., 1988) that is initiated by introduction of a site-specific cut on both recombining segments at the border of signal and coding sequences. Bases may then be removed from the coding but not signal sequence ends and inserted de novo between either set of ends, which are then ligated to form respective "coding" and "signal" joins. Intervening DNA is generally deleted during joining between H chain variable region gene segments, which are oriented with their signal sequences facing each other. However, many murine and human kappa (κ) L chain variable region gene segments are oriented with their signal sequences pointed in the same direction, so that joining between them occurs by inversion and the signal join is retained on the chromosome (reviewed in Alt, et al, 1987).

Current Topics in Microbiology and Immunology, Vol. 152

During B lymphocyte differentiation H chain genes are assembled before L chain genes -- it has been proposed that initiation of L chain gene assembly is signalled by production of the membrane-bound form of μ H chain (μ_m) (Reth, et al., 1987; reviewed in Blackwell and Alt, 1988). Assembly of Ig and the related T cell antigen receptor (TCR) variable region genes appears to be controlled by modulation of the recombinational "accessibility" of these loci (Yancopoulos, et al., 1986; Blackwell, et al., 1986), which may be related either directly or indirectly to their transcriptional competence (reviewed in Blackwell and Alt, 1988). For example, early pre-B cells that join V_H and DJ_H segments but have not begun L chain gene assembly express germline V_H segments of the J558 family (Yancopoulos and Alt, 1985), but do not transcribe the germline C_κ region at high levels (Nelson, et al., 1985). However, more mature B lineage cells produce high levels of germline C_κ transcripts (Van Ness, et al., 1981; Nelson, et al., 1985; Kelley, et al., 1988).

The <u>scid</u> mutation impairs V(D)J recombinase activity (Schuler, et al., 1986), so that homozygous <u>scid</u> mice do not generate functional B or T lymphocytes (Bosma, et al., 1983; Dorshkind, et al., 1984). A-MuLV transformation of <u>scid</u> bone marrow generates pre-B cell lines that actively form aberrant rearrangements of their Ig H chain locus (Malynn, et al., 1988). These rearrangement events involve appropriate gene segments, but generally result either in joining of one segment to sequences that are distal to the other, or in recombination between sequences distal to both recombining partners (Schuler, et al., 1986; Hendrickson, et al., 1988; Kim, et al., 1988; Malynn, et al., 1988; Okazaki, et al., 1988). These findings suggested that the <u>scid</u> V(D)J recombinase can recognize and cut normally at variable region gene segments, but is unable to join coding segment ends together (Hendrickson, et al., 1988; Malynn, et al., 1988).

The μ-signalling hypothesis predicts that because <u>scid</u> pre-B cells do not correctly assemble H chain genes, they should be analogous to normal early pre-B cells that have not initiated L chain gene assembly. We have therefore assayed <u>scid</u> A-MuLV transformants for expression of several RNA transcripts that are produced by such early pre-B cells. We have also analyzed formation of coding joins between endogenous Ig variable region gene segments and within an introduced recombination substrate, and have assayed for expression and rearrangement of the J_κ-C_κ locus in <u>scid</u> A-MuLV transformants that either lack μ expression or that express μ_m chains from an introduced vector. In these experiments we have identified two <u>scid</u> A-MuLV transformants that attempt V_κ-to-J_κ rearrangement during culture, thus allowing us to study the effect of the <u>scid</u> mutation on signal join formation.

<u>RESULTS</u>

<u>Most scid A-MuLV Transformants Express RNA Transcripts Characteristic of Early Pre-B Cells</u>

We have compared the levels at which A-MuLV transformants derived from either <u>scid</u> or normal C.B-17 mice (Malynn, et al., 1987) express RNA transcripts that are produced by pre-B cells. These cell lines express similar levels of "sterile" (non-coding) μ H chain constant region (C_μ) transcripts (from 1.9 to 3.0 kb; figure 1), which are produced by early pre-B cells but may also be expressed later in the B-lineage (reviewed in Blackwell and Alt, 1988). With the exception of the <u>scid</u> transformant SC11A, all of these lines express significant levels of germline V_HJ558 transcripts (figure 1), which are normally produced by pre-B cells that undergo V_H-to-DJ_H rearrangement, but not by later B-lineage cells (Yancopoulos and Alt, 1985). Significantly, SC11A is also the only cell line analyzed that does not express a high level of transcripts that encode terminal deoxynucleotidyl transferase (TdT). TdT is thought to introduce extra bases (N regions) into V_HDJ_H and TCR but generally not V_LJ_L rearrangements, and is expressed in pre-B cells that assemble H chain genes but apparently not in cells that assemble L chain genes (reviewed in Alt, et al., 1987). All of these A-MuLV transformants express the N-<u>myc</u> gene, which is expressed in all pre-B stages (Zimmerman et al., 1986). Thus, with the exception of SC11A, the <u>scid</u> A-MuLV transformants express a set of RNA transcripts that are characteristically produced by pre-B cells that assemble V_HDJ_H rearrangements. The finding that SC11A does not express germline V_HJ558 or TdT transcripts raises the possibility that it might represent a later-stage pre-B cell that has targeted L chain gene segments for rearrangement.

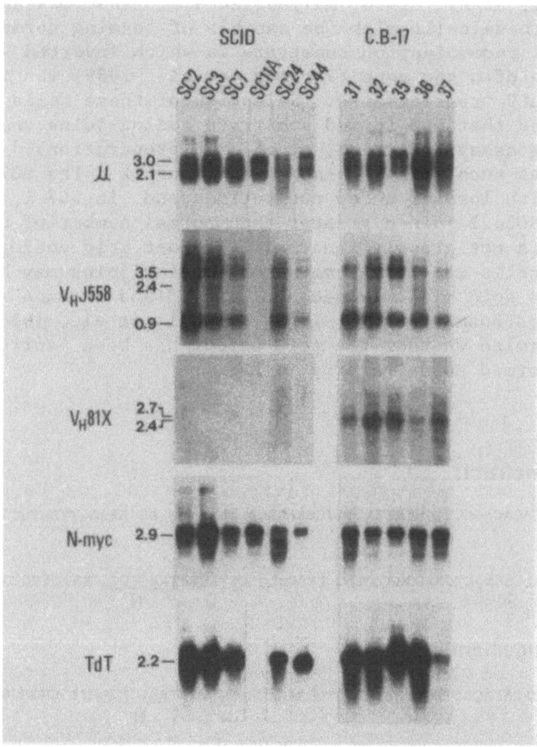

<u>Figure 1. Expression in scid and normal A-MuLV transformants of genes that are expressed in pre-B cells</u>. Total RNA (10μg) from C.B-17 <u>scid</u> (left panel) and from normal C.B-17 (right panel) cell lines was assayed by Northern blotting for hybridization to the indicated ^{32}P-labelled probes. Sizes are in kb. Many C_μ transcripts hybridize to J_H-C_μ intron probes (not shown), thus identifying them as germline C_μ transcripts that initiate within this region (reviewed in Blackwell and Alt, 1988).

Most of the <u>scid</u> A-MuLV transformants actively rearrange H chain gene segments (Malynn, et al., 1987). V_HDJ_H rearrangements that occur during culture of A-MuLV transformants can be detected as H chain mRNA sequences that hybridize with V_H-specific probes--even when they are not detectable by Southern blotting (Yancopoulos et al., 1984). C.B-17 has the IgH^b H chain locus (Bosma et al., 1983), in which the V_H7183 family is the most J_H-proximal and is preferentially utilized in V_H to DJ_H rearrangements (Yancopoulos et al., 1984; Yancopoulos, Malynn, and Alt, 1988). Although the

C.B-17 A-MuLV transformants express substantial levels of V_H7183-hybridizing μ mRNA, the <u>scid</u> A-MuLV transformants do not express detectable amounts of μ mRNA sequences that hybridize to V_H81X or several other V_H probes (figure 1; not shown), implying that assembly of normal V_HDJ_H rearrangements is a very infrequent event in <u>scid</u> pre-B cells.

<u>Formation of Coding Joins in scid Pre-B Cells</u>

In <u>scid</u> pre-B cells attempts at coding join formation generally result in deletion of one or both recombining segments (Schuler, et al., 1986; Hendrickson, et al., 1988; Kim, et al., 1988; Malynn, et al., 1988; Okazaki, et al., 1988). To test whether these cells might be capable of forming normal coding joins at all, a retroviral recombination substrate in which inverted V_H-to-DJ_H joining allows expression of a <u>gpt</u> gene (Ferrier, et al., 1989) was introduced into several <u>scid</u> A-MuLV transformants. Selection of these cells for <u>gpt</u> expression isolated subclones that had formed substrate coding joins that appear normal by Southern blotting assays (Ferrier, et al., in preparation). The nucleotide sequences of three such joins are shown in figure 2. The SC7-1 and SC24-1 joins appear normal, with loss of a few nucleotides and, in SC7.1, presence of an N region. In the SC24.2 join a greater than normal number of bases have been deleted, but it is not grossly abnormal like most <u>scid</u> coding joins. The finding that <u>scid</u> pre-B cells can form some normal coding joins may be relevant to understanding how <u>scid</u> mice may become "leaky," and produce an oligoclonal B and T cell repertoire (Bosma, et al., 1988; Carroll, et al., 1988). Studies are underway to determine whether the <u>scid</u> defect has been "corrected" in the <u>scid</u> subclones that formed normal coding joins.

A: VH-to-DJH Construct:

```
VH81X  -- TCTGAGGACACAGCCTTGTTATTACTGTGCAAGACA cacaatg AGCAAAAGTTACTGTGAGCTCAA actaaaacc --
                         VH81X                   7                                9

DJH3       -- gattttgt CAAGGATCTAC tactgtg CCTACTATAAGCCCCCTTACTGGGGCCAAGGGACTCTGGTCACTG --
                  9                   7       D      N                JH3
```

B: VHDJH Rearrangements:

```
SC7-1  -- TCTGAGGACACAGCCTTGTTATTACTGTGCAAGACA TG CTATAAGCCCCCTTACTGGGGCCAAGGGACTCTGGTCACTG --
                         VH81X                N      N                JH3

SC24-1 -- TCTGAGGACACAGCCTTGTTATTACTGTGCAAGAC - A - CTATAAGCCCCCTTACTGGGGCCAAGGGACTCTGGTCACTG --
                         VH81X                       N                JH3

SC24-2 -- TCTGAGGACACA ---------------------- GCC --------------------- AGGGACTCTGGTCACTG --
              VH81X                                                          JH3
```

<u>Figure 2: "Normal V-to-DJ_H Rearrangements in scid Pre-B Cell Lines.</u> A. Partial nucleotide sequences of the V_H and DJ_H segments within the retroviral construct. Signal heptamer and nonamer sequences are indicated by lower case letters, and an N region is underlined. B. Nucleotide sequences of rearrangements within the substrate in <u>gpt</u>-resistant SC7 and SC24 subclones, as isolated by the polymerase chain reaction (Saiki, et al., 1988). The joins are aligned with the sequences in A so that base loss is apparent. Bases that could be derived from either partner are indicated between dashes, and an N region is shown as in A.

During culture <u>scid</u> A-MuLV transformants attempt to join endogenous D and

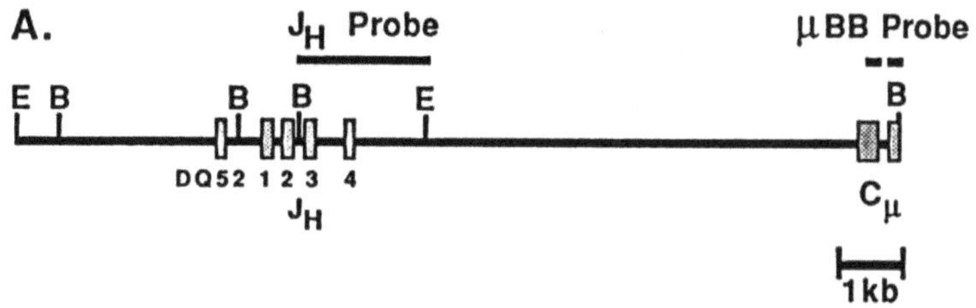

J_H segments, or can attempt to join a V_H segment to an aberrant "DJ_H" complex if the recognition signals upstream of the D segment are intact (Malynn, et al., 1988). Rearrangements within the J_H locus can be identified as novel EcoR1 fragments that hybridize with the J_H probe (figure 3A), or as novel BamH1 fragments that hybridize with the μB-B probe (figure 3A). Many <u>scid</u> A-MuLV transformants are characterized by multiple J_H-hybridizing EcoR1 fragments, indicating ongoing rearrangement (Malynn, et al., 1988). However, during passage these lines often rapidly delete the J_H region, so that in this respect they then resemble many <u>scid</u> A-MuLV transformants described previously (Schuler, et al., 1986). To investigate the basis for these deletions, we have analyzed J_H-associated rearrangements in multiple subclones of <u>scid</u> A-MuLV transformants by Southern blotting (not shown). A representative <u>scid</u> transformant, SC7 (figure 3B, lane 2), has many J_H-hybridizing Eco R1 fragments, but in nine of eleven SC7 subclones that have J_H-hybridizing sequences these rearrangements seem identical to those present in SC7.1 (figure 3B, lanes 3, 7, 11, and 15; table 1). Most SC7 subclones have deleted the J_H region, and of these that were tested nearly all are identical to SC7.2 (figure 3B, lanes 4, 8, 12, and 16; table 1). Most D segments are present in SC7.2 (figure 3B, lanes 8 and 12), but SC7.1 retains only a 10 kb DFL16-positive Eco R1 fragment (figure 3B, lanes 7 and 11) that does not contain a functional D segment (Alt, et al., 1984). Thus, SC7.1- and SC7.2-like subclones have completely deleted either their D or J_H loci and accordingly do not undergo further rearrangement (not shown). Significantly, such subclones are derived from a small minority of SC7 cells (figure 3B, lanes 2, 3, and 14-16), suggesting that <u>scid</u> pre-B cells that cease V(D)J recombination may have a survival advantage.

B:

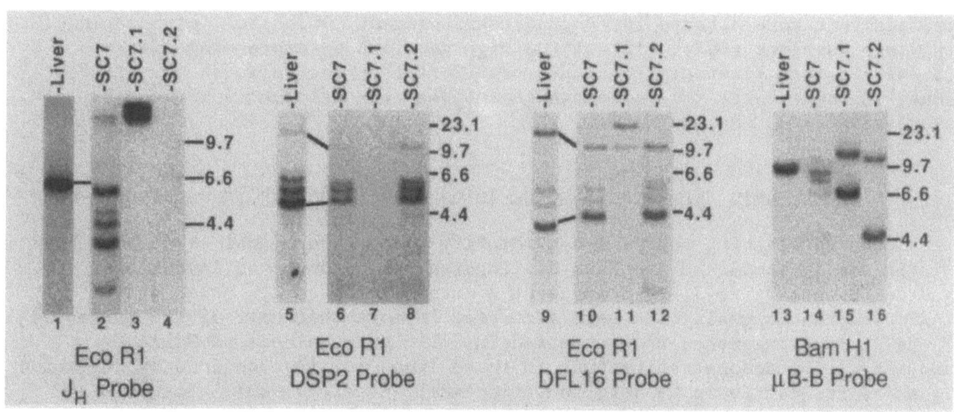

<u>Figure 3. J_H-associated rearrangement in SC7 subclones.</u> A. Partial restriction map of the J_H-C_μ region (reviewed in Tonegawa, 1983). DQ52, J_H, $C_\mu 1$, and part of $C_\mu 2$ coding regions (not to scale) are indicated. E: Eco R1; B: Bam H1. B. 10 μg of genomic DNA from <u>scid</u> liver or the indicated cell lines was digested with Eco R1 or Bam H1 and analyzed by Southern blotting for hybridization to the indicated ^{32}P-labelled probes. Sizes are in kb. Lines indicate identical sizes.

Table 1

		1	2	Total
SC7 Subclones analyzed:		24	17	41
J_H Hybridizing Sequences Present	SC7.1:	7	2	9
	Other:	2	0	2
J_H Hybridizing Sequences Deleted	SC7.2:	2	15	17
	Other:	1	0	1
	Uncharacterized:	12	0	12

Table 1. Characterization of SC7 Subclones. SC7 cellular subclones from two successive experiments (denoted 1 and 2) are classified based on the assays shown in figure 3.

Expression and Rearrangement of Kappa Light Chain Genes in scid Transformants

Five of six analyzed scid A-MuLV transformants express germline C_κ transcripts at low-to-undetectable levels, but SC11A expresses them at a strikingly high level (Blackwell, et al., 1989). J_κ-associated rearrangement was not detected by Southern blotting in cellular subclones of scid A-MuLV transformants that do not express the C_κ region at high levels, but products of multiple κ gene rearrangement events were apparent in SC11A and its cellular subclones (Blackwell, et al., 1989). Although SC11A underwent κ gene rearrangement, it does not produce detectable levels of μ H chains and has not assembled a V_HDJ_H rearrangement (Blackwell, et al., 1989). To examine the role of μ_m expression in induction of κ gene rearrangement, a μ_m-expression vector was introduced into two scid A-MuLV transformants (SC24 and SC44) that rearrange H chain genes only (Blackwell, et al., 1989). J_κ-associated rearrangement was not detected in any of fourteen independent μ_m-expressing transfectants, among which germline C_κ transcripts were expressed at levels ranging from undetectable to nearly as high as SC11A Cellular subclones of both a high and a low germline C_κ-expressing transfectant were assayed for κ gene rearrangement, which was detected in only one subclone (SC24.1.13) of the high germline C_κ-expressing line (SC24.1) and in secondary subclones of SC24.1.13 (Blackwell, et al., 1989). Thus, in these cells κ gene rearrangement was not efficiently induced by expression of μ_m chains.

κ L Chain Signal and Coding Joins in scid A-MuLV Transformants

Southern blotting analyses can identify coding and signal joins that occur within the J_κ locus and can thus distinguish the products of inversional and deletional V_κ-to-J_κ rearrangement events. These assays have indicated that signal joins were generated in more that half of the attempted V_κ-to-J_κ rearrangements that occurred in cellular subclones of SC11A and SC24.1.13, demonstrating that in these lines joining occurred by inversion at least as frequently as by deletion (Blackwell, et al., 1989). Several J_κ-associated rearrangements, most of which could be identified as putative coding and signal joins, have been molecularly cloned from these cell lines (Blackwell, et al., 1989). The scid κ coding joins are aberrant and analogous in structure to scid H chain coding joins. In contrast, five of six predicted scid signal joins consist of J_κ recognition sequences (with 23 bp spacers) joined back-to-back with V_κ recognition sequences (with 12 bp spacers) -- only one of these joins is aberrant and can not be definitively classified as a signal join. In one of the <u>scid</u> signal joins 2 bp have been lost from the $J_\kappa 1$ heptamer, but in the other four joins the heptamers are joined precisely. None of these signal joins contain N regions, as would be predicted from the observation that SC11A (figure 1) and SC24.1 (not shown) do not express TdT transcripts. These findings demonstrate the <u>scid</u> V(D)J recombinase can form normal signal joins at high frequency (Blackwell, et al., 1989). The V_κ sequences flanking these <u>scid</u> signal joins are highly similar to each other and to the 3' flank of a member of the $V_\kappa 5$ subgroup, suggesting that this family of V_κ segments may be preferentially utilized in inverted V_κ-to-J_κ rearrangements or in secondary signal joins that may follow (Blackwell, et al., 1989).

DISCUSSION

Regulation of Ig Variable Region Gene Rearrangement in scid Pre-B Cells

Most analyzed <u>scid</u> A-MuLV transformants express germline V_H but not germline C_κ transcripts and rearrange H chain gene segments only; in contrast, the <u>scid</u> cell lines that rearrange κ genes have previously rearranged their H chain locus and express germline C_κ but not germline V_H transcripts at high levels (figure 1; not shown; Malynn, et al., 1988; Blackwell, et al., 1989). These findings are consistent with an ordered program of Ig H and L chain gene assembly, and with the notion that transcriptional competence is a prerequisite for rearrangement of these loci (reviewed in Blackwell and Alt, 1988). In this context, it is of interest to determine whether pre-B cells that undergo V_κ-to-J_κ rearrangement transcribe germline V_κ segments. Consistent with the notion that pre-B cells that undergo κ gene rearrangement generally do not express TdT (Alt, et al., 1982), neither SC11A (figure 1) nor SC24.1 (not shown) express significant levels of TdT transcripts, suggesting that during pre-B cell differentiation TdT expression might be terminated by a signal that is involved in initiation of κ gene assembly.

Some of our results appear inconsistent with the proposal that the μ H chain signals initiation of L chain gene assembly (Reth, et al., 1985, 1987; Blackwell and Alt, 1988). SC11A rearranges κ gene segments but has not assembled a $V_H DJ_H$ rearrangement and does not express μ protein, and κ gene rearrangement is not induced efficiently in numerous <u>scid</u> (and normal) A-MuLV transformants by expression of μ_m chains from an introduced vector (Blackwell, et al., 1989). Nevertheless, most <u>scid</u> A-MuLV transformants do not rearrange their κ genes (Blackwell et al., 1989), and evidence from several experimental systems supports the μ-signalling model (reviewed in Blackwell and Alt, 1988). It is possible that μ_m expression, perhaps together with other factors, may normally signal initiation of κ gene assembly, and that aspects of A-MuLV transformation or of growth of these cell lines outside of their <u>in vivo</u> environment may interfere with their capacity to differentiate. In this regard, it has been suggested that only cells that have assembled endogenous $V_H DJ_H$ rearrangements may efficiently initiate L chain gene assembly (Reth, et al., 1987).

scid A-MuLV Transformants Form Normal Signal and Abnormal Coding Joins

Analysis of <u>scid</u> H chain coding joins led to the proposal that the <u>scid</u> V(D)J recombinase can bind to, juxtapose, and cut at Ig variable region gene segments with normal specificity, but is essentially unable to join the coding sequences to each other (Malynn, et al., 1988). Aberrant <u>scid</u> coding joins therefore derive from "rescue" of this otherwise lethal double-strand break by illegitimate recombination (Malynn, et al., 1988). Consistent with this model, nearly all surviving SC7 subclones derive from cells that have deleted either their D or J_H loci and no longer undergo V(D)J recombination (figure 3; table 1). Thus, although the scid V(D)J recombinase can form structurally normal coding joins at low frequency (figure 2), in most cases the chromosomal break between the coding segments that is generated by the V(D)J recombination event is unrepaired and therefore lethal (Malynn, et al., 1988). In contrast, in SC11A and SC24.1.13 the scid V(D)J recombinase forms generally normal signal joins at an approximately normal efficiency (Blackwell, et al., 1989). These findings demonstrate that the V(D)J recombinase functions that recognize, cut at, and join signal sequences are largely unaffected by the scid mutation. Experiments involving extrachromosomal recombination substrates have led to similar conclusions (Lieber, et al., 1988a). Our results further demonstrate that the scid V(D)J recombinase can juxtapose and join signal sequences that are separated by large distances on the chromosome (Blackwell, et al., 1989). Therefore, coding and signal joins are formed by processes that are mechanistically distinct.

Although scid mice appear to have an approximately normal number of early B cell precursors (Fulop et al., 1988) and express germline Ig transcripts in their lymphopoietic organs (Schuler, et al., 1988), these organs do not contain readily detectable numbers of B or T cells that have undergone rearrangement events (Schuler, et al., 1986). This observation may suggest that mechanisms exist that quickly eliminate precursor lymphocytes that do not assemble functional antigen receptor genes (Schuler, et al., 1986). However, it is also possible that the scid defect itself may be lethal to developing lymphocytes. Developing B and T cells in vivo must undergo V(D)J recombination events in rapid succession, which in scid mice might result in overwhelming cell death due to unrepaired chromosomal breaks. Perhaps abnormal coding join formation can be observed during culture of scid A-MuLV transformants only because of a much lower rate of V(D)J recombination in vitro versus in vivo, thus allowing expansion of a cell that has "rescued" an abortive V(D)J recombination event by illegitimate recombination (Malynn, et al., 1988).

Acknowledgements

T.K.B. and F.F. are fellows of the Howard Hughes Medical Institute, B.A.M. and R.R.P. are fellows of the Leukemia Society of America, L.R.C. is a recipient of a Cancer Research Institute fellowship, and G.D.Y. is a Lucille P. Markey Scholar. This work was supported by the Howard Hughes Medical Institute, grants from the NIH (AI-20047 and CA-40427), and grants from the Medical Research Council and the National Cancer Institute of Canada to R.A.P.

REFERENCES

Alt F, Rosenberg N, Lewis S, Thomas E, Baltimore D (1981) Organization and reorganization of immunoglobulin genes in A-MuLV-transformed cells: Rearrangement of heavy but not light chain genes. Cell 27: 381-390

Alt FW, Baltimore D (1982) Joining of immunoglobulin heavy chain gene segments: Implications from a chromosome with evidence of three D-J_H fusions. Proc Natl Acad Sci USA 79: 4118-4122

Alt FW, Yancopoulos GD, Blackwell TK, Wood C, Thomas E, Boss M, Coffman R, Rosenberg N, Tonegawa S, Baltimore D (1984) Ordered rearrangement of immunoglobulin heavy chain variable region gene segments. EMBO J 3: 1209-1219

Alt FW, Blackwell TK, Yancopoulos GD (1987) Development of the primary antibody repertoire. Science 238: 1079-1087

Blackwell TK, Alt FW (1988) Immunoglobulin genes. In: Hames BD, Glover D (eds) Molecular immunology. IRL Press, Oxford, pp 1-60

Blackwell TK, Moore MW, Yancopoulos GD, Suh H, Lutzker S, Selsing E, Alt FW (1986) Recombination between immunoglobulin variable region segments is enhanced by transcription. Nature 324: 585-589
Blackwell TK, Malynn BA, Pollock RR, Ferrier P, Covey L, Fulop GM, Phillips RA, Yancopoulos GD, Alt FW (1989) Isolation of scid pre-B cells that rearrange kappa light chain genes: Formation of normal signal and abnormal coding joins. EMBO J 8: 735-742
Bosma GC, Custer RP, and Bosma MJ A severe combined immunodeficiency mutation in the mouse. (1983) Nature 301: 527-530
Bosma GC, Fried M, Custer RP, Carroll A, Gibson D, and Bosma MJ (1988) Evidence of functional lymphocytes in some (leaky) scid mice. J Exp Med 167: 1016-1033
Carroll AM, Bosma MJ (1988) Detection and characterization of functional cells in mice with severe combined immune deficiency. Eur J Immunol 18: 1965-1971
Dorshkind K, Keller GM, Phillips RA, Miller RG, Bosma GC, O'Toole M, and Bosma MJ (1984) Functional status of cells from lymphoid and myeloid tissues in mice with severe combined immune deficiency disease. J Immunol 132, 1804-1808
Durdik J, Moore MW, Selsing E (1984) Novel κ light-chain gene rearrangements in mouse λ light chain-producing B lymphocytes. Nature 307: 749-752
Ferrier P, Covey L, Suh H, Winoto A, Hood L, Alt FW (1989) T cell receptor DJ but not VDJ rearrangement within a recombination substrate introduced into a pre-B cell line. Int Immunol 1, 66-74
Fulop GM, Bosma GC, Bosma MJ, Phillips RA (1988) Early B-cell precursors in scid mice: Normal numbers of cells transformable with Abelson murine leukemia virus (A-MuLV). Cell Immunol 113: 192-201
Hendrickson EA, Schatz DG, Weaver DT (1988) The scid gene encodes a trans-acting factor that mediates the rejoining event of Ig gene rearrangement. Genes and Dev 2: 817-829
Kelley DE, Pollok BA, Atchinson ML, Perry RP (1988) The coupling between enhancer activity and hypomethylation of κ immunoglobulin genes is developmentally regulated. Mol Cell Biol 8: 930-937
Kim M-G, Schuler W, Bosma MJ, Marcu K (1988) Abnormal recombination of Igh D and J gene segments in transformed pre-B cells of scid mice. J Immunol 141: 1341-1347
Lieber MR, Hesse JE, Lewis S, Bosma GC, Rosenberg N, Mizuuchi K, Bosma MJ, Gellert M (1988a) The defect in murine severe combined immune deficiency: Joining of signal sequences but not coding segments in V(D)J recombination. Cell 55: 7-16
Lieber MR, Hesse JE, Mizuuchi K, Gellert M (1988b) Lymphoid V(D)J recombination: Nucleotide insertion at signal joints as well as coding joints. Proc Natl Acad Sci USA 85: 8588-8592
Malynn BA, Blackwell TK, Fulop GM, Rathbun G, Furley AJW, Ferrier P, Heinke LB, Phillips RA, Yancopoulos GD, Alt FW (1988) The scid defect affects the final step of the immunoglobulin VDJ recombinase mechanism. Cell 54: 453-460
Nelson KJ, Kelley DE, Perry RP (1985) Inducible transcription of the unrearranged C_{κ} locus is a common feature of pre-B cells and does not require DNA or protein synthesis. Proc Natl Acad Sci USA 82: 5305-5309
Okazaki K, Nishikawa S-I, Sakano H (1988) Aberrant immunoglobulin gene rearrangement in scid mouse bone marrow cells. J Immunol 141: 1348-1352
Reth M, Ammirati P, Jackson S, Alt FW (1985) Regulated progression of a cultured pre-B-cell line to the B-cell stage. Nature 317: 353-355
Reth M, Petrac E, Wiese P, Lobel L, Alt FW (1987) Activation of V_{κ} gene rearrangement in pre-B cells follows the expression of membrane-bound immunoglobulin heavy chains. EMBO J 6: 3299-3305
Saiki RK, Gelfand DH, Stoffel S, Scharf SJ, Higuchi R, Horn GT, Mullis KB, Erlich HA (1988) Primer-directed enzymatic amplification of DNA with a thermostable DNA polymerase. Science 239: 487-491
Schuler W, Weller IJ, Schuler A, Phillips RA, Rosenberg N, Mak TW, Kearney JF, Perry RP, Bosma MJ (1986) Rearrangement of antigen-receptor genes is defective in mice with severe combined immune deficiency. Cell 46: 963-972
Schuler W, Schuler A, Lennon GG, Bosma G, Bosma MJ (1988) Transcription of unrearranged antigen receptor genes in scid mice. EMBO J 7: 2019-2024

Tonegawa S (1983) Somatic generation of antibody diversity. Nature 302: 575-581
Van Ness BG, Weigert M, Coleclough C, Mather EL, Kelley DE, Perry RP (1981) Transcription of the unrearranged mouse C_κ locus: Sequence of the initiation region and comparison of activity with a rearranged V_κ-C_κ gene. Cell, 27: 593-602
Yancopoulos GD, Alt FW (1985) Developmentally controlled and tissue specific expression of unrearranged V_H gene segments. Cell 40: 271-281
Yancopoulos GD, Desiderio SV, Paskind M, Kearney JF, Baltimore D, Alt FW (1984) Preferential utilization of the most J_H-proximal V_H gene segments in pre-B-cell lines. Nature 311: 727-733
Yancopoulos GD, Blackwell TK, Suh H, Hood L, Alt FW (1986) Introduced T cell receptor variable region gene segments recombine in pre-B cells: Evidence that B and T cells use a common recombinase. Cell 44: 251-259
Yancopoulos GD, Malynn BA, and Alt FW (1988) Developmentally regulated and strain-specific expression of murine V_H gene families. J Exp Med 168: 417-435
Zimmerman KA, Yancopoulos GD, Collum RG, Smith RK, Kohl NE, Denis KA, Nau MM, Witte ON, Toran-Allerand D, Gee CE, Minna JD, Alt FW (1986) Differential expression of myc family genes during murine development. Nature 319, 780-783

III. Introduction of Functional Ig or TCR Transgenes

The Utilization of the *scid* Mutation in the Study of T Cell Development

Harald von Boehmer[1], Horst Blüthmann[2], Hung Sia Teh[3] and Bernadette Scott[1]

Basel Institute for Immunology, Grenzacherstrasse 487
4058 Basel, Switzerland

Introduction

It is believed that in scid mice the rearrangement process required for the expression of immunoglobulin (Ig) and T cell receptors (TCR) genes is defective (Bosma et al., 1983, Schuler et al., 1986). The ability to exploit this defect would provide unique opportunities to study the control of lymphocyte development by receptors encoded by rearranging gene segments.

To this end we have crossed TCR transgenic mice, containing productively rearranged α and β TCR genes, with scid mice and then backcrossed transgenic offspring of the F1 generation to scid mice. From the second cross we studied both transgene positive and transgene negative animals homozygous for the scid mutation (serum Ig negative animals). A comparison of the lymphoid cell composition of these mice indicated that the arrest in T lymphocyte development in scid mice can be overcome by introducing rearranged TCR genes, i.e. that the lack of T cell development in scid mice is solely a consequence of the defective rearrangment process (Scott et al. 1989).

Because we had introduced a TCR of defined specificity into these mice (the receptor is specific for the male [HY] antigen in the context of class I H-2D^{b} major histocompatibility complex [MHC] antigens) and the scid/scid offspring of the first backcross generation differed in their MHC antigens (H-2$^{b/d}$ heterozygous or H-2$^{d/d}$ homozygous) we could also analyse the impact of TCR specificity as well as MHC haplotype on the development of lymphocytes. We conclude from these studies that in female mice (i.e. in the absence of nominal antigen) a specific interaction of the TCR with thymic MHC antigens is necessary for the development of mature T cells (Scott et al. 1989).

The presence of rearranged TCR genes also allowed us to address the question of whether the 'leakiness' of the immune deficiency in scid mice is due to a general reversal of the genetic defect (back mutation) or simply reflects occasional productive rearrangements in the genetically defective mice. These studies led to the conclusion that productive rearrangements of TCR α chain gene segments can occur in scid mice in the absence of significant numbers of B lymphocytes indicating the rescue of rare productive rearrangements by the presence of the transgenic β chain (Scott et al. 1989).

[2] Central Research Units, Hoffmann-La Roche & Co. Ltd, 4002 Basel, Switzerland

[3] Department of Microbiology
University of British Columbia
Vancouver, Canada V6T IW5

Current Topics in Microbiology and Immunology, Vol. 152

Finally, the analysis of male αβ transgenic scid mice showed that autospecific T cells can be negatively selected by deletion at an early stage of T cell development and that the deletion is MHC restricted.

The development of immature $CD4^{+}8^{+}$ thymocytes requires productive TCR rearrangement(s).

The thymus of nontransgenic scid mice contained low numbers of lymphocytes (2-3 x 10^{6}), all of which were Thy-1 positive but lacked CD4 and CD8 surface markers. None of these cells expressed TCR associated CD3 molecules and they resembled $CD4^{-}8^{-}$ thymocytes found early in development (14 days of gestation) in normal mice (Kisielow et al. 1984) before significant productive rearrangement of α and β TCR genes (Snodgrass et al. 1985ab; Raulet et al. 1985). This picture of the thymus of scid mice was drastically changed after the introduction of productively rearranged α and β TCR genes into this mouse strain: thymocyte numbers fell into a normal range (5-8 x 10^{7}), the majority of lymphocytes expressed CD4 as well as CD8 antigens and practically all cells expressed the transgenic αβ TCR on the cell surface. These changes occurred irrespective of whether the αβ transgenic scid offsprings expressed $H\text{-}2^{b}$ MHC antigens or not (Fig. 1). Thus, the introduction of productively rearranged TCR genes into scid mice had overcome at least one block in T cell development and allowed the transition of $CD4^{-}8^{-}$ precursors into $CD4^{+}8^{+}$ progeny. This indicates that it is only the defective rearrangement process which arrests T lymphocyte development in these mice. At present it is not clear whether β or β and α rearrangements of TCR loci are required for this step. It is clear however that differentiation of $CD4^{-}8^{-}$ into $CD4^{+}8^{+}$ progeny does not depend on the interaction of the TCR with MHC ligands in the thymus as it also occurs in $H\text{-}2^{d}$ αβ transgenic scid mice lacking the restricting MHC determinants recognized by the introduced TCR.

The development of mature $CD4^{+}8^{-}$ and $CD4^{-}8^{+}$ thymocytes requires an interaction of the TCR with thymic MHC antigens

$H\text{-}2^{b/d}$ heterozygous and $H\text{-}2^{d}$ homozygous αβ transgenic scid differed with regard to the development of mature single positive $CD4^{+}8^{-}$ and $CD4^{-}8^{+}$ T cells in that the thymus of $H\text{-}2^{d}$ animals contained practically no mature T cells whereas a significant portion (~20%) of thymocytes in $H\text{-}2^{b/d}$ animals was of the $CD4^{-}8^{+}$ phenotype (Fig. 1). These experiments corroborate earlier observations indicating that the specific interaction of the αβ TCR with thymic MHC antigens determines the CD4/CD8 phenotype of mature T cells (Teh et al. 1988; Kisielow et al. 1988b; von Boehmer et al. 1989). In addition, they show for the first time that an interaction of the TCR with thymic MHC antigens is required for the development of mature T cells (Teh et al. 1988; Kisielow et al. 1988a/1988b). Experiments in non-scid mice, as well as earlier experiments in χ-irradiation chimeras (Bevan 1977; Zinkernagel et al. 1978; von Boehmer et al. 1978; Kappler et al. 1987), could not address the question of whether, in the thymus, T cells expressing certain receptors were just preferred (or others suppressed, Matzinger 1981), or whether the interaction of the TCR with thymic MHC antigens played an essential role in T cell development.

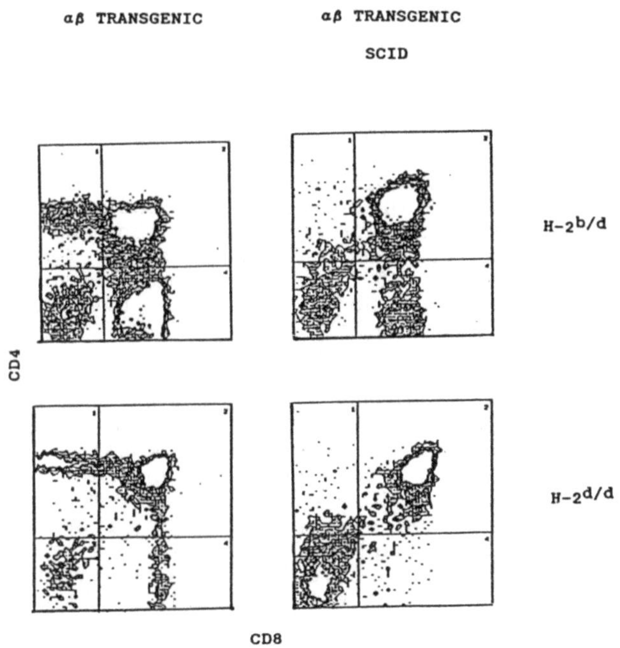

Fig. 1: Staining of thymocytes from female αβ transgenic mice (left) and female αβ transgenic scid mice (right) with CD4 (y axis) and CD8 (x axis) antibodies.

In contrast, the experiments in the 'quasi monoclonal' (see below) αβ transgenic scid mice, showing the virtual absence of single positive T cells in $H\text{-}2^d$ but not $H\text{-}2^{b/d}$ animals, rule out a role for regulatory T cells in the selection process and show that a TCR-MHC interaction is essential for obtaining single positive T cells. Additionally, these and previous experiments (Teh et al. 1988; Kisielow et al. 1988b) show that the intrathymic TCR-MHC interaction, in the absence of nominal (HY) antigen in female mice, determines the CD4/CD8 phenotype of mature T cells. The differentiation into single positive cells may be achieved in one of two ways: either the interaction of the TCR on $CD4^+8^+$ thymocytes with thymic MHC antigens instructs these cells to produce only one type of accessory molecules. Alternatively $CD4^+8^+$ cells may randomly lose CD4 or CD8 accessory molecules and the TCR-MHC interaction, for instance of a class I restricted receptor with thymic class I MHC antigens, can only select $CD4^-8^+$ T cells for further maturation. If the loss of CD4 and CD8 molecules were random, the lifespan of these single positive T cells must be extremely short because we do not detect significant numbers of these cells expressing the transgenic TCR in the thymus or periphery of $H\text{-}2^d$ αβ transgenic scid mice. Thus the final CD4/CD8 phenotype of single positive, mature T cells is determined by either instruction or selection through the specificity of the αβ TCR.

Infrequent productive rearrangements in immunodeficient scid mice

The analysis of peripheral lymphoid organs in αβ transgenic scid mice expressing either $H\text{-}2^{b/d}$ or $H\text{-}2^d$ MHC antigens revealed two facts: first, as expected, from the analysis of thymocytes, only $H\text{-}2^{b/d}$ animals contained substantial numbers of $CD4^-8^+$ lymphocytes expressing high levels of both transgenes in spleen and lymph nodes. Second, the lymph nodes of all αβ transgenic scid mice contained T lymphocytes, mostly of the $CD4^+8^-$ phenotype, which expressed, in addition to the transgenic β chain, endogenous α chains. Nevertheless, the peripheral lymphoid tissue of these mice was underdeveloped (about 10% of the size of that from conventional mice), so that the total number of T lymphocytes expressing endogenous α chains was small. In nontransgenic scid mice, to express a TCR, three independent productive joinings would have to occur in a given clone of T cells, the probability of which is very low. We believe that productive rearrangements of the TCR α locus are easily observed in transgenic scid mice because only a single productive rearrangement is required to become apparent in a cell which already contains a productively rearranged TCR β gene. The cell is then subjected to selection in the thymus. These data then indicate that productive rearrangements occur in scid mice. We propose that the peripheral T cells expressing endogenous α chains in the αβ transgenic scid mice represent an oligoclonal population formed by some thymic immigrants which expand, to some extent, in peripheral lymphoid tissue. This appears, at present, the best interpretation of the results showing that though very few mature $CD4^+8^-$ T cells are found in the thymus of these mice, such cells exist in significant numbers in the periphery.

Immature $CD4^+8^+$ thymocytes are deleted in male $H\text{-}2^{d/b}$ but not $H\text{-}2^d$ αβ transgenic scid mice

We have shown in previous experiments in αβ transgenic mice that tolerance to self peptides, presented by class I MHC antigens, can be achieved by deletion of immature $CD4^+8^+$ thymocytes expressing an autospecific transgenic receptor as well as high levels of CD8 accessory molecules (Kisielow et al. 1988a; Teh et al. 1989).

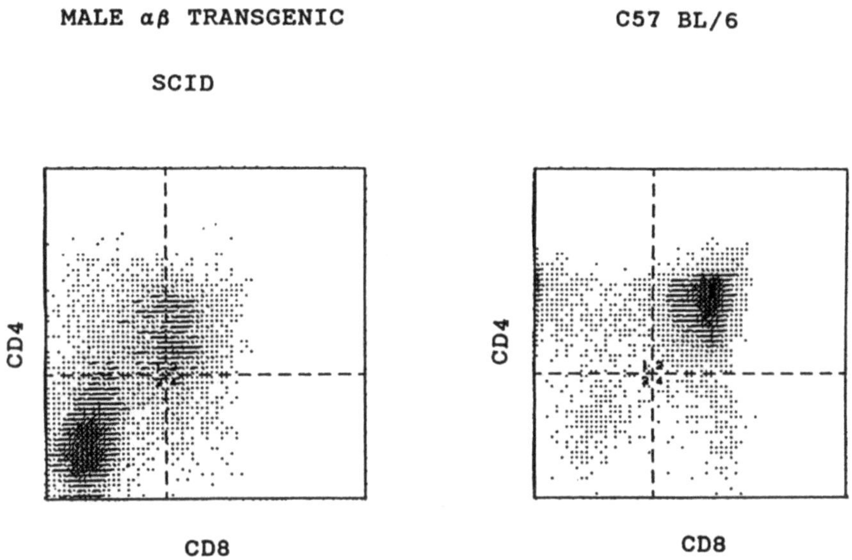

Fig. 2: Staining of thymocytes from male αβ transgenic scid mice (left) and C57 BL/6 mice (right) with CD4 (y axis) and CD8 (x axis) antibodies.

The same phenomenon is seen in αβ transgenic scid mice (Fig. 2), which also demonstrates that the deletion is independent of other regulatory T cells, i.e. the "quasi monoclonal" nature of the αβ transgenic scid thymus precludes the existence of T cells other than those expressing the transgene. As this deletion does not occur in male $H\text{-}2^d$ αβ transgenic scid mice one can conclude that deletion, just like T cell activation, is MHC restricted, a conclusion which has been drawn previously from more indirect experiments (Rammensee and Bevan 1984; Matzinger 1984). In addition, the data show that T cell deletion as well as T cell activation depends heavily on CD4/CD8 accessory molecules: $CD4^-8^-$ or $CD4^-8^{+(low)}$ T cells expressing high levels of the transgenic receptor are neither deleted in male αβ transgenic mice nor can they be activated by male D^b cells, though they can easily be activated by Concanavalin A (Kisielow et al. 1988a).

Discussion

The introduction of TCR transgenes into scid mice has provided new evidence that early T cell development is dependent on the productive rearrangement of TCR genes, and at a later stage, on the specific interaction of the αβ TCR with thymic MHC antigens. The first TCR dependent differentiation step is the transition of $CD4^-8^-$, TCR negative cells into $CD4^+8^+$ thymocytes, which begin to express the TCR on the cell surface.

It is important to point out that the early $CD4^-8^-$ cells with precursor potential for all other thymocyte subsets are distinct from those $CD4^-8^-$ cells which express γδ or αβ receptors on their surface and do not yield progeny after intrathymic injection (Kisielow et al. 1984; Crispe et al. 1987; Bluestone et al. 1987). Because our TCR transgenic scid mice lack γδ T cells, it appears unlikely that γδ cells play an essential role in the development of αβ T cells once these cells have rearranged their α and β genes. It can, however, not be excluded that they play some role in the rearrangement process of αβ cells. The role of $CD4^-8^-$ αβ TCR positive T cells is likewise mysterious: these cells appear late in ontogeny in normal mice (Crispe et al. 1987), preferentially utilise $V_\beta8$ genes and, like γδ cells, can be induced to become lymphokine producing cells or cytolytic cells. In the transgenic mice we detect $CD4^-8^-$ TCR^+ cells early in ontogeny before $CD4^+8^+$ thymocytes, but at present, it is not clear whether these early $CD4^-8^-$, TCR^+ cells correspond to those found in normal mice. Some $CD4^-8^-$ αβ TCR positive cells, however, leave the thymus of transgenic mice and accumulate in the periphery where they can be detected as functional T cells (Kisielow et al. 1988a; Teh et al. 1989). The migration of $CD4^-8^-$ $\alpha\beta^+$ cells is observed in H-$2^{b/d}$ as well as H-2^d αβ transgenic scid mice, indicating that they do not have to undergo positive selection to leave the thymus as functional T cells.

Though the role and lineage relationship of the $CD4^-8^-$ γδ and αβ TCR positive cells remains unclear, the data obtained in the αβ transgenic scid mice indicate that the transition from $CD4^-8^-$ TCR^- cells into $CD4^+8^+$ TCR^+ cells requires productive TCR rearrangement(s).

The second differentiation step dependent on the αβ TCR is the change in phenotype of immature $CD4^+8^+$ thymocytes such that they become single positive $CD4^+8^-$ and $CD4^-8^+$ cells. The evidence that $CD4^+8^+$ thymocytes are the precursors of the single positive $CD4^+8^-$ and $CD4^-8^+$ mature T cells stems from recent experiments concerned with tolerance of class I and class II MHC restricted T cells, showing that the intrathymic deletion of autospecific T cells affects $CD4^+8^+$ precursors of single positive T cells (Kisielow et al. 1988a; Fowlkes et al. 1988; MacDonald et al. 1988). According to the data obtained here, and in αβ transgenic mice on the positive selection of T cells (Teh et al. 1988; Kisielow et al. 1988b), one has to assume that the change in phenotype associated with positive selection occurs late during the lifespan of $CD4^+8^+$ thymocytes: the first cells acquiring the $CD4^+8^+$ phenotype are cycling blasts, which thereafter become resting $CD4^+8^+$ cells. The lifespan of $CD4^+8^+$ cells is no longer than three days (Shortman and Jackson, 1974). As the ratio of cycling to resting $CD4^+8^+$ is not significantly altered in female αβ transgenic mice (von Boehmer et al. 1989) as compared to normal mice, it is unlikely that the conversion of the $CD4^+8^+$ into the $CD4^-8^+$ phenotype happens at the blast stage or even at an early stage during the lifespan of resting $CD4^+8^+$ thymocytes. This leaves open the question at which stage of T cell development the critical interaction of the TCR with thymic MHC antigen, required for positive selection and change of phenotype, occurs: the signal provided by this interaction may have immediate or late consequences resulting in the phenotypic change.

According to our data, the consequences of negative selection become apparent earlier than those of positive selection. The former affects $CD4^+8^+$ cells such that even the earliest $CD4^+8^+$ blasts are diminished while the latter does not appear to affect the pool of $CD4^+8^+$ thymocytes to any significant extent. Thus, in terms of phenotypic changes, negative selection precedes positive selection. The experiments in αβ transgenic scid mice rule out the possibility that regulatory T cells are essential for any of the two selection steps, indicating that negative as well as positive selection is strictly dependant upon the interaction of αβ TCRs with antigen in the thymus.

Occasional in frame joinings may be the reason for the observed leakiness in scid mice: a series of several productive rearrangements may occur infrequently in clones of T or B cells in normal scid mice. Such clones may be selected and expanded by internal and external antigens which may, in turn, lead to a significant accumulation of T or B cells in some scid mice. In the TCR transgenic scid mice which have been screened for the absence of Ig secretion the detection of a productive rearrangement of the TCR α locus has been greatly facilitated by the introduction of an already rearranged TCR β gene. Our results demonstrate that in frame joinings occur frequently in scid mice but not as frequently as in normal mice.

References

Bevan MJ (1977) In a radiation chimaera, host H-2 antigens determine immune responsiveness of donor cytotoxic cells. Nature 269:417

Bluestone J, Pardoll DM, Sharrow SO, Fowlkes BJ (1987) Characterization of murine thymocytes with CD3-associated T-cell receptor structures. Nature 326:82

Bosma GC, Custer PR, Bosma MJ (1983) A severe combined immunodeficiency mutation in the mouse. Nature 301:527

Crispe IN, Moore MW, Husmann LA, Smith L, Bevan MJ, Simenkavitz RP (1987) Differantiation potential of subsets of CD4-CD8- thymocytes. Nature 329:336

Fowlkes BJ, Schwartz RH, Pardoll DM (1988) Deletion of self-reactive thymocytes occurs at a CD4+CD8+ precursor stage. Nature 334:620

Kappler JW, Wade T, White J, Kushnir E, Blackman M, Bill J, Roehm N, Marrack PC (1987) A T cell receptor Vβ segment that imparts reactivity to a class II major histocompatibility complex product. Cell 49:263

Kisielow P, Blüthmann H, Staerz UD, Steinmetz M, von Boehmer H (1988a) Tolerance in T cell receptor transgenic mice involves deletion of nonmature CD4+8- thymocytes. Nature 333:742

Kisielow P, Leiserson W, von Boehmer H (1984) Differentiation of thymocytes in fetal organ culture : analysis of phenotypic changes accompanying the appearance of cytolytic and interleukin 2 producing cells. J.Immunol. 133:1117

Kisielow P, Teh HS, Blüthmann H, von Boehmer H (1988b) Positive selection of antigen-specific T cells in thymus by restricting MHC molecules. Nature 335:730

MacDonald HR, Hemgartner H, Pedrazzini T (1988) Intrathymic deletion of self-reactive cells prevented by neonatal anti CD4 antibody treatment. Nature 335:174

Matzinger P (1981) A one receptor view of T-cell behavior. Nature 292:497

Matzinger P, Zamoyska R, Waldmann H (1984) Self-tolerance is H-2 restricted. Nature 308:738

Rammensee HG, Bevan MJ (1984) Evidence from in vitro studies that tolerance to self antigens is MHC-restricted. Nature 308:741

Raulet DH, Garman RD, Saito H, Tonegawa S (1985) Developmental regulation of T-cell receptor gene expression. Nature 314:232

Schuler W, Weiler IJ, Schuler A, Phillips RA, Rosenberg N, Mak TW, Kearney JF, Perry RP, Bosma MJ (1986) Rearrangement of antigen receptor genes is defective in mice with severe combined immune deficiency (SCID). Cell 46:963

Scott B, Blüthmann H, Teh HS, von Boehmer H (1989) The generation of mature T cells requires an interaction of the αβ T cell receptor with major histocompatibility antigens. Nature in press:

Shortman K, Jackson H (1974) The differatiation of T lymphocytes I. Proliferation kinetics and interrelationships of subpopulations of mouse thymus cells. Cell.Immunol. 12:230

Snodgrass R, Kisielow P, Kiefer H, Steinmetz M, von Boehmer H (1985) Ontogeny of the T-cell antigen receptor within the thymus. Nature 313:592

Snodgrass RH, Dembic Z, Steinmetz M, von Boehmer H (1985) Expression of T cell antigen receptor genes during fetal development in the thymus. Nature 315:232

Teh HS, Kishi H, Scott B, von Boehmer H (1989) Deletion of autospecific T cells in T cell receptor transgenic mice spares cells with normal TCR levels and low levels of CD8 molecules. J.Exp.Med. in press:

Teh HS, Kisielow P, Scott B, Kishi H, Uematsu Y, Blüthmann H, von Boehmer H (1988) Thymic major histocompatibility complex antigens and the αβ T cell receptor determine the CD4/CD8 phenotype of T cells. Nature 335:229

von Boehmer H, Haas W, Jerne NK (1978) Major histocompatibility complex linked immune-responsiveness is acquired by lymphocytes of low-responder mice differentiating in thymus of high-responder mice. Proc.Natl.Acad.Sci.USA 75:2439

von Boehmer H, Teh HS, Kisielow P (1989) The thymus selects the useful, neglects the useless and destroys the harmful. Immunol.Today 10:57

Zinkernagel RM, Callahan GN, Althage A, Cooper S, Klein PA, Klein J (1978) On the thymus in the differentiation of "H-2 self-recognition" by T cells : Evidence for dual recognition ?. J.Exp.Med. 147:882

Transgenic Scid Mice With a Functionally Rearranged Immunoglobulin Heavy Chain Gene

Michal Fried, Richard R. Hardy and Melvin J. Bosma

Institute for Cancer Research, Fox Chase Cancer Center, Philadelphia, PA 19111, USA

INTRODUCTION

As reported and cited in this volume, there is now considerable evidence in support of the hypothesis (Schuler et al. 1986) that V(D)J recombination is severely impaired in developing B and T lymphocytes of mice homozygous for the scid mutation (scid mice). Consequently, early scid lymphocytes fail to mature and presumably die prematurely. This would explain, for example, the absence of detectable pre-B cells in the bone marrow of scid mice (Dorshkind et al. 1984; Schuler et al. 1988). However, on rare occasions, a given recombinase-active scid lymphocyte may succeed in making functional V(D)J rearrangements at two critical alleles and give rise to a clone of B or T lymphocytes; mice with one or more such clones are referred to as leaky scid mice (Bosma et al. 1988; Carroll and Bosma 1988).

In order to test the above hypothesis and possibly gain further insight into the basis of leaky scid mice, a functionally rearranged IgM heavy chain (μ) gene was introduced into the scid mouse genome. Our first and foremost interest was whether the resulting μ transgenic scid mice would develop normal numbers of pre-B cells in their bone marrow, and second, whether such mice would show a higher incidence of leaky B cells than non-transgenic scid mice. The second possibility could be expected if leaky B lymphocytes were simply to reflect the occurrence of rare functional V(D)J rearrangements in recombinase-impaired B cell precursors. All such precursors in μ transgenic scid mice would contain a functional VDJ heavy chain gene as opposed to few, if any, functional VDJ rearrangements in the early B cells of non-transgenic scid mice.

RESULTS AND DISCUSSION

Construction of μ-transgenic scid mice

The introduction of a functionally rearranged μ gene into the scid mouse genome was accomplished by selective genetic crosses with an established transgenic mouse strain (M54). The M54 strain carries a functional μ gene whose variable region is derived from the NP (4-hydroxy-3-nitrophenyl)-specific BALB/c hybridoma (17.2.25); M54 was constructed and described by Grosschedl et al. (1984) and kindly supplied to us by T. Imanishi-Kari (Tufts University Medical School, Boston, MA). To construct μ transgenic scid mice we took advantage of the known linkage of scid to mahoganoid (md), a recessive coat color gene, and to Igl-1, the Ig lambda light chain locus (Bosma et al. 1989). All three loci are closely linked and map to the centromeric end of chromosome 16 (Epstein et al. 1984, 1986; Davisson 1985; Bosma et al. 1989).

We first crossed the μ transgene of the M54 strain (μ-NP) onto a constructed mouse stock derived by Epstein et al. (1986) and kindly supplied by M. T. Davisson of the Jackson Laboratories (Bar

Current Topics in Microbiology and Immunology, Vol. 152

Harbor, ME). This stock is homozygous for md and Igl-1^b, an allelic variant of Igl-1. Selected μ-NP/+ offspring of the first cross were backcrossed to the same stock in order to obtain μ-NP/+ mice that were homozygous for md and Igl-1^b. These mice were then used as the genetic donors of the μ-NP transgene in subsequent crosses to scid mice; in all crosses the μ-NP transgene segregated independently of Igl-1 and scid. Scid mice are wild-type at the md locus and carry the Igl-1^a allele which is clearly distinguishable from Igl-1^b on the basis of a restriction fragment length polymorphism (Arp et al. 1982; Epstein et al. 1986). Therefore, by using Igl-1^a as an independent marker for scid it was possible in subsequent genetic crosses to identify unambigously mice that were homozygous and heterozygous for scid. Four experimental classes of mice were generated by backcrossing μ-NP/+ scid heterozygotes (scid Igl-1^a/+ Igl-1^b) to scid mice (scid Igl-1^a/scid Igl-1^a): (1) μ transgenic scid mice (μ-NP/+, scid Igl-1^a/scid Igl-1^a), (2) non-transgenic scid mice (+/+, scid Igl-1^a/scid Igl-1^a), (3) μ transgenic heterozygous scid mice (μ-NP/+, scid Igl-1^a/+ Igl-1^b) and (4) non-transgenic heterozygous scid mice (+/+, scid Igl-1^a/+ Igl-1^b), where classes 2,3 and 4 served as controls for class 1. The results reported below were obtained with mice from backcross generations 1-4.

Fluorescent activated cell sorter (FACS) analysis of μ-transgenic scid bone marrow and spleen

The surface phenotype (B220, Thy-1.2, IgM) of bone marrow and splenic cells from transgenic scid mice was examined by FACS analysis to ascertain the effect of the μ transgene on B cell development. The expression of the B220 cell surface antigen is specific for both immature and mature (IgM^+) B cells (Coffman and Weissman 1981a, b). Prior to expression of surface IgM, early $B220^+$ progenitors of B cells express low levels of Thy-1.2 (Thy-1.2^{lo}), and presumably give rise to phenotypically distinct pre-B cells ($B220^+$ Thy-1.2^-) (Müller-Sieburg et al. 1989; Tidmarsh et al. 1989). Thus, by using monoclonal antibody to B220 in conjunction with anti-Thy-1.2 or anti-IgM, we could identify and distinguish $B220^+$ Thy-1.2^{lo} and $B220^+$ Thy-1.2^- progenitor B cells, and also $B220^+$ IgM^+ mature B cells.

As indicated in Table 1, normal percentages of early B cell progenitors ($B220^+$ Thy-1.2^{lo} IgM^-) were detected in bone marrow of scid mice (non-transgenic scid controls). This direct demonstration of early B cell progenitors confirms previous findings indicating the presence of such cells (Schuler et al. 1986; Fulop et al. 1988; Schuler et al. 1988). Though the progression of these progenitors to a later stage ($B220^+$ Thy-1.2^- IgM^-) apparently did not occur in scid mice, it did occur in transgenic scid mice as these mice showed a normal percentage of $B220^+$ Thy-1.2^- IgM^- cells in their bone marrow. However, similar to scid mice, transgenic scid mice failed to develop mature $B220^+$ Thy-1.2^- IgM^+ B cells in either bone marrow or spleen.

FACS profiles of Thy-1.2 expression on $B220^+$ bone marrow cells of transgenic and non-transgenic scid mice are shown in Figure 1. As can be seen, the majority of $B220^+$ cells were Thy-1.2^{lo} in the bone marrow of the scid controls. In transgenic scid bone marrow, on the other hand, the presence of $B220^+$ Thy-1.2^- cells is clearly apparent in addition to $B220^+$ Thy-1.2^{lo} cells. Using fluorescinated anti-IgM, we found that ~50% of the FACS sorted $B220^+$ cells from transgenic mice stained positive for cytoplasmic μ chains ($C\mu^+$) (Figure 2); in contrast, no $C\mu^+$ cells were observed in the $B220^+$ fraction from scid control bone marrow.

Table 1. B lineage cell populations in the bone marrow and spleen of μ-transgenic scid and control mice

Mouse Genotype	bone marrow B220+ Thy-1.2lo IgM-	bone marrow B220+ Thy-1.2- IgM-	bone marrow B220+ Thy-1.2- IgM+	spleen B220+ Thy-1.2- IgM+
scid	3.3 ± 0.4	0.3 ± 0.2	--	--
transgenic scid	3.6 ± 0.3	5.7 ± 0.9	--	--
heterozygous scid	2.5 ±0.4	4.3 ± 0.8	5.1 ± 0.8	43.5 ± 3.5
transgenic heterozygous scid	2.6 ±0.3	6.4 ± 0.6	3.9 ± 0.6	23.6 ± 2.2

Table 1. The surface phenotype (B220, Thy-1.2, IgM) of the cell populations shown was deduced from simultaneous two-color FACS analysis. Single cell suspensions were prepared from bone marrow and spleen of 2-to-4-month-old mice and stained as previously described (Hayakawa et al. 1985) using anti-B220, RA3-6B2/phycocyanin with either anti-IgM, 331.12/fluorescein or anti-Thy-1.2, 30-H12/fluorescein (Coffman and Weissman, 1983; Kincade et al. 1981; Ledbetter and Herzenberg, 1979). Dead cells were excluded from the analysis by propidium iodide staining. Data were collected from analyses of four to seven individual mice of each genotype. Values are presented as average percentage of total viable cells examined (30,000 or 100,000 cells per analysis) ± SEM.

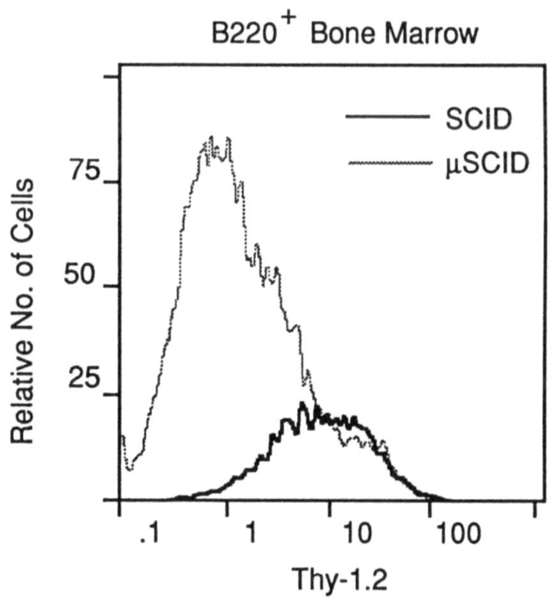

Figure 1. Expression of Thy-1.2 on the surface of B220+ cells from transgenic and non-transgenic scid bone marrow. Cell preparation, staining and FACS analyses were carried out as described for Table 1. Histograms are representative of data obtained from three experiments.

A C B D

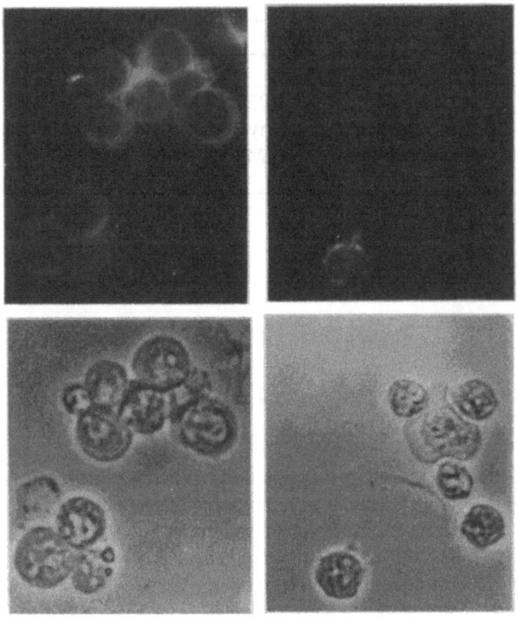

Figure 2. Detection of Cμ chains by fluorescence microscopy in B220+ cells sorted from transgenic scid bone marrow. Fields of positively (A) and negatively (C) stained cells were taken from the same slide; phase contrast of the same fields under white light illumination is shown in (B) and (D). B220+ cells were isolated by cell sorting and subsequently were spun down onto microscope slides by cytocentrifuge at a concentration of ~2 x 10^5 per slide. Ethanol/acetic acid-fixed cytospots were incubated with affinity-purified goat-anti mouse IgM/fluorescein (Southern Biotechnology), washed and mounted as previously described (Stall et al. 1988). Proportions of cells positive for cytoplasmic staining of μ chains were determined by counting at least 300 cells per slide.

Taken together, the above results directly demonstrate that scid mice contain normal numbers of early B cell progenitors ($B220^+$ $Thy\text{-}1.2^{lo}$); further, that the introduction of a functional μ gene into the scid mouse genome enables these early B cell progenitors to differentiate into pre-B cells. The latter finding lends strong support to the hypothesis that early scid B cells cannot progress beyond the stage at which Ig heavy chain (Igh) rearrangements normally occur (Schuler et al. 1988), because these cells are unable to make functional Igh rearrangements. Whether the pre-B cells that arise in μ transgenic scid mice ever reach the stage at which Ig light chain (Igl) rearrangements normally occur is not yet clear. If Igl rearrangements were to occur, they presumably would be abnormal and prevent further maturation of pre-B cells into mature ($B220^+$ IgM^+) B cells.

Leaky μ transgenic scid mice

Although mature B cells ($B220^+$ IgM^+) were not observed in the bone marrow or spleen of μ transgenic scid mice, some of these mice did apparently develop limited numbers of Ig-producing plasma cells. About 26% of the transgenic mice (29/111) contained ≥0.1 mg/ml of serum Ig(κ) as measured by ELISA (assay described in Bosma et al. 1988). Serum and spleen cell lysates from these leaky mice showed both μ and κ chains on Western blots in contrast to μ chains only in most transgenic mice (Figure 3). Sixty percent of the leaky mice expressed IgM only, and in these cases, all of the detected IgM was of the BALB/c or transgene allotype (Igh-6a). About 40% of the leaky mice produced one or more IgG isotypes together with Igh-6a and Igh-6b or Igh-6b only (Igh-6b is the endogenous C.B-17 IgM allotype). The allotyping results (not shown) were obtained by dot blot analysis of serum samples using ^{125}I-labeled anti-Igh-6a (Schüppel et al. 1987) and ^{125}I-labeled anti-Igh-6b (Stall and Loken, 1984) monoclonal antibodies. Affinity purified anti-Igh-6a was generously supplied by Drs. Schüppel and Weiler of the University of Konstanz (Konstanz, Germany).

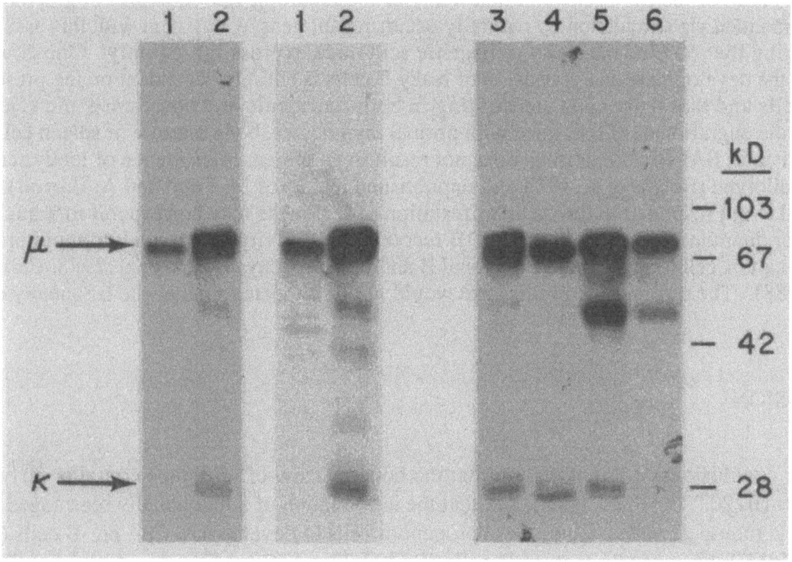

Figure 3. SDS-PAGE (10%) Western blots of reduced cell lysates (40 x 10^6 cells/ml) from spleen of Ig^- transgenic scid (1) and transgenic heterozygous scid (2). The blot in the left panel was exposed for 6 hours at -70°C. An overnight exposure of the same blot (middle panel) served to confirm the absence of κ chains in the spleen of Ig^- transgenic scid mice. The right panel shows the presence of both μ and κ chains in spleen (4) and serum (6) of Ig^+ transgenic scid mice compared to spleen (3) and serum (5) from transgenic heterozygous scid mice. The blots were first sequentially overlaid with affinity-purified anti-mouse IgM (15 μg/ml, Southern Biotechnology) and ^{125}I-IgM of MOPC-104E to reveal μ chains. A second sequential overlay of affinity-purified anti-mouse κ (15 μg/ml, Southern Biotechnology) and ^{125}I-IgG-κ of MOPC 31-C was carried out to detect κ chains. The positions of the molecular mass standards are indicated. Serum samples were diluted 1:200. Cell lysate preparation, electrophoresis and Western blotting procedure were performed as previously described (Marks and Bosma, 1985). Proteins were radiolabeled with ^{125}I (Amersham Corp.) using the iodination method of Salacinski (1981).

Two conclusions can be drawn from these results: First, that early scid B cells expressing the μ transgene can differentiate into plasma cells provided the appropriate intracellular events occur, including a functional Igl rearrangement. Second, the suppression of endogenous heavy chain expression by the μ transgene (allelic exclusion) is incomplete since 40% of the leaky transgenic mice contained endogenous serum IgM and IgG. The expression of endogenous Igh genes was also previously observed in the M54 μ transgenic strain (Weaver et al. 1986; Stall et al. 1988).

A final point of interest is that serum Ig+ transgenic scid mice (29 out of 111 mice) were no more frequent than serum Ig+ mice in the non-transgenic scid control group (20 out of 86 mice). This result appears inconsistent with the notion that rare, chance productive rearrangements are responsible for the appearance of leaky scid lymphocytes. If early scid B cells with severely impaired V(D)J recombinase activity have a low chance of making two functional Ig gene rearrangements, then one might expect a higher incidence of detectable Ig in transgenic vs. non-transgenic scid mice, because all developing B cells in the transgenic mice already contain a functional μ gene. This expectation assumes that different endogenous Igl chains will associate with the μ chain of the transgene and that many of the resulting Igh-6a-expressing cells will be clonally expanded via stimulation by naturally occurring antigens. Consistent with this assumption is our finding that 60% of the leaky transgenic scid mice express Igh-6a only. One could still argue that the development and expansion of leaky B cells is entirely dependent on the presence of leaky T cells and that these cells are limiting in both transgenic and non-transgenic scid mice. However, the engraftment of scid mice with normal thymocytes, bone marrow or spleen cells from allotype-distinct BALB/c donor mice does not result in an increased incidence of leaky scid mice with host allotype (Bosma et al. 1983 and unpublished results of M. Fried and A. Carroll). Thus, the critical event responsible for the manifestation of leakiness may correspond to a qualitative (genetic) or quantitative change in the V(D)J recombinase activity of a given lymphoid precursor cell enabling this cell to give rise to functional B and/or T progeny (Bosma et al. 1988; Carroll and Bosma 1988). The μ transgene in this case would not be expected to alter the frequency of such events.

CONCLUSION

The major conclusions of this study are that the bone marrow of scid mice contains early B cell progenitors ($B220^+$ $Thy\text{-}1.2^{lo}$ IgM^-), and that the introduction of a functionally rearranged μ gene into the scid mouse genome enables these progenitor cells to develop into $C\mu^+$ pre-B cells ($B220^+$ $Thy\text{-}1.2^-$ IgM^-), but not into mature B cells ($B220^+$ $Thy\text{-}1.2^-$ IgM^+). The latter possibility is presumably precluded by the inability of μ transgenic scid mice to make functional light chain gene rearrangements.

ACKNOWLDGEMENTS

We thank T. Imanshi-Kari for M54 transgenic mice, M. T. Davisson for the constructed mouse stock and G. C. Bosma for supervision of the mouse breedings. We also thank A. M. Carroll for helpful discussion, E. Cunningham and S. Shinton for technical help, and M. Piatek and A. Capriotti for typing this manuscript.

This work was supported by NIH grants AI-13323, CA-04946, CA-06927 and RR-05539 and by an appropriation from the Commonwealth of Pennsylvania.

REFERENCES

Arp B, McMullen MD, Storb U (1982) Sequence of immunoglobulin λ1 genes in a λ1 defective mouse strain. Nature 298: 184-186

Bosma GC, Fried M, Custer RP, Carroll AM, Gibson DM, Bosma MJ (1988) Evidence of functional lymphocytes in some (leaky) scid mice. J Exp Med 167: 1016-1033

Bosma GC, Davisson MT, Ruetsch NR, Sweet HO, Schultz LD, Bosma MJ (1989) The mouse mutation severe combined immune deficiency (scid) is on chromosome 16. Immunogenetics 29: 54-57

Carroll AM, Bosma MJ (1988) Detection and characterization of functional T cells in mice with severe combined immune deficiency. Eur J Immunol 18: 1965-1971

Coffman RL, Weissman IL (1981a) B220: a B cell-specific member of the T200 glycoprotein family. Nature 289: 681-683

Coffman RL, Weissman IL (1981b) A monoclonal antibody that recognizes B cells and B cell precursors in mice. J Exp Med 153: 269-279

Coffman RL, Weissman IL (1983) Immunoglobulin gene rearrangement during pre-B cell differentiation. J Mol Cell Immunol 1: 31-38

Davisson MT (1985) The Jackson Laboratory. 56th Annual Report 1984-1985. p 40

Dorshkind K, Keller GM, Phillips RA, Miller RG, Bosma GC, O'Toole M, Bosma MJ (1984) Functional status of cells from lymphoid and myeloid tissues in mice with severe combined immunodeficiency disease. J Immunol 132: 1804-1808

Epstein R, Lehmann K, Cohn M, Buckler C, Rowe W, Davisson M (1984) Linkage of the Igl-1 structural and regulatory genes to Akv-2 on chromosome 16. Immunogenetics 19: 527-537

Epstein R, Davisson M, Lehmann K, Akeson EC, Cohn M (1986) Position of Igl-1 md, and Bst Loci on chromosome 16 of the mouse. Immunogenetics 23: 78-83

Fulop GM, Bosma GC, Bosma MJ, Phillips RA (1988) Early B cell precursors in scid mice - normal numbers of cells transformable with Abelson murine leukemia virus (A-MuLV). Cell Immunol 113: 193-201.

Grosschedl R, Weaver D, Baltimore D, Costantini F (1984) Introduction of a μ immunoglobulin gene into the mouse germ line: specific expression in lymphoid cells and synthesis of functional antibody. Cell 38: 647-658

Hayakawa K, Hardy RR, Herzenberg LA, Herzenberg LA (1985) Progenitors for Ly-1 B cells are distinct from progenitors for other B cells. J Exp Med 161: 1554-1568

Kincade PW, Lee G, Sun L, Watanabe T (1981) Monoclonal rat antibodies to murine IgM determinants. J Immunol Methods 42: 17-26

Ledbetter JA, Herzenberg LA (1979) Xenogenic monoclonal antibodies to mouse lymphoid differentiation antigens. Immunol Rev 47: 63-90

Marks R, Bosma MJ (1985) Truncated μ (μ') chains in murine IgM. Evidence that μ' chains lack variable regions. J Exp Med 162: 1862-1877

Müller-Sieburg CE, Tidmarsh GF, Weissman IL, Spangrude GJ (1989) Maturation of hematolymphoid cells that express Thy-1. In: Reif AF, Schlesinger M (eds) Cell surface antigen Thy-1, Vol. 45, Immunology. Marcel Dekker, New York, p 289

Salacinski PRP (1981) Iodination of proteins glycoproteins and peptides using a solid phase oxidizing agent, 1,3,4,6-Tetrachloro-3α,6α-diphenyl glycoluril (Iodogen). Analytical Biochem 117: 136-146

Schuler W, Weiler IJ, Schuler A, Phillips RA, Rosenberg N, Mak TW, Kearney JF, Perry RP, Bosma MJ (1986) Rearrangement of antigen receptor genes is defective in mice with severe combined immune deficiency. Cell 46: 963-972

Schuler W, Schuler A, Lennon GG, Bosma GC, Bosma MJ (1988) Transcription of unrearranged antigen receptor genes in scid mice. EMBO J 7: 2019-2024

Schüppel R, Wilke J, Weiler E (1987) Monoclonal anti-allotype antibody towards BALB/c IgM. Analysis of specificity and site of a V-C crossover in recombinant strain BALB-Igh-V^a/Igh-C^b. Eur J Immunol 17: 739-741

Stall AM, Loken MK (1984) Allotypic specificities of murine IgD and IgM recognized by monoclonal antibodies. J Immunol 132: 787-795

Stall AM, Kroese FGM, Gadus FT, Sieckmann DG, Herzenberg LA, Herzenberg LA (1988) Rearrangement and expression of endogenous immunoglobulin genes occur in many murine B cells expressing transgenic membrane IgM. Proc Natl Acad Sci USA 85: 3546-3550

Tidmarsh GF, Heimfeld S, Whitlock CA, Weissman IL, Müller-Sieburg CE (1989) Identificaton of a novel bone marrow derived B cell progenitor population which coexpresses B220 and Thy-1 and is highly enriched for Abelson leukemia virus targets. Mol Cell Biol (in press)

Weaver D, Reis MH, Albanese C, Costantini F, Baltimore D, Imanishi-Kari T (1986) Altered repertoire of endogenous immunoglobulin gene expression in transgenic mice containing a rearranged mu heavy chain gene. Cell 45: 247-259

IV. Leaky Phenotypes

T Cell Leakiness in Scid Mice

Ann M. Carroll, Richard R. Hardy, John Petrini and Melvin J. Bosma

Institute for Cancer Research, Fox Chase Cancer Center, Philadelphia, PA 19111, USA

INTRODUCTION

The leaky scid phenotype

A variable percentage (~10-20%) of young adult scid mice produce detectable serum Ig, and have been designated "leaky", or scid Ig^+ mice (Bosma et al. 1988). Leakiness appears to be due to a somatic event as scid mice cannot be selectively bred for this phenotype (G. Bosma and M. Bosma, unpublished data). The frequency of scid Ig^+ mice increases with age, and mice housed under non-specific-pathogen free (SPF) conditions show a higher incidence of leakiness than SPF mice, suggesting that naturally-occurring antigens play an important role in the expansion of leaky B cell clones. Scid Ig^+ mice generally fail to express all of the major serum Ig classes (i.e., IgM, IgG3, IgG1, IgG2a, IgG2b and IgA), and isoelectric focusing of serum Ig light chains indicates only one to seven Ig-producing clones in most mice (Bosma et al. 1988; Gibson et al. 1989).

Scid mice that are leaky for B cells may also be leaky for T cells. This was first indicated by the apparent selective ability of scid Ig^+ mice to reject skin allografts (Bosma et al. 1988), a classic assay for the presence of functional T cells. We recently obtained alloresponsive T cell clones from scid Ig^+ mice, confirming the occurrence of leaky T lymphocytes (Carroll and Bosma 1988). In the present report, we show that T cell leakiness is not restricted to scid Ig^+ mice; in fact, $CD3^+$ T cells are detectable in almost all old scid mice. However, the number of $CD3^+$ T cells is low (generally $<10^5$ per tissue), and Southern blot analysis of the number of T cell receptor (TCR) gene rearrangements detected in polyclonally stimulated scid T cell cultures indicates that leaky T cell populations, like leaky B cell populations, are pauciclonal.

We postulated two alternative (but not mutually exclusive) hypotheses to account for the origin(s) of leaky B and T cells (Bosma et al. 1988; Carroll and Bosma 1988), based on evidence that the scid mutation impairs the VDJ recombinase activity in developing pre-lymphocytes (Schuler et al. 1986, and this volume). According to the first hypothesis, normal VDJ joining occurs at a very low frequency in developing scid lymphocytes despite their defective VDJ recombinase activity, and a rare cell succeeds in making a productive V(D)J join at two critical alleles, resulting in production of a heterodimeric receptor. In the second hypothesis, the generation of leaky cells requires a qualitative change in the VDJ recombinase activity of one or more scid lymphocyte progenitors during development. This model entails normalization of the defective recombinase activity, resulting from either genetic or epigenetic events. As the normal process of V-J and V-D-J recombination results in a high frequency of nonproductive (out-of-frame) rearrangements, we reasoned that analysis of nonproductively rearranged antigen receptor genes in leaky clones might allow us to distinguish between these hypotheses. If leaky cells were generated by a defective recombinase activity, their nonproductively recombined alleles should contain abnormal J-associated deletions as characteristically seen in transformed scid lymphocytes. Alternatively,

Current Topics in Microbiology and Immunology, Vol. 152

leaky lymphocytes that derive from progenitors with normalized recombinase activity should contain nonproductive rearrangements indistinguishable from those of normal cells. Our results indicate that most rearranged TCR genes in leaky scid T cells are indistinguishable from rearranged TCR genes in normal cells.

RESULTS AND DISCUSSION

Incidence and frequency of leaky scid T cells

Lymphoid tissue suspensions of individual scid mice were examined by multiparameter fluorescence activated cell sorter (FACS) analysis for detection of mature T lineage cells. Thymic and peripheral [spleen and peritoneal cavity (PerC)] lymphoid tissue suspensions from mice of three age groups (I: 4 wks; II: 12 wks; and III: >52 wks) were triple stained with antibodies to CD3, CD4, and CD8. Table 1 shows the percentages of cells expressing CD3, a T cell surface complex which shows obligate co-expression with the surface TCR heterodimer.

Table 1. Incidence of $CD3^+$ cells in lymphoid tissues of scid mice[a]

Group/Mouse #		[Ig(κ)]	% $CD3^+$ Cells Thymus	Spleen	PerC
I	1	-	-	-	-
4 wks	2	-	-	-	-
	3	-	-	-	-
	4	-	-	-	0.2
	5	-	-	-	-
	6	-	1.3	-	0.4
II	7	-	-	-	-
12 wks	8	-	1.0	-	-
	9	-	0.3	-	-
	10	+	1.3	0.6	2.2
	11	-	-	-	-
	12				
III	13	-	9.8	-	0.1
>52wks	14	-	2.9	-	0.4
	15	-	-	-	-
	16	+	1.0	-	0.3
	17	+	24.0	0.5	0.1
	18	+	0.3	0.1	-
	19	-	-	-	0.1
C.B-17+/+ (12 wks)		+	41.8	28.7	3.2

[a] Lymphoid tissue suspensions of individual scid mice and a normal C.B-17+/+ mouse were stained using phycoerythrin labeled anti-CD3 (2C11-145, Leo et al. 1987), allophycocyanin labeled anti-CD4 (GK-15; Dialynas et al. 1983) and fluorescein labeled anti-CD8 (53-6, Ledbetter and Herzenberg 1979). Dead cells were excluded from analysis by propidium iodide staining. Triple staining provided evidence for different functional and differentiation-stage specific T cell subsets (data not shown). Serum Ig(κ) concentrations were determined by Elisa; values <0.05 mg/ml are indicated with a (-).

In both 4 wk and 12 wk old scid mice, $CD3^+$ cells were rare, and only two mice (#6, #10) contained detectable T cells in more than one tissue. In all these young mice, the numbers of $CD3^+$ cells were very low ($<10^5$ per tissue). The incidence of leaky T cells did increase with age; almost all scid mice over one year of age contained detectable $CD3^+$ cells, regardless of serum Ig positivity. Despite this increased incidence, the overall numbers of leaky T cells in individual old scid mice was still very low. The percentages of thymic $CD3^+$ cells was sometimes high (up to 24.0%), but thymuses of old scid mice characteristically contained $<5 \times 10^5$ cells, so even higher percentages represent low numbers (10^3-10^5) of cells. $CD3^+$ thymocytes of old scid mice were of medullary phenotypes with respect to expression of the T subset markers CD4 and CD8 ($CD3^+4^+8^-$, $CD3^+4^-8^+$ and $CD3^+4^-8^-$). In individual scid mice, however, subsets of both thymic and peripheral $CD3^+$ populations showed considerable skewing from normal ratios. Peripheral T cells in some mice, for example, were comprised of $CD3^+4^+8^-$ or $CD3^+4^-8^+$ cells only. This finding suggested limited clonality of leaky T cell populations.

Pauciclonality of T cell populations in leaky scid mice

To evaluate the apparent limited clonality of leaky T cell populations, cells from spleen and lymph nodes of individual old scid mice were cultured short-term in vitro in the presence of IL-2 and Con A, a classic mitogen for mature $CD3^+$ cells, or in the presence of phorbol ester (PMA) and ionomycin, which induce proliferation of T lineage cells independent of surface CD3/TCR expression. DNA was prepared from these cultures and examined by Southern blot analysis with a series of TCR gene probes. Results are shown in Figure 1.

Examination of TCRβ and TCRδ loci with appropriate probes (Figure 1A and B) can distinguish alleles in germline and rearranged configuration. By examining the number of detectable rearranged alleles at both TCR loci, it is possible to make a minimal estimation of the number of stimulated clones per culture. The pJβ2 probe (Fig. 1A) detected a 2 kb germline fragment in scid liver and a complex set of rearrangement fragments in the normal spleen culture. By contrast, the stimulated scid cultures (8706, 7881, 7938) showed 1-4 non-germline bands, indicating pauciclonality. When the same filter was reprobed with p3'Jδ1 (Fig. 1B), no major fragments were detected in Con A stimulated normal spleen cells. This result is consistent with expection as most mature spleen T cells delete their TCRδ alleles by subsequent rearrangement of the TCRα locus (Chien et al. 1987a). However, some non-germline bands were detected in the individual scid cultures. An assessment of the number of non-germline TCRδ and TCRβ fragments detected in DNA from cultures of 12 scid mice has indicated the presence of 1-5 stimulatable T lineage clones in the peripheral lymphoid tissues of individual leaky scid mice (Carroll et al. 1989).

Status of rearranged TCRβ and TCRγ alleles in leaky T cells

As shown in Figure 1, hybridization of pJβ2 to DNA from the Con A stimulated culture, #8706, showed only a single detectable Jβ2 allele. The failure to detect a second hybridizing fragment suggested that this culture consisted of one T cell clone in which the Jβ2 sequence of the second TCRβ allele had been deleted as an apparent result of defective scid recombinase activity. This interpretation was supported by hybridization of the same filter with a probe (pCγ2) to the TCR γ locus (Fig. 1C). pCγ2 hybridizes to γ1, 2 and 3 genes. In germline configuration these genes are

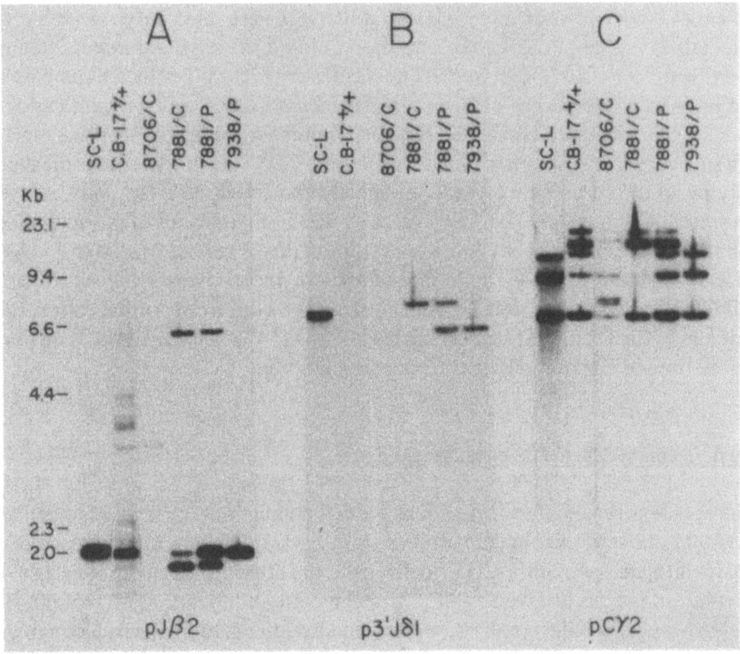

Figure 1. Southern blot analysis of TCR gene rearrangements in Eco-R1 restricted DNA of scid liver (SC-L), Con A stimulated C.B-17+/+ spleen, and individual scid peripheral lymphoid tissue suspensions [indicated by mouse identification numbers (8706, 7881, 7938)] stimulated with IL-2 and Con A (C) or PMA and ionomycin (P). The same blot was sequentially hybridized with probes to TCRβ (A, pJβ2; Malissen et al. 1984), TCRδ (B, p3'Jδ1; Chien et al. 1987b), and TCRγ (C, pCγ2, Iwamoto et al. 1986) region genes.

represented by Eco R1 fragments of 13.4, 10.5 and 7.5 kb (see scid liver in panel C) whereas rearranged TCRγ alleles appear as 16, 17 or 22 kb fragments (see C.B-17+/+ spleen in panel C). In contrast, 8706 DNA showed two hybridizing fragments of abnormal size (~8-9 kb), indicative of defective scid recombinase activity. Thus, this $CD3^+$ Con A stimulated clone appeared to contain three grossly defective TCR gene rearrangements characteristic of scid T cell lymphomas (Schuler et al. 1986). This phenotype is consistent with generation of a leaky T cell by a defective scid recombinase, as postulated in our first hypothesis. It should be noted, however, that other scid cultures (7881p and c, 7938c) contained only conventional sized recombination fragments. Indeed, when a total of 12 cultures were so examined, eight showed detectable TCRγ gene rearrangements, and in seven of these eight cases all rearrangements appeared conventional (Carroll et al. 1989). These results are therefore consistent with generation of leaky T cells from precursors with normalized VDJ recombinase activity.

To ascertain the status of non-productively rearranged TCR alleles in single leaky T cell clones, five alloresponsive T cell clones were isolated from individual leaky scid mice. The TCRβ and TCRγ alleles in these clones were examined by Southern blot analysis (Carroll and Bosma 1988), and a summary of the results is shown in Table 2.

Table 2. TCR gene status of T cell clones from scid Ig^+ mice[a]

T clone	TCRβ		TCRγ		
	Jβ1	Jβ2	JCγ1	JCγ2	JCγ4
scid 1374	G,G	G,R	G,G	G,G	G,G
scid 1233	R,_	R,R*	G,G	R,R*	R,R*
scid 1280	G,R	G,R	G,G	R,R	ND
scid 1287	_,_	R,R	G,G	G,R	ND
scid 7167	G,_	R,R	G,R	R,R	ND

[a] DNA was prepared from individual scid T clones, restricted with Eco R1 and/or Pvu II and hybridized with probes to TCRβ (pJβ1, pCβ1A, pCβ1A-5′, pJβ2, pCβ2, pCβ2-5′) and TCRγ (pCγ2, pCγ4, pVγ1.2) as detailed elsewhere (Carroll and Bosma 1988). Configuration for both alleles of presented loci (Jβ1, Jβ2, Cγ1, Cγ2, Cγ4) is indicated as follows: G, germline; R, rearranged; R* indicates a grossly aberrant rearrangement, resulting in deletion of a Jβ2 coding allele (S1233) or generation of a novel sized pCγ2, pCγ4 hybridizing fragment (S1233); (-) indicates that Jβ1-Cβ1 coding sequence has been deleted by conventional recombination of an upstream Vβ or Dβ1 element to the downstream Jβ2; ND indicates not determined.

The first clone, S1374, contained a single detectable Jβ2-associated rearrangement (the other allele was in germline configuration) and no TCRγ rearrangements. Retention of so many alleles in germline configuration is unusual, but the lack of nonproductively rearranged alleles rendered this clone useless for the present analysis. The other four clones showed multiple rearrangements of TCRβ and TCRγ genes. Clone S1233 contained two abnormally rearranged TCR genes: One TCRβ allele incurred a deletion of the Jβ2 region and one of the TCRγ alleles yielded an Eco R1 fragment of abnormal size which hybridized to both pCγ2 and pCγ4. Thus, this clone may have derived from a progenitor which contained a defective scid recombinase activity. In contrast to 1233, the remaining scid clones (S1280, S1287 and S7167) all contained multiple TCR gene rearrangements (4 for 1280, 3 for 1287, 5 for 7167) with no evidence of gross aberrancy. TCRγ rearrangements in these clones resulted in conventional sized (17 kb, 22 kb) hybridizing fragments, and no deletion of the Jβ2 region was detected. Thus, the phenotypic profile of these three clones was indistinguishable from that of normal T cell clones and notably different from that of scid T cell lymphomas, which typically contain rearranged TCR alleles with abnormal J-associated deletions of >1 kb (Schuler et al. 1986). Therefore, S1280, S1287 and S7167 appear to have derived from progenitors containing normalized recombinase activity.

To test whether S1280, S1287 and S7167 may have abnormal TCR gene deletions too small to be detected by Southern blot analysis (up to a few hundred nucleotides), we are currently amplifying selected TCR recombination joints for sequencing analysis. We recently obtained sequences for three rearranged TCRγ2 genes, one in S1287 and two in S7167. All sequenced recombination joints are normal, and the two sequenced TCRγ gene rearrangements of S7167 (Vγ1.2-Jγ2) have been confirmed to be nonproductive (Petrini, Carroll and Bosma, in preparation). The occurrence of at least five independent normal recombination events (three productive rearrangements, Vα-Jα,

Dβ-Jβ and Vβ-Dβ; and two nonproductive TCRγ rearrangements) argues strongly for a normalized scid recombinase activity in the progenitor of the S7167 clone.

SUMMARY

FACS analysis showed that the incidence of leaky T cells increases with age, such that virtually all old scid mice (>1 year) contain detectable $CD3^+$ cells. The number of detectable T cells remained very low; individual old scid mice generally contained $<10^5$ $CD3^+$ cells. When $CD3^+$ populations in individual leaky mice were analyzed for expression of the T cell subset markers, CD4 and CD8, the ratios of CD4/CD8 were found to be markedly skewed relative to normal mice. This suggested the presence of very few T cell clones. Indeed, the analysis of TCR gene rearrangements in polyclonally stimulated T cell cultures revealed only 1-5 clones in the pooled spleen and lymph nodes of individual old scid mice. These studies also indicated that TCR gene rearrangements in the majority of the stimulated T cell cultures did not contain abnormal J-associated deletions that are characteristic of antigen receptor genes of scid lymphomas. Four of five alloreactive T cell clones from leaky scid mice also apparently lacked abnormal J-associated deletions in their rearranged TCR alleles. Therefore, most leaky lymphocytes appear to derive from progenitors with normal or near-normal scid recombinase activity. However, one of five leaky T cell clones (S1233) and one Con A stimulated monoclonal culture (8706) contained both normally and abnormally rearranged TCR genes. The configuration of TCR loci in such clones may reflect the ability of the defective scid recombinase to mediate normal rearrangements at a low frequency.

ACKNOWLEDGEMENTS

We thank N. Ghobadi and S. Shinton for technical help, and Y. Chien for p3'Jδ1. We also thank M. Fried for critical review of the manuscript, and A. Capriotti and M. Piatek for typing this manuscript. This work was supported by NIH grants AI-13323, CA-04946, CA-06927 and RR-05539 and by an appropriation from the Commonwealth of Pennsylvania.

REFERENCES

Bosma GC, Fried M, Custer RP, Carroll AM, Gibson DM, Bosma MJ (1988) Evidence of functional lymphocytes in some (leaky) scid mice. J Exp Med 167: 1016-1033

Carroll AM, Bosma MJ (1988) Detection and characterization of functional T cells in mice with severe combined immune deficiency. Eur J Immunol 18: 1965-1971

Carroll AM, Hardy RR, Bosma MJ (1989) Occurrence of mature B (IgM^+, $B220^+$) and T ($CD3^+$) lymphocytes in scid mice. J Immunol (in press).

Chien Y, Iwashima M, Kaplan KB, Elliott JF, Davis MM (1987a) A new T-cell receptor gene located within the alpha locus and expressed early in T cell differentiation. Nature 327: 677-682

Chien, Y, Iwashima M, Wettstein DA, Kaplan KB, Elliott JF, Born W, Davis MM (1987b) T-cell receptor δ gene rearrangements in early thymocytes. Nature 330: 722-727

Dialynas DB, Wilde P, Marrack A, Pierres KA, Wall W, Hairan G, Otten MR, Loken M, Pierres J, Kappler J, Fitch FW (1983) Characterization of the murine antigenic determinant, designated

L3T4a, recognized by monoclonal antibody GK1.5: Expression of L3T4a by functional T cell clones appears to correlate primarily with class II MHC antigen reactivity. Immunol Rev 74: 29-56

Gibson D, Bosma GC, Bosma MJ (1989) Limited clonal diversity of serum immunoglobulin in leaky scid mice, this volume

Iwamoto A, Rupp F, Ohashi PS, Walker CL, Pircher H, Joho R, Hengartner H, Mak TW (1986) T cell specific γ genes in C57BL/10 mice. J Exp Med 163: 1203-1212

Ledbetter JA, Herzenberg LA (1979) Xenogeneic monoclonal antibodies to mouse lymphoid differentiation antigens. Immunol Rev 47: 63-90

Leo OM, Too DH, Sachs D, Samelson L, Bluestone JA (1987) Identification of a monoclonal antibody specific for a murine T3 polypeptide. Proc Natl Acad Sci 84: 1374-1378

Malissen M, Minard K, Mjolsness S, Kronenberg M, Goverman J, Hunkapillar T, Prystowsky MB, Yoshikai Y, Fitch F, Mak TW, Hood L (1984) Mouse T cell antigen receptor: structure and organization of constant and joining gene segments encoding the β polypeptide. Cell 37: 1101-1110

Schuler W, Weiler IJ, Phillips RA, Rosenberg N, Mak TW, Kearney JF, Perry RP, Bosma MJ (1986) Rearrangement of antigen receptor genes is defective in mice with severe combined immune deficiency. Cell 46: 963-972

Limited Clonal Diversity of Serum Immunoglobulin in Leaky Scid Mice

David M. Gibson†, Gayle C. Bosma* and Melvin J. Bosma*

†Département de biochimie Université de Sherbrooke, Sherbrooke Qué., Canada J1H 5N4. *Institute for Cancer Research, Fox Chase Cancer Center, 7701 Burholme Avenue, Philadelphia, Pennsylvania 19111, USA

Mice homozygous for the *scid* mutation are characterized by the absence of functional B and T cells (Bosma et al. 1983; Dorshkind et al. 1984; Custer et al. 1985). Most scid mice do not possess detectable serum immunoglobulin (Bosma et al. 1983). The exact nature of the scid defect is not yet known but considerable evidence suggests that it may be due to a defective recombinase system which precludes the functional rearrangement of antigen receptor genes for B and T cells (Schuler et al. 1986; Kim et al. 1988; Malynn et al. 1988; Okazaki et al. 1988).

An interesting observation is that approximately 15% of young adult scid mice produce serum immunoglobulin and invariably these mice also display some T-cell function, as seen in their ability to reject skin allografts (Bosma et al. 1988). These mice have been designated "leaky" or scid Ig+. It is of considerable interest to know the extent of immunoglobulin (or T-cell receptor) diversity in scid Ig+ mice for this could provide an indication of the potential for diversification of the cells which become leaky.

In order to shed some light on the nature of the leaky scid phenotype, we have assessed the heterogeneity of serum immunoglobulin in a series of scid Ig+ mice. A relatively simple but sensitive analysis for Ig heterogeneity is based on the isoelectric focusing (IEF) of serum Ig light (Igl) chains (Gibson 1984). The analysis of Igl chains using this technique can reliably identify the clonality of the Ig and an assessment can be made of changes in clonality with time. As previously reported, the diversity of Ig present in the serum of scid Ig+ mice is extremely limited, often consisting of as few as 1-3 clonotypes (Bosma et al. 1988).

In this study we present a more detailed description of the nature of Igl chain heterogeneity in scid Ig+ mice and discuss these findings in terms of our understanding of the ontogeny of B cell diversification.

Materials and Methods

Mice and serum: C.B-17, BALB/c and C.AL-20 scid mice were maintained at the ICR, Philadelphia. The majority of scid mice possessed undetectable levels of Ig(κ) (<0.01 mg/ml). Sera from mice possessing Ig(κ) concentrations of >0.05 mg/ml were further analyzed for Igl chain heterogeneity.

Current Topics in Microbiology and Immunology, Vol. 152

Immunoglobulin and Light Chain Isolation: Serum immunoglobulin was isolated from 0.015 to 0.100 ml of serum by absorption onto 1 ml columns of protein-A Sepharose (Pharmacia Canada Ltd, Dorval, Qué.). Non-absorbed proteins were eluted with PBS and the Ig fraction was eluted in 8 M urea, 0.05 M tris-HCl, pH 8.0 (reduction and alkylation buffer). The Ig was then reduced and alkylated with [^{14}C] iodoacetamide as previously described (Gibson 1984).

Isoelectric Focusing of Light Chains: Procedures for separation of light chains on urea-formate gels (pH 3.0) and isoelectric focusing have been described (Gibson 1984). Isoelectric focusing gels contained 2% ampholines, pH range 3-10 (LKB) and 6.6 M urea.

Results

Heterogeneity of Scid Ig+ Light Chains

Individual light chains, as found in myelomas, generally give rise to 2-3 evenly spaced bands in isoelectric focusing, each band representing a light chain differing by a single charge (Gibson 1984). The presence of doublets or triplets in the case of myelomas has been interpreted as being due to post-synthetic deamidation of the light chain (Awdeh et al. 1970; Williamson 1971). In contrast to myelomas, normal Ig light chain patterns are highly complex, consisting of >65 bands in addition to a diffuse general background in parts of the pH gradient (Gibson et al. 1983).

The isoelectric focusing profiles of light chains of 94 scid Ig+ mice have been analyzed in the present study. One obvious feature of the Igl chain patterns of scid Ig+ mice is that without exception they are highly restricted compared with patterns found in normal mice. Considerable variation in complexity is seen in different individual scid Ig+ mice however. While many scid Ig+ mice displayed a virtually monoclonal light chain pattern (Fig. 1, mouse 58; Fig.4, mouse 3228), others displayed increased levels of diversity, up to 10 or more clonotypes (Fig. 4, mouse 6342). Examples of Igl chain IEF spectra found in a variety of scid Ig+ mice are illustrated in Figures 1,2, 4 and 5.

In estimating the number of clonotypes present in scid Ig+ mice, short (1 day) and long (>6 days) autoradiographic exposures were examined. In most cases, bands corresponding to major clonotypes were clearly evident in 1 day autoradiograms of the gels. In many cases, a number of minor clonotypes (<10% intensity of major bands) could be identified in longer exposures. Figure 2 illustrates the effect of longer exposure on the light chain IEF patterns. In Fig. 2(a) the dried gel was exposed for 4 days while in Fig. 2(b) the same gel was exposed for 3 weeks. As can be seen in this example, relatively few new bands appear in the longer exposure. In practice, exposures greater than 8 days have not been found to add any reliable

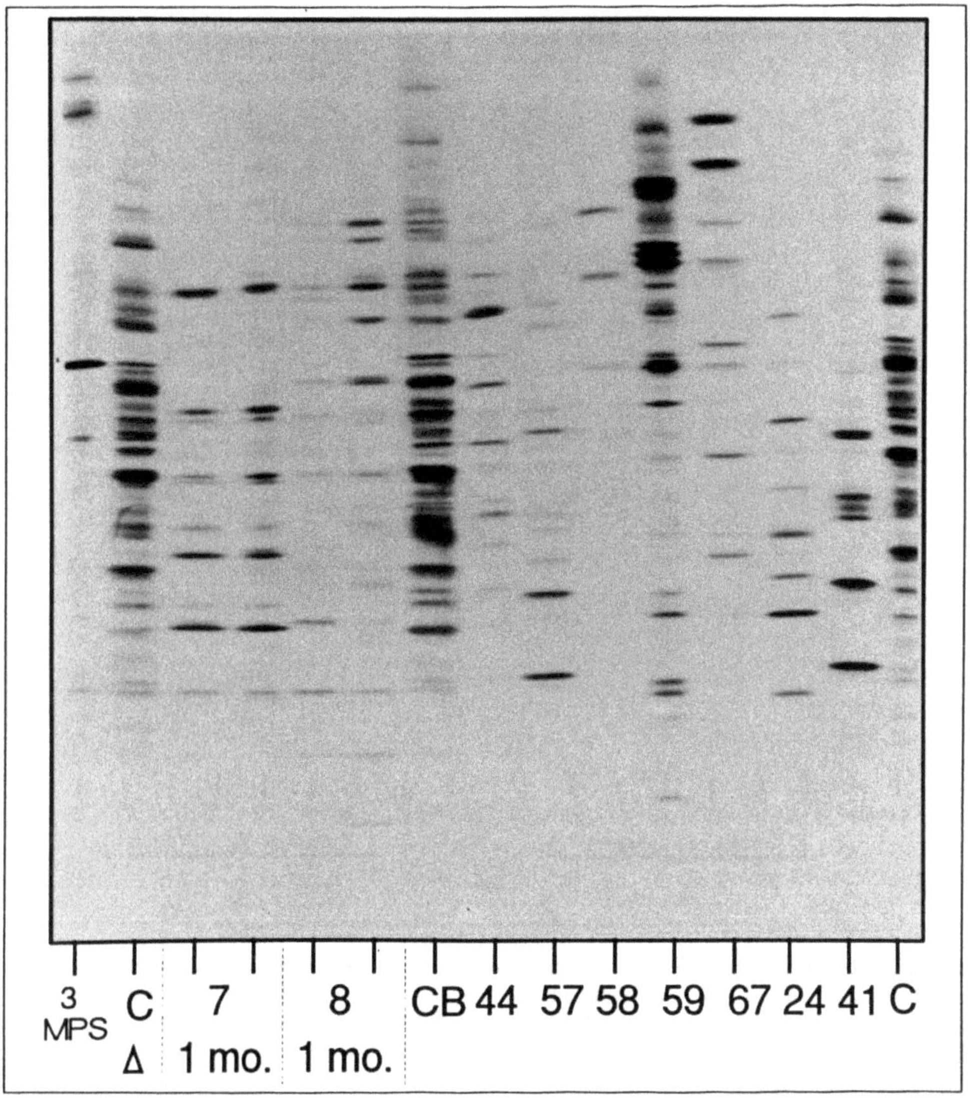

Fig.1 Representative examples of individual scid Ig+ light chain isoelectric focusing profiles. The anode is at the top of the gel. The first lane (3 MPS) includes the light chains of myelomas PC-7043 (NZB Vκ-21D)(top), HOPC-1 (Vλ-1)(middle) and FLOPC-1 (Vκ-1)(lower). Lanes labelled C and CB are normal serum Ig light chains of BALB/c and C.B-17 respectively. Numbers below individual scid Ig+ samples indicate animal number.

information to the analysis. In fact, minor degradation products may begin to appear in longer exposures, as is evident in the lane containing the three myeloma light chains. These degradation products may arise as a result of carbamylation of lysine residues or even partial hydrolysis during the electrophoresis step. Figure 3

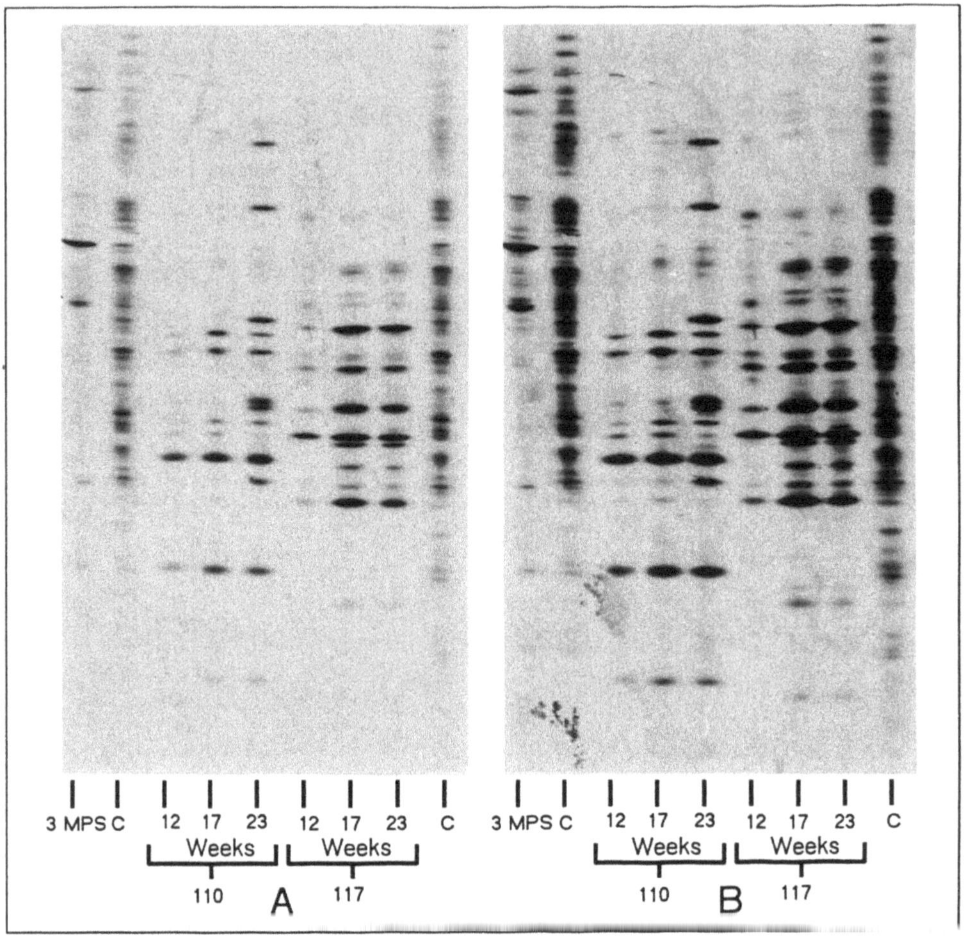

Fig.2 Illustration of the effect of longer exposure times on Igl chain banding patterns. The autoradiograph on the left was exposed for 4 days while on the right the same gel was autoradiographed for 3 weeks. Standards are as in Fig. 1. Individual scid Ig+ samples are indicated by animal number.

summarizes the clonality estimates made on Igl chain IEF patterns of 94 scid Ig+ mice. Close to 25% of the patterns consist of 3 or fewer clones. The largest group includes mice expressing between 4 and 7 clonotypes, with the remainder showing even more complex patterns. Even the most complex of the leaky patterns was relatively simple compared with normal Igl chain spectra.

The variation in complexity of different individual scid patterns did not appear to bear any correlation with total Ig(k) serum concentration, which itself shows wide variation (Bosma et al. 1988). Examples of both simple and complex patterns were

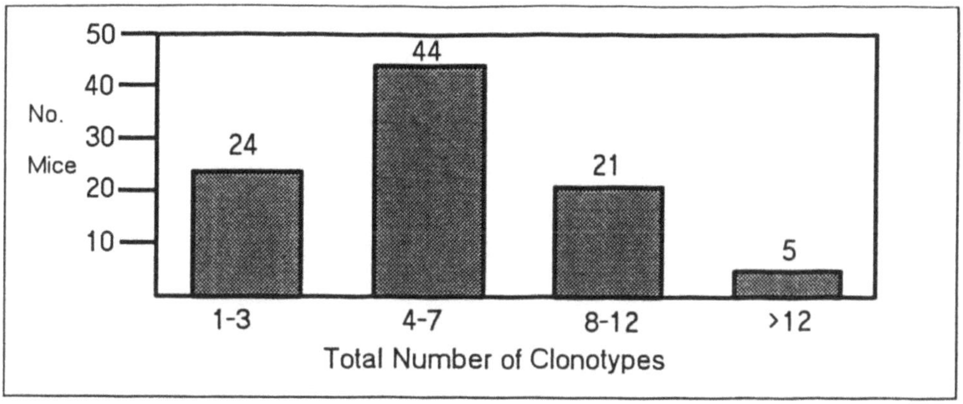

Fig.3 Frequency distribution of Igl chain clonotypes found in 94 scid Ig+ mice.

observed in mice with low (≈0.05 mg/ml) and high (≈5 mg/ml) serum Ig(k) concentrations (Fig. 4).

Absence of Bias in Igl Chain Utilization

A question which can be addressed by this analysis is whether Igl chain subgroups utilized in leaky scid mice may represent a select subset. In the present studies, we find no evidence for bias towards the expression of given Igl chain subgroups. The pattern of Igl chain(s) expressed by each of the mice was unique. Some Igl chain subgroups were identified in more than one mouse, but the repeated occurrence of given Igl chain spectrotypes was not more frequent than previously found in screening of large numbers of randomly selected myelomas (Gibson 1984). Based on their characteristic isoelectric focusing position and band spacing, light chains belonging to some kappa subgroups can be directly identified in IEF profiles (Gibson 1984). Using this criteria, light chains belonging to the kappa subgroups Vκ-21 and Vκ-1 have been identified in several of the leaky scid patterns. As the Vκ genes encoding these two kappa groups have been mapped to Jκ- proximal and distal positions, respectively (D'Hoostelaere et al. 1988; D'Hoostelaere and Gibson 1986), this would indicate that Vκ gene usage in these animals is not restricted with respect to chromosomal location.

As previously noted, relatively few scid Ig+ mice express lambda chains (Bosma et al. 1988). Out of the 94 scid (Ig+) patterns analyzed, candidates for λ1 chains were seen in only 4 mice. The identification of (un-mutated) λ1 light chains is reliable on the IEF gels as this light chain has a unique isoelectric point (the most prominent myeloma light chain in Fig. 1 lane 1 is the λ1 chain of HOPC-1).

Persistence of Clonotypes

One of the most striking findings of the present study is the stability of the

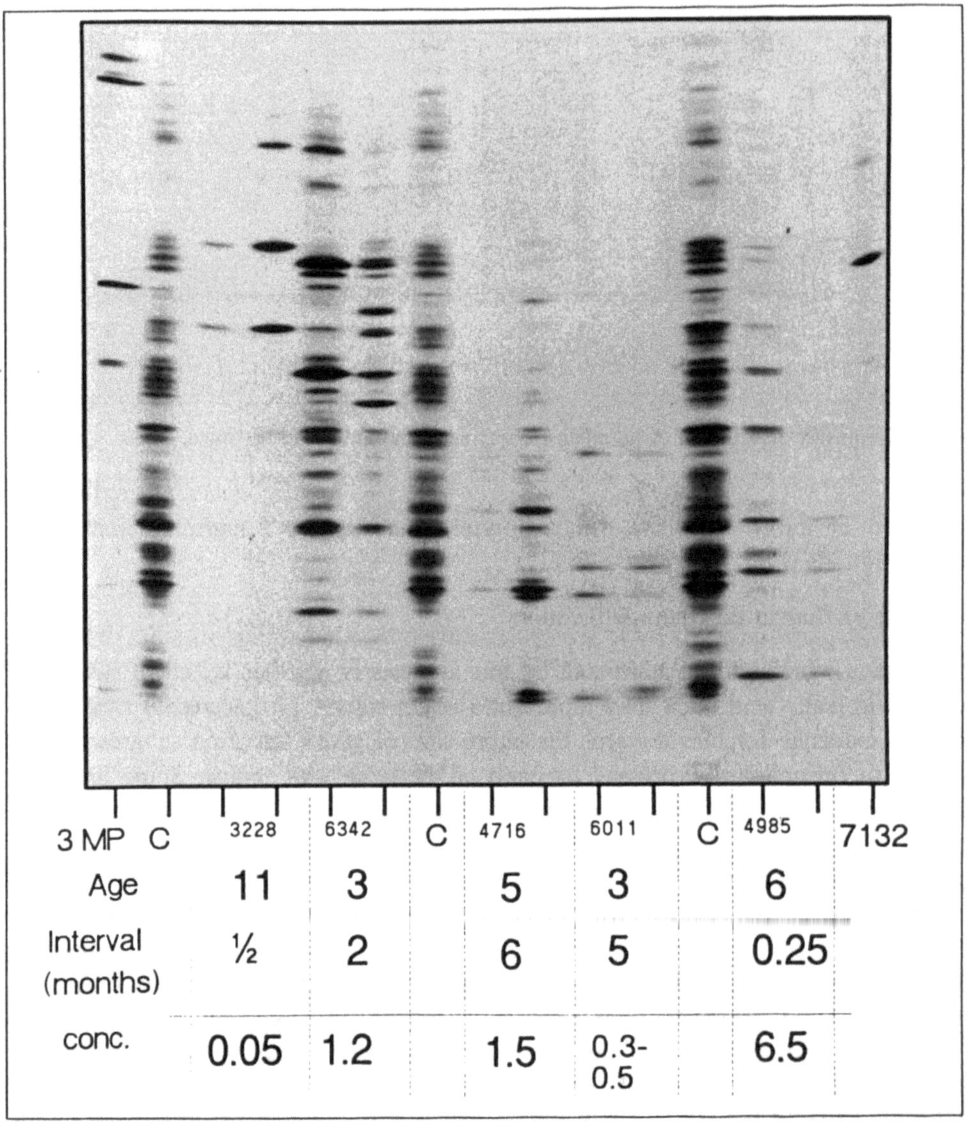

Fig.4 Examples of scid Ig+ light chain IEF profiles of individual animals sampled twice. The age at the initial bleed is in months. The serum Ig concentration given is in mg/ml (kappa). Abbreviations as in Fig. 1. The last lane contains the light chain of the myeloma PC 7132 (NZB Vκ-21A).

clonotype patterns over long periods of time. Examples of the stability of clonotype expression are shown in Figs. 1, 2, 4 and 5. Other examples of this were reported earlier (Bosma et al. 1988).

While the majority of scid Ig+ patterns were found to be remarkably stable,

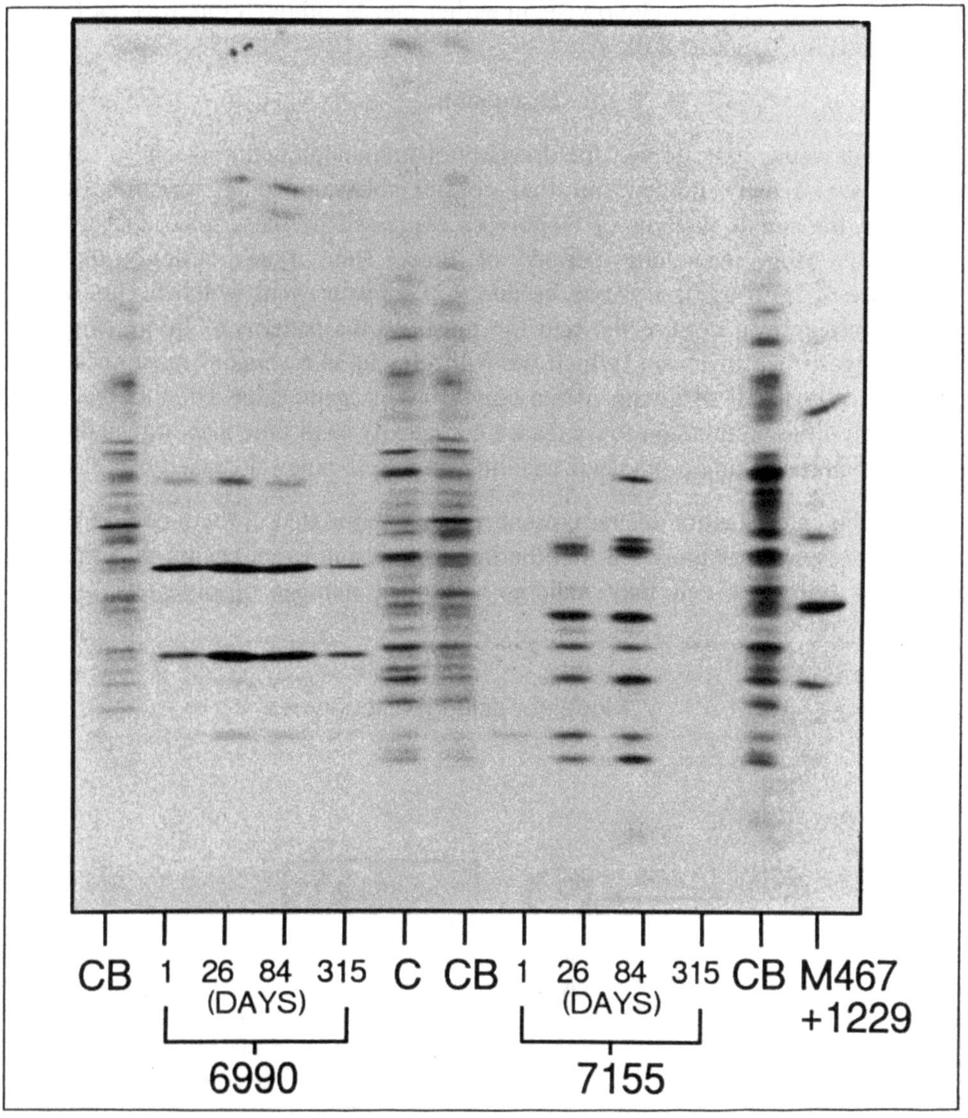

Fig. 5 Examples of analyses of Igl chain IEF spectra of scid Ig+ animals over longer periods of time. Two animals (6990 and 7155) are illustrated. Standards are as in Fig. 1. The last lane contains the Igl chains of NZB PC-1229 (Vκ-21A) (upper) and MOPC-467 (Vκ-1C) (lower) .

examples of patterns showing both quantitative and qualitative changes were also noted in serial bleed samples. An example of a scid Ig+ pattern showing the appearance of a new Igl chain clonotype is seen in Fig. 2 (mouse 110). In Figure 5, animal 7155 appears to have lost most of the major clonotypes between the 84 day and 315 day bleeds. In the majority of instances, clonotypes which increased in time were also present as minor clonotypes in earlier bleeds although this was not true in every case (Fig. 2, mouse 110).

Discussion

The present results indicate that the diversity of immunoglobulin present in scid Ig+ animals is extremely limited but that considerable individual variation occurs. Moreover, the simple patterns of clonotypes expressed by these mice appear to be remarkably stable over long periods of time. This argues strongly that the appearance of Ig in scid mice may be due to a somatic event which occurs rarely, perhaps only once in most of the scid Ig+ mice. As the patterns of Ig are simple, it would suggest that this event (which we will refer to as reversion) must occur in a precursor cell capable of giving rise to only a few progeny cells. The fact that there is relatively little progression to increased complexity with time also argues that the cell in which reversion occurs itself has little or no self-renewal capacity.

One possible interpretation of the present results is presented in the model in Figure 6. In this scheme, it is proposed that the initial reversion event occurs in a single B progenitor cell. This cell may undergo a limited number of divisions prior to

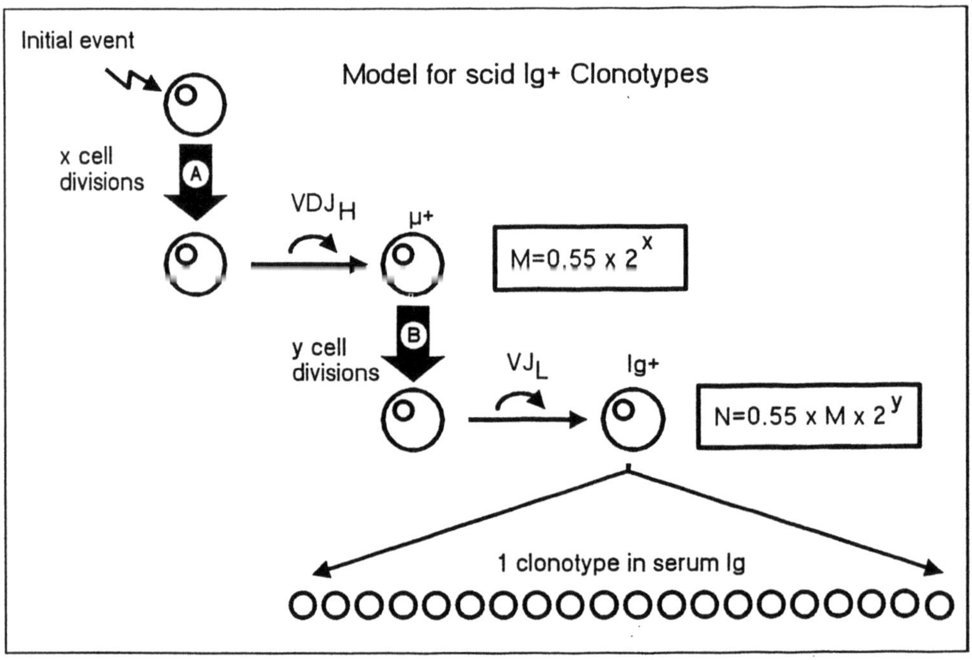

Fig.6 Model for origin of scid Ig+ cells. A single reversion in a B cell progenitor could give rise to multiple clonotypes of serum Igl chains if the revertant cell underwent cell division prior to VDJ_H rearrangement (step A) and VJ_L rearrangement (step B).

Table I Number of Clonotypes From Single Revertant Cell*

Number of cell divisions prior to VDJ rearrangement (X); Number of cell divisions prior to L. Chain gene rearrangement (Y)

Y \ X	0	1	2	3	4
0	0.3	0.6	1.2	2.4	4.8
1	0.6	1.2	2.4	4.8	9.6
2	1.2	2.4	4.8	9.6	19.2
3	2.4	4.8	9.6	19.2	38.4
4	4.8	9.6	19.2	38.4	76.8

* Assuming probability of selection=1.0

undergoing VDJ rearrangements (Alt et al. 1984; Yancoupoulos and Alt 1986; Alt et al. 1987). Once a successful VDJ rearrangement occurs for the heavy chain, the cell may again multiply to a limited extent and its progeny go on to rearrange a variety of kappa chains. The total number of clonotypes present in the animal would then depend on the number of cell divisions preceding VDJ_H rearrangement and VJ_L rearrangement (steps A and B; Fig. 6) and the probability of successful VDJ_H and VJ_L rearrangements (1/3 for each allele (Alt et al. 1984)). The number of clones that would go on to express detectable levels of serum Ig (in particular IgG) would then depend on antigenic stimulation (Askonas and Williamson 1972).

From work with cultures of early (15 day) fetal liver cells, Sauter and Paige (1988) have identified early B cell progenitors which undergo division prior to VDJ_H commitment. These cells are capable of giving rise to single colonies of B cells in agar culture in which more than one Igh chain is expressed. If the somatic reversion occurred in this early cell, a variety of clonotypes possessing different Igh chains could result. In addition, single colonies of pre-B cells derived from later stage fetal liver cultures express a single heavy chain but have been shown to give rise to progeny producing both lambda and kappa chains (Sauter and Paige 1987). This also indicates that the cells must have undergone at least one cell division between VDJ_H and VJ_L rearrangement. Although the exact number of cell divisions occurring at steps A and B (Figure 6) is not known, the present results argue that the number of divisions cannot be large. Table I gives a simple calculation of the number of potential clonotypes generated from a single revertant cell for different numbers of cell divisions at steps A and B. As can be seen from this table, if the number of cell divisions were even as large as 4, the number of clonotypes which could be generated would be as large as 76. In the present case, as the majority of scid Ig+ mice express fewer than 8 clonotypes it would suggest that the number of cell divisions at these steps must be 3 or less.

The considerable variation in numbers of clonotypes in different scid Ig+ mice could result partly from variation in number of cell divisions and successful rearrangements in the clonal progeny of a single cell. Additional variation could arise as a result of differences in the ability of a particular VDJ_H to pair successfully with different Igl chains. It seems likely that certain VDJ_H combinations may possess an intrinsically narrower range of specificities even when combined with different Igl chains.

The above model based on somatic reversion of the *scid* mutation in a single B progenitor cell can explain the appearance in some scid mice of a group of clonotypes of Ig and the stability of the clonotype patterns with time. Other results consistent with rare somatic reversion or normalization of scid lymphoid cells have been reported previously (Bosma et al. 1988; Carroll and Bosma 1988). It is important to note that our data are not easily explained by an alternative model (Bosma et al. 1988) which postulates that most scid mice possess a small population of B cells with functional Igh and Igl chain genes that have arisen by chance correct rearrangements at both Igh and Igl chain loci. Such models would predict a gradual increase in complexity of the Ig clonotype patterns with time, a prediction not born out by the present results.

Acknowledgements: The authors appreciate the skilled technical assistance of Eva Cunningham in collecting serum samples and Diane Cloutier and Sheila MacLean in doing the IEF analysis. This work was supported by the Medical Research Council of Canada, and the National Cancer Institute of Canada, grants AI-13323, CA-04946, and CA-06927 and by an appropriation from the Commonwealth of Pennsylvania.

References

Alt FW, Yancopoulos GD, Blackwell TK, Wood C, Thomas E, Boss M, Coffman R, Rosenberg N, Tonegawa S and Baltimore D (1984) Ordered rearrangement of immunoglobulin heavy chain variable region segments EMBO J 3:1209-1219

Alt FW, Blackwell TK, Yancopoulos GD (1987) Development of the primary antibody repertoire. Science 238:1079-1087

Askonas BA, Williamson AR (1972) Factors affecting the propagation of a B cell clone forming antibody to the 2,4-dinitrophenyl group. Eur J Immunol 2: 487-493

Awdeh ZL, Williamson AR, Askonas BA (1970) One-cell-one immunoglobulin. Origin of limited heterogeneity of myeloma proteins. Biochem J 116:241-248

Bosma GC, Custer RP, Bosma MJ (1983) A severe combined immunodeficiency mutation in the mouse. Nature 301: 527-530

Bosma GC, Fried M, Custer RP, Carroll A, Gibson DM, Bosma MJ (1988) Evidence of functional lymphocytes in some (leaky) scid mice. J Exp Med 167:1016-1033

Carroll AM, Bosma MJ (1988) Detection and characterization of functional T-cells in mice with severe combined immune deficiency. Eur J Immunol 18: 1965-1971

Custer RP, Bosma GC, Bosma MJ (1985) Severe combined immunodeficiency (SCID) in the mouse. Am J Pathol 120: 464-477

D'Hoostelaere LA, Gibson DM (1986) The organization of immunoglobulin variable kappa chain genes on mouse chromosome 6. Immunogenetics 23:260-265

D'Hoostelaere LA, Huppi K, Mock B, Mallett C, Potter M (1988) The Ig kappa L chain allelic groups among the Ig kappa haplotypes and Ig kappa crossover populations suggest a gene order. J Immunol 141:652-661

Dorshkind K, Keller GM, Phillips RA, Miller RG, Bosma GC, O'Toole M, Bosma MJ (1984) Functional status of cells from lymphoid and myeloid tissues in mice with severe combined immunodeficiency disease. J Immunol 132: 1804-1808.

Gibson DM, Maclean SJ, Cherry M (1983) Recombination between kappa chain genetic markers and the Lyt-3 locus. Immunogenetics 18: 111-116

Gibson DM (1984) Evidence for 65 electrophoretically distinct groups of light chains in BALB/c and NZB myelomas. Mol Immunol 21: 421-432

Kim MG, Schuler W, Bosma MJ, Marcu KB (1988) Aberrant Igh locus rearrangements in A-MuLV pre B lines of scid mice: evidence for deregulated D-J recombination. Curr Top Microbiol Immunol 137: 203-210

Malynn BA, Blackwell TK, Fulop GM, Rathbun GA, Furley AJ, Ferrier P, Heinke LB, Phillips RA, Yancopoulos GD, Alt FW (1988) The scid defect affects the final step of the immunoglobulin VDJ recombinase mechanism. Cell 54: 453-460

Okazaki K, Nishikawa S, Sakano H (1988) Aberrant immunoglobulin gene rearrangement in scid mouse bone marrow cells. J Immunol. 141: 1348-1352

Sauter H, Paige CJ (1987) Detection of normal B-cell precursors that give rise to colonies producing both kappa and lambda light immunoglobulin chains. Proc Natl Acad Sci USA 84: 4989-4993

Sauter H, Paige CJ (1988) B cell progenitors have different growth requirements before and after immunoglobulin heavy chain commitment. J Exp Med 168: 1511-1516

Schuler W, Weiler IJ, Schuler A, Phillips RA, Rosenberg N, Mak TW, Kearney JF, Perry RP, Bosma MJ (1986) Rearrangement of antigen receptor genes is defective in mice with severe combined immune deficiency. Cell 46: 963-972

Williamson AR (1971) Antibody isoelectric focusing spectra-analysis of the heterogeneity of antibody molecules in serum by isoelectric focusing in gel and specific detection with hapten. Eur J Immunol 1: 390-394

Yancoupoulos GD, Alt FW (1986) Regulation of the assembly and expression of variable region genes. Ann Rev Immunol 4: 339-368

Paucicional B Cell Involvement in Production of Immunoglobulin in Scid Ig+ Mice

John F. Kearney, Nanette Solvason, Robert Stohrer, Jie Ma, Victor Van Cleave, Agnes Leheun, Gabrielle Fulop* and Michal Fried^

Department of Microbiology, Division of Developmental and Clinical Immunology, The Comprehensive Cancer Center, University of Alabama at Birmingham, Birmingham, Alabama 35294

INTRODUCTION

Although the scid defect is complete in the majority of young and adult mice, ~41% of those that are kept for nine months or more, develop some form of T and B cell involvement. The nature of the T cells that develop in these scid mice is described by Carroll et al. this volume. B cell involvement is characterized by the presence of large amounts of serum immunoglobulin in mice which were previously immunoglobulin negative. Mice which have high serum antibody titres have been called "leaky" and will be referred to as scid Ig^+ in this text. Scid Ig^+ mice have expressed μ, γ_1, γ_{2b}, γ_3 and κ; so far no γ_{2a}, α or λ production has been detected. We have isolated hybridomas from completely deficient and scid Ig^+ mice. In independent fusions from scid mice no Ig producing hybridomas were obtained; however, large numbers of Ig producing hybridomas have been isolated from fusion of spleen cells of scid Ig^+ mice. Analysis of the hybridomas isolated from scid Ig^+ mice has revealed a restricted heterogeneity with respect to the κ light chains and the V_H regions expressed in the collection of hybridomas obtained from each individual mouse. Furthermore, idiotypes appear to be shared by immunoglobulins derived from some individual scid Ig^+ mice analyzed. Preliminary studies suggested that in scid Ig^+ mice the defect involved in the failure to produce normal B cells is overcome on rare occasions. The paucicional B cells that emerge proliferate in the absence of normal autoregulatory mechanisms perhaps being activated by auto- or environmental antigens. That both stimuli may be involved is suggested by our finding that a proportion of antibodies from these scid Ig^+ mice have specificity for nuclear antigens and antigens expressed by certain strains of *Enterobacter* and *Serratia*. As suggested by previous studies there appears to be a relationship between the T and B cells that appear in scid Ig^+ mice. We show here that certain T cell populations can be expanded by administration of particular immunoglobulins isolated from hybridomas derived from scid Ig^+ mice.

RESULTS

Isotype profiles of hybridomas

A number of fusions were made with spleen and bone marrow from scid and scid Ig^+ mice, however it was only from the latter that immunoglobulin producing hybridomas were obtained and then only from the spleen. No μ only pre-B cell like hybridomas were found in spleen or bone marrow. Prior to cervical dislocation and subsequent fusion of spleen and bone marrow tissues each mouse was bled and serum assayed for the expression of immunoglobulin isotypes. As shown in Table 1, $\mu\kappa$ immunoglobulin was far in excess of the other isotypes present in

---

*The Hospital for Sick Children, Toronto, Canada
^The Fox Chase Cancer Center, Philadelphia, Pennsylvania

TABLE 1 SERUM IMMUNOGLOBULIN LEVELS IN SCID Ig+ MICE

	4545[a]		6468		6407		6448		5500		5849		6409	
	Amt In Sera mg/ml	#Hybrids	Amt In Sera mg/ml	#Hybrids	Amt In Sera mg/ml	#Hybrids	Amt In Sera mg/ml	#Hybrids	Amt In Sera mg/ml	#Hybrids	Amt In Sera mg/ml	#Hybrids	Amt In Sera mg/ml	#Hybrids
μ	-	14	3.0	6	4.76	13	0.15	0	0.54	0	1.2	0	2.27	-
γ1	-	0	0.2	1	0.1	0	0.034	0	<0.1	0	<0.1	0	<0.1	-
γ2a	-	0	<0.1	0	<0.1	<0.1	0.00	0	<0.1	0	<0.1	0	0	-
γ2b	-	5	<0.1	0	0.1	3	0.07	0	0.08	0	11.7	19	0	-
γ3	-	0	1.3	0	0.1	0	0.06	0	0.04	0	0.11	0	0	-

[a]Serum was not available from this mouse.

the serum of each mouse with the exception of mouse 5849 in which $\gamma_{2b}\kappa$ antibodies predominated. The serum from mouse 4545 was not available for analysis.It can also be seen that the isotype distribution of hybridomas collected from a particular mouse generally reflects the relative isotype distribution in sera. No hybridomas synthesizing λ light chains were detected and no λ^+ Ig was found in the sera.

Diversity of immunoglobulins synthesized by scid Ig+ mice

We next wished to determine whether the apparent breakthrough of B cells in scid Ig^+ mice was multi- or pauci-clonal in nature. Initially we examined clonal diversity in these mice by subjecting hybridoma derived immunoglobulins to isoelectric focussing (IEF) of their light chains. IEF analysis of κ chains produced by the separate panel of hybridomas from an individual scid Ig^+ mouse demonstrates the marked restriction in the κ chain repertoire (Fig. 1A), while a comparable number of κ chains examined from adult mice expressed 12 different patterns widely differing in their isoelectric points (Fig. 1B). Despite the restriction in the κ chain repertoire expressed by the scid Ig^+ mice as determined by IEF, no pattern was repeated between mice, i.e., each pattern was present in only one mouse.

Idiotype restriction in leaky scid derived hybridomas

Monoclonal anti-idiotype antibodies were prepared by immunizing a Lewis rat with two of the purified immunoglobulins from mouse 45 which exhibited distinct κ chain focussing patterns. The two antibodies chosen for the immunization were 45-33-2 (γ_{2b},κ) and 45-29-1 ($\gamma_{2b}\kappa$) (Table 2).

Five monoclonal anti-idiotype antibodies were found that recognized either 45-29-1 or 45-33-2 or both. These were further screened against a panel of purified non-related $\gamma_{2b}\kappa$ and $\gamma_{2b}\lambda$ antibodies from mice of the Igh6b allotype. Four of the five antibodies recognized either 45-29-1 or 45-33-2 and probably detect a private idiotope. One of the five named B1-1, however, recognized both 45-29-1 and 45-33-2. B1-1 as indicated in Table 2 recognizes a common determinant found on at least half of the members from three of the four different κ chain groups derived from mouse 45. B1-1, however, did not recognize any member of any κ chain group present in mouse 58. We then wished to determine whether this idiotope which is frequently expressed on the panel of hybridomas from mouse 4545, is represented in the normal plasma cell populations of C57B1/6 spleen and bone marrow. Cytocentrifuge smears were made from these tissues, followed by staining with FITC anti-Mig and then costaining with B1-1 to determine the frequency of B1-1 expression. In addition, single cell suspensions from seven other independent leaky scid mice were stained. The results in Table 3 show that this idiotope is expressed on about 2-3 percent of plasma cells in spleens of normal C57B1/6 mice. In more than half of the scid Ig^+ mice, despite the small numbers of plasma cells available for scoring, it can be seen that there is a tendency for this idiotope to be expressed at a 5- to 10-fold higher frequency than in normal mice, suggesting that there may be a common V_H or V_L expressed frequently in scid Ig^+ mice. The high degree of reactivity within the 45 panel again supports the pauci-clonality of these hybridomas. The apparent non-reactive clones may have mutations which cause loss of the B1-1 idiotope.

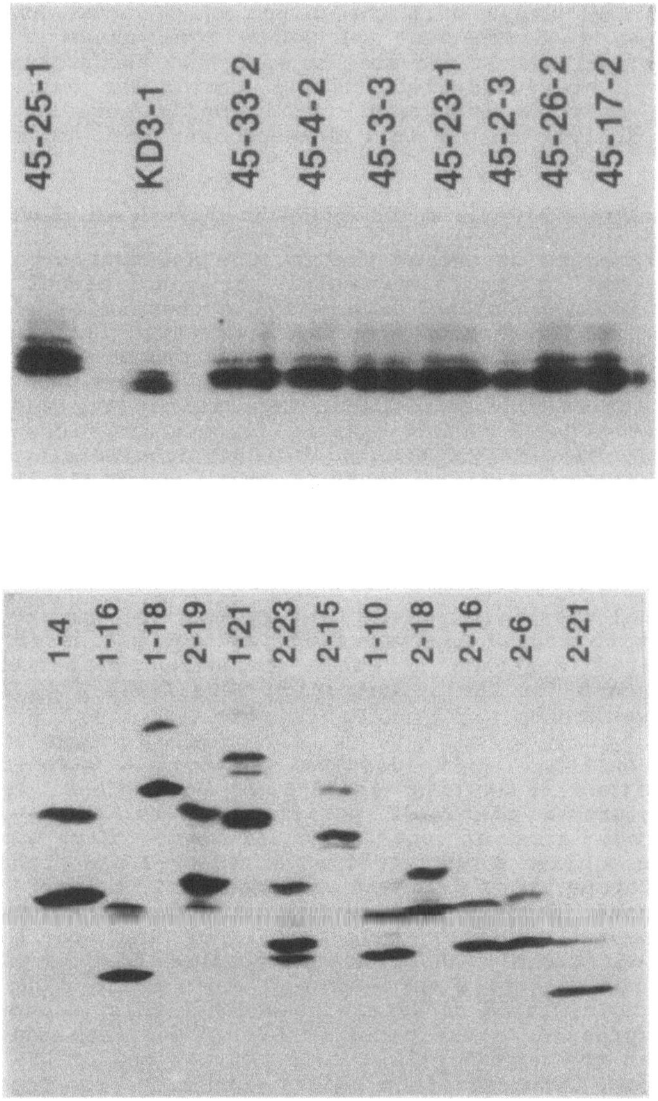

FIGURE 1A,B. *Secreted immunoglobulins were biosynthetically labeled by incubating 10^6 hybridoma cells in 1 ml of methionine free medium containing 50 μCi (^{35}S) methionine for 18 hr. Labeled antibodies were purified from supernatants by incubation with goat anti-mouse isotype antibodies followed by precipitation with a Staphylococcus aureus protein A preparation. Precipitates were reduced and run on a preparative SDS-PAGE gel to separate the light and heavy chains. Light chains were cut from the gel and analyzed by isoelectric focussing. Figure 1A (top) shows IEF patterns of light chains from scid mouse 4545 and KD3-1 which is a selected fetal liver derived hybridoma. Figure 1B (bottom) shows the IEF patterns of light chains from randomly picked hybridomas of normal adult mouse spleen.*

TABLE 2. CLONAL ANALYSIS OF HYBRIDOMAS FROM MOUSE 45

IEF LC Pattern	Hybridoma	Isotype	Anti-Nuclear	B1-1	*S. marcescens*	*E. cloacae*
A	23	$\mu\kappa$	+	+	+	+
	26	$\mu\kappa$	+	+	+	+
	9	$\mu\kappa$	+	+	+	-
	4	$\mu\kappa$	+	+	-	-
	3	$\mu\kappa$	+	-	-	-
	2	$\mu\kappa$	+	-	-	-
	33*	$\gamma_{2b}\kappa$	-	+	-	-
	17	$\mu\kappa$	nd	nd	nd	nd
	7	$\mu\kappa$	nd	nd	nd	nd
B	10	$\mu\kappa$	+	+	+	-
	20	$\mu\kappa$	+	+	-	+
	25	$\mu\kappa$	+	+	-	-
	27	$\mu\kappa$	+	+	-	-
	29*	$\gamma_{2b}\kappa$	+	+	-	-
	32	$\gamma_{2b}\kappa$	+	+	-	-
	40	$\gamma_{2b}\kappa$	+	+	-	-
	39	$\gamma_{2b}\kappa$	nd	nd	nd	nd
C	28	$\mu\kappa$	+	+	-	+
	5	$\mu\kappa$	+	-	-	-
D	6	$\mu\kappa$	-	-	-	-
	22	$\mu\kappa$	nd	nd	nd	nd

*These antibodies were used as immunogens to construct the anti-idiotypic antibody, B1-1.

TABLE 3. DISTRIBUTION OF B1-1 IDIOTOPE IN NORMAL C57Bl/6 AND SCID Ig^+ MICE LYMPHOID TISSUES

Cell Source	#B1-1+	Total Plasma Cells	%B1-1+
C57Bl/6 Spleen 1	15	526	2.8
C57Bl/6 Spleen 2	6	341	1.8
C57Bl/6 Bone Marrow 1	1	72	1.4
C57Bl/6 Bone Marrow 2	0	20	0.0
Scid Ig^+ 6468 Spleen	12	589	2.0
6468 BM	2	15	13.0
6407 Spleen	17	170	10.0
6407 BM	2	15	20.0
6448 Spleen	1	13	7.6
6448 BM	0	0	0.0
2931 Spleen	0	35	0.0
2931 BM	0	2	0.0
5500 Spleen	1	41	2.4
5500 BM	0	0	0.0
5859 Spleen	0	20	0.0
5859 BM	2	17	11.7

Specificity analyses of individual hybridomas

To search for clues for a mechanism to explain apparent large scale clonal expansion of the few B cells that arise in scid Ig^{+} mice, we screened the hybridoma antibodies against a large panel of antigens including homologous defined plasmacytoma and hybridoma proteins (for elements of idiotypic networks), polysaccharide and haptenic antigens and a large panel of enteric organisms described previously (Kearney et al. 1985; Kearney et al.1987). During this search we found that some but not all 45 series antibodies reacted by agglutination tests and FACS analysis with *Serratia liquefaciens* and *Enterobacter cloacae* (Fig. 2).

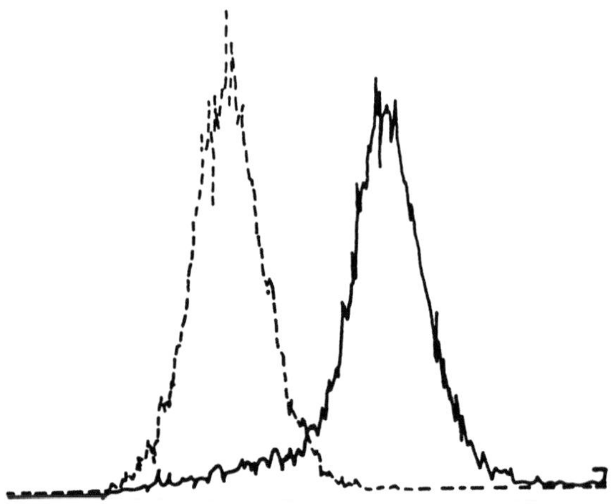

FIGURE 2. *Each hybridoma antibody was tested for specificity on a large panel of different strains of bacteria. Heat killed bacterial were stained with 10 microliters of hybridoma supernatant, washed and then stained with FITC anti-mouse Ig. The samples were then analyzed by flow cytometry. These profiles show staining of Serratia marcescens with an IgM antibody from mouse 4545 (solid lines) and staining with an irrelevant control antibody (slashed line).*

Staining of fixed autologous cells revealed that many 45 series hybridomas stained the nucleus intensely (Fig. 3) suggesting potential autoreactivity. The results of nuclear staining, bacterial agglutination and B1-1 idiotype expression are summarized in Table 2. It can be seen that while most of the 45 series react with nuclear material and are B1-1 Id positive, fewer react with the bacteria described. These results would suggest that this pauci-clonal B cell expansion may be driven by self reactivity and/or a response to environmental bacteria. The production of large amounts of immunoglobulin in this mouse resulted from the clonal expansion of progeny derived from one or a very few B cell clones that successfully rearranged heavy and light chain immunoglobulin genes.

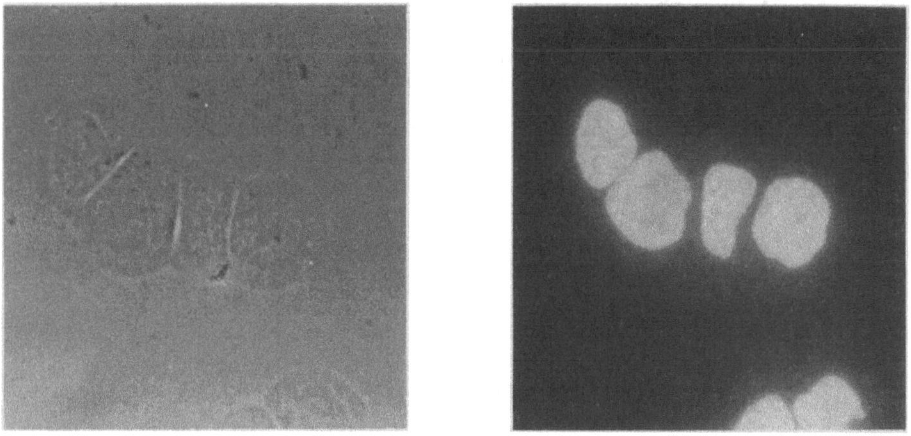

FIGURE 3. *Immunofluorescence (right) and phase contrast (left) micrograph showing binding of hybridoma antibody 45-33 to nucleus. Cytocentrifuge preparations were made of the Ag8.653 myeloma cell line. The preps were fixed in acid alcohol and subsequently stained with supernatant from hybridomas derived from mouse 45 followed by FITC anti-mouse Ig.*

V_H and V_L nucleotide sequences

Nucleotide sequences from hybridomas derived from scid Ig^+ mice have been used as a further means of analyzing the clonality of these cells.

Table 4 contains a preliminary description of the V_HDJ_H and V_LJ_L genes used in hybridomas derived from 5 independent fusions of spleen cells from scid Ig^+ mice. As can be seen, four different V_H gene families, and five $V\kappa$ families are represented. There does not seem to be an over-representation of 3' V_H genes as appears during ontogeny. Four different D genes appear to be used while all hybridomas except those from mouse 58 used $J\kappa_1$.

Although many of the hybridomas derived from a single mouse use the same V, D and J gene segments, a closer examination of the sequence data clearly demonstrates that there are differences in the N region diversity between these hybridomas (data not shown). The possibility exists that these N segments may be a result of somatic mutation. Preliminary Southern blot analysis of DNA from these hybridomas using a J_H4 probe showed that each individual hybridoma contained not only a common fragment representing the expressed heavy chain allele (non-fusion partner derived) but also showed a separate set of bands representing unique fragments from the non-expressed allele. The latter set of bands differed in size between each individual hybridoma. In contrast when DNA was of serum antibodies and expansion of T cells in secondary lymphoid organs. Because the IgM antibody injected was of the same IgM allotype it is not possible to exclude that IgM present in the serum was that which was passively administered. However, since only IgM was administered any IgG isotypes detected must be derived by endogenous production. Furthermore, λ antibody was detected in some mice which has not yet been observed in naturally occurring scid Ig^+ mice. Although our initial efforts focussed on B cell leakiness, it became apparent that certain of these antibodies were efficient in promoting T cell leakiness. This was evidenced by the presence of enlarged lymph nodes and substantial numbers of T cells present in these nodes. As shown in the profiles in Fig. 4 these cells are T3 positive and present a variety of phenotypes.

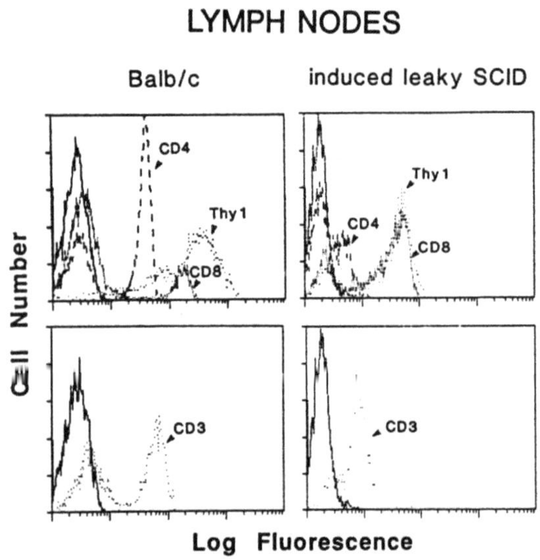

FIGURE 4. *Flow cytometric analysis of lymphocytes from lymph nodes of normal BALB/c (left) or "induced" scid Ig^+ mice (right). Lymph nodes were pooled, gently teased apart and a single cell suspension was made. Aliquots were stained with either FITC anti-CD4, anti-CD8, anti-Thy-1, or anti-CD3 as indicated. Negative control staining is indicated in each plot as a solid line.*

Some mice treated in this way also have thymomas. Examination of the thymus, however, in these mice showed little evidence for extensive T cell development (Fig. 5). These results suggest that administration of these antibodies may induce an expansion of rare CD3+ T cells that break through the scid defect and by some as yet undetermined means are expanded in peripheral lymph nodes.

hybridized with the J_κ probe the same restriction fragments were observed in all hybridomas analyzed. Although it is not clear at the moment how to interpret these results, several events during the development of a single clone may result in the observed phenotype and rearrangement status of immunoglobulin genes that have been observed in these hybridomas. Firstly, a B cell precursor could have successfully rearranged a $V_\kappa J_\kappa$ initially and then in the progeny of that cell $V_H DJ_H$ rearrangements occurred all involving the same genes. This would be unlikely but if strong antigenic selection were to occur (as is likely in this clone) it is a possibility. Alternatively, the distinct rearrangement patterns of the unexpressed alleles may be the result of the abnormal recombination process in scid mice. More extensive sequence analysis of hybridoma clones from these mice will be revealing.

TABLE 4. $V_H DJ_H$ AND $V_L J_L$ OF HYBRIDOMAS FROM FIVE DIFFERENT SCID Ig+ MICE.

Scid Hybridoma	Isotype	H Chain V	H Chain D	H Chain J	L Chain V	L Chain J
45-5	$\mu\kappa$	3609	DSP2.8-C DFL16.1	JH2	-	JK1
45-10	$\mu\kappa$	"	"	"	K-VI	"
45-20	$\mu\kappa$	"	"	"	-	-
45-26	$\mu\kappa$	"	"	"	K-VI	JK1
45-27	$\mu\kappa$	"	"	"	-	-
45-28	$\mu\kappa$	"	"	"	K-VI	JK1
45-29	$\gamma_{2b}\kappa$	"	"	"	"	"
45-33	$\gamma_{2b}\kappa$	"	"	"	"	"
A-2	$\mu\kappa$	7183	DQ52	JH4	K-IV	JKI
F-23	$\mu\kappa$	"	"	"	"	"
F-14	$\mu\kappa$	"	"	"	-	-
F-18	$\gamma_{2b}\kappa$	J558	DFL16.2	"	K-III	JKI
58-45	$\gamma_{2b}\kappa$	7183	DQ52	JH4	-	-
58-51	$\gamma_{2b}\kappa$	"	"	"	K-II	JK2
58-59	$\gamma_{2b}\kappa$	-	-	-	"	"
67-36	$\gamma_1\kappa$	-	-	-	K-V	JK1
67-53	$\gamma_1\kappa$	-	-	-	-	"
67-75a	$\gamma_3\kappa$	-	-	JH2	K-V	"
35-15-C1	$\mu\kappa$	S107	DSP2.3	JH4	K-V	JK1

Leakiness in scid mice may involve both T and B cells

Observations by Carroll et al. (1989) have shown that T and B cell leakiness was strongly correlated, suggesting a possible interaction between the two cell types which break through the defective rearrangement of receptor genes. To investigate this we purified immunoglobulins from the scid Ig^+ mice described above and re-injected them into litters of scid mice from the day of birth. The regimen used involved the tri-weekly injection of 50-100 μg of antibody for 6-8 weeks. It was found that certain IgM antibodies but not IgG induced leakiness in these mice both in the production

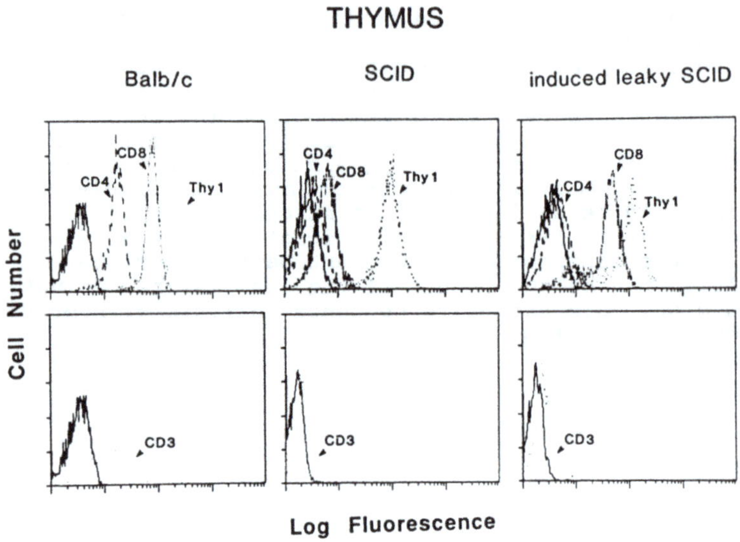

FIGURE 5. *Flow cytometric analysis of thymocytes from normal BALB/c (left), scid (middle) and induced scid Ig$^+$ mice (right) Thymic lobes were removed, teased apart and single cell suspensions were made. Aliquots were stained with either FITC anti-CD4, anti-CD8, anti-Thy-1 or anti-CD3 as indicated. Negative control staining is indicated in each plot as a solid line.*

DISCUSSION

The block in normal differentiation of lymphoid cells results in almost complete absence of functional T and B cells in scid mice. While the mechanism resulting in this block is not known it is thought that defective mechanisms involved in T and B cell receptor gene rearrangements may be the prime cause. Mechanisms involved in this defect are discussed elsewhere in this volume. Nevertheless from observations made in scid Ig$^+$ mice this block is sometimes overcome and a few B or T cells may develop. In the case of T cell involvement, sometimes functional T cells develop and/or development of thymomas occurs (Carroll et al., in press). It has previously been noted that leakiness tends to occur as the animals age and that a non-germfree environment promotes the development of T and B cells in scid populations. These and other results suggest that the breakthrough in B cell differentiation is a rare event and that environmental antigens may be involved in the expansion and growth of these B cells eventually resulting in the high serum immunoglobulin levels in these leaky mice. We have shown that this B cell expansion is pauci- or oligoclonal and involves an environmental or a self antigen.

The IEF and sequence analyses produced evidence that in each individual mouse a single light chain or a family of very closely related light chains was used. No λ chains were detected in any experiments. There was evidence described previously that this selection may be strongly driven by external or self antigens. The sequence data for the heavy chain variable region is more difficult to interpret since from a single mouse many of the hybridomas use the same V, D and J segments suggesting that the hybridomas were clonally related. However, analysis of DNA fragments by Southern analysis indicates that there are unique rearrangements on the non-expressed alleles indicating that each clone arose independently. The apparent non-biased usage of the V_H repertoire with respect to chromosomal location also suggests that at least within the panel of scid Ig^+ mice analyzed so far there does not appear to be a recapitulation of ontogeny with respect to the preferential use of the most 3' proximal V_H genes.

Finally, we have shown that chronic treatment from birth with hybridoma derived IgM antibodies promoted leakiness in both T and B cell compartments. This was evidenced by expansion of serum immunoglobulins of multiple isotypes and the presence of T cells in peripheral lymph nodes but not in the thymus.

ACKNOWLEDGEMENTS

We wish to thank Ann Brookshire for her patience and excellent secreterial assistance and Ann Marie Hamilton for assistance in the preparation of this manuscript. This work was supported by NIH grants CA 16673, CA 13148 and AI 14782.

REFERENCES

Carroll AM, Hardy RR, Bosma M (in press) Occurrence of mature B (IgM^+, $B220^+$) and T ($CD3^+$) lymphocytes in scid mice. J. Immunol.

Kearney JF, McCarthy MTM, Stohrer R, Benjamin WH, Briles DE (1985) Induction of germline anti-α-3 dextran antibody responses in mice by members of the enterobacteriaceae family. J Immunol 135:3468

Kearney J, Vakil M, Nicholson N (1987) Non-random V_H gene expression and idiotype anti-idiotype expression in early B cells. In: Kelsoe G, Schulze D (eds) Evolution and Vertebrate Immunity: The antigen receptor and MHC families, Texas University (Austin), pp 175-180

V. Potential Uses: Engraftment with Syngeneic Mouse Cells

Reconstitution of Scid Mice by Injection of Varying Numbers of Normal Fetal Liver Cells into Scid Neonates

Gayle C. Bosma,[1] David M. Gibson,[2] R. Phillip Custer[1†] and Melvin J. Bosma[1]

1 Institute for Cancer Research, Fox Chase Cancer Center, Philadelphia, PA 19111, USA
2 Université de Sherbrooke, Sherbrooke, Québec, Canada J1H 5N4

INTRODUCTION

In our original report (Bosma et al. 1983) we showed that young adult C.B-17scid/scid mice (scid mice) could be successfully engrafted with bone marrow cells of normal BALB/c donors. The reconstituted recipients readily produced serum immunoglobulin heavy (Igh) chain allotype of the donor cells, indicating that they could support the differentiation of normal lymphocytes. However, the extent of reconstitution in these and other similarly reconstituted scid mice was variable and often incomplete (Custer et al. 1985). Furthermore, the extent of reconstitution did not correlate with the number of donor cells injected (1-5 x 10^6 bone marrow cells) nor with the elapsed time between injection and analysis. In agreement with these findings, Fulop and Phillips (1986) reported that injection of variable numbers of normal bone marrow cells into young adult scid mice seldom resulted in normal numbers of lymphoid cells, surface Ig^+ cells or colony-forming B cells. To ensure complete reconstitution, they found it necessary to irradiate scid recipients prior to cell transfer (Fulop and Phillips 1986).

In view of the above findings, we were interested in reconstituting scid mice with donor lymphocytes in a dose-dependent manner without the use of irradiation, and in a way that would closely mimic normal lymphoid development. In an attempt to satisfy these objectives we initiated reconstitution in newborn scid mice by injecting them with varying numbers of fetal liver cells from normal, allotype-distinct donors. The recipients were analyzed at 12-14 weeks of age. We specifically wanted to know: 1) Whether the frequency of recipients with donor allotype would follow a Poisson distribution at limiting cell doses, and thus enable us to estimate the average number of injected Ig progenitor cells; 2) would the recipients repopulated by progeny of one Ig progenitor cell show complete or partial reconstitution, and if the latter were true, would such recipients resemble leaky scid mice; and 3) would recipients show complete reconstitution at non-limiting cell doses containing multiple Ig progenitor cells.

RESULTS

Dose response analysis of reconstitution

In order to reconstitute scid mice in a cell dose-dependent manner, varying numbers of fetal liver cells from 13-14 day-old embryos of normal BALB/c or C.AL-20 mice were injected

† deceased May 3, 1989

Current Topics in Microbiology and Immunology, Vol. 152

intraperitoneally into C.B-17scid/scid neonates within 24 hours of birth. The BALB/c, C.AL-20 and C.B-17 allotype congenic strains share the same genetic background (BALB/c), but express different serum Igh chain allotypes; i.e., Igh^a (BALB/c), Igh^d (C.AL-20) and Igh^b (C.B-17). Thus, the presence of donor allotype in scid recipients at 12-14 weeks of age served as an indication of reconstitution.

When relatively low numbers of fetal liver cells (0.20-1.0 x 10^5) were injected into scid neonates, some recipients became reconstituted, others did not. The probability of reconstituting or not reconstituting a scid recipient was an exponential function of the number of fetal liver cells injected. This is indicated in the semi-logarithmic plots of Figure 1, where the percentage of non-reconstituted recipients is shown to be a linear function of the number of injected donor cells. This kind of relationship conforms to a distribution (Poisson) descriptive of the frequencies of rare, discreet events (see legend of Figure 1). The events in this case would correspond to rare lymphocyte progenitor cells in the donor fetal liver cell population that give rise to allotype producing progeny. We will refer to these precursors as Ig progenitor cells though this is not meant to imply an exclusive commitment to the B cell lineage; such cells may also give rise to T cells. Estimates of the Ig progenitor cell frequencies were derived from the slope (λ) of the linear regression lines in Figure 1. As can be seen, the estimated frequency of Ig progenitor cells was 1.47 x 10^{-5} in BALB/c and 0.76 x 10^{-5} in C.AL-20.

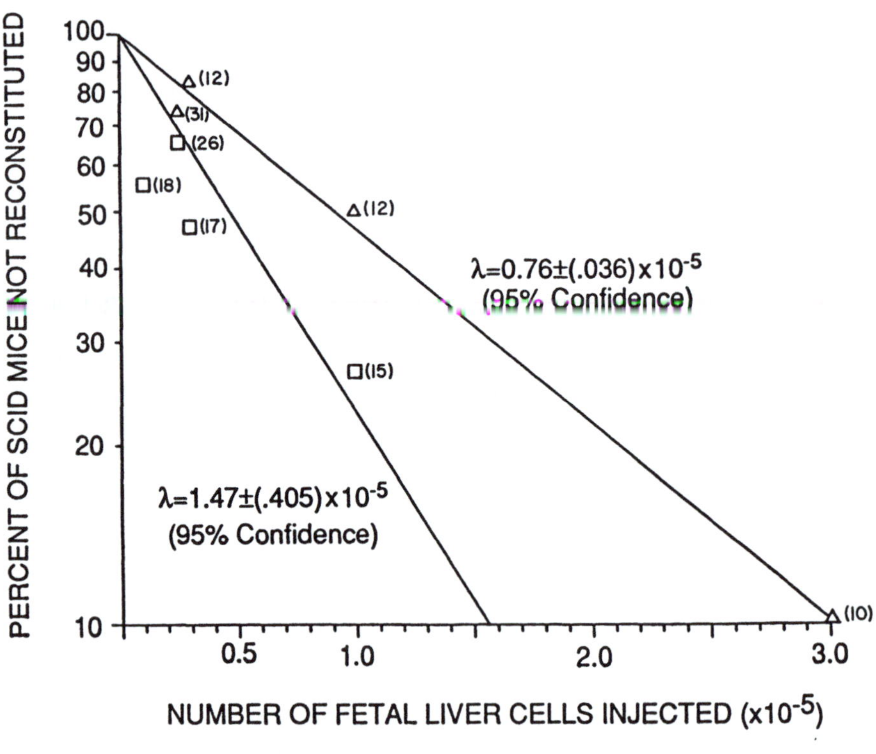

Figure 1. All-or-none reconstitution of scid mice. Neonatal scid mice were injected with varying numbers of fetal liver cells from 13-14 day-old normal BALB/c (Δ) and C.AL-20 (□) embryos and tested for donor

Igh allotype 12-14 weeks later. The number of mice examined per cell dose is indicated in parentheses. The percentage of scid recipients not reconstituted by donor allotype-producing cells is shown on a logarithmic scale as a function of the number of donor cells injected. The line of best fit was determined by linear regression through the ordinate at unity and is described by the equation: $\log_{10}Pr(0) = -\lambda d/2.3$; where Pr(0) is the probability of no reconstitution, λ, the negative of the slope of the line, is the frequency of Ig progenitor cells in the fetal liver cell inocula and d is the dose or number of fetal liver cells injected. The same relationship can be expressed as $Pr(0) = e^{-\lambda d}$ which describes a Poisson distribution. The presence or absence of donor allotype (BALB/c, Igh^a and C.AL-20, Igh^d) was scored by immunodiffusion in micro-Ouchterlony using an antiserum crossreactive to Igh^a and Igh^d (anti-$Igh^{a/d}$) and, in addition, antisera specific for Igh^a only and Igh^d only.

To validate the above estimates of Ig progenitor cell frequencies, three different limiting cell doses (0.25×10^5, 0.50×10^5 and 1.0×10^5) of a 1:1 mixture of BALB/c and C.AL-20 fetal liver cells were injected into scid neonates. The number of recipients that became reconstituted with allotype-producing cells of only one donor or both donors was then compared to the number expected based

Table 1. Injection of scid neonates with a 1:1 mixture of BALB/c and C.AL-20 fetal liver cells: observed vs. expected number of recipients with allotype-producing cells of only one donor or both donors.

	Average number of donor Ig progenitor cells/recipient[o]			C.B-17 scid Recipients: Number Observed/Number Expected*			
Gr.	From BALB/c	From C.AL20	Number injected	Number reconstituted+	Igh^a only	Igh^d only	Igh^a Igh^d
A	0.18	0.10	69	29/17.07	15/10.83	4/5.16	6/1.07
B	0.37	0.19	21	9/ 8.99	7/ 5.42	1/2.46	1/1.11
C	0.74	0.38	24	19/16.16	10/ 8.69	1/3.54	5/3.93

o The average number of BALB/c and C.AL-20 Ig progenitor cells in a 1:1 mixture of fetal liver cells from both mouse strains was determined from the data of Figure 1. The total number of cells was 0.25×10^5 in group A, 0.5×10^5 in group B and 1.0×10^5 in group C.

* A probabilistic model (Litwin 1988) was used to calculate the expected number of scid recipients that would express Igh^a only, Igh^d only and both Igh^a and Igh^d. The terms used included λ (number of Ig progenitor cells seeded/cell inoculum), d (cell dose injected), p (fraction of cell dose from the Igh^a donor), Pr(0) (probability of seeding no Ig progenitor cells=$e^{-\lambda d}$) and Pr (probability of seeding at least one Ig progenitor cell). The values of Pr were determined from the following relationships: Pr of Igh^a only=$e^{-\lambda d}(e^{\lambda dp}-1)$; Pr of Igh^d only=$e^{-\lambda d}(e^{\lambda d(1-p)}-1)$; and the Pr of both Igh^a and Igh^d=1-Pr (Igh^a only)-Pr (Igh^d only)-Pr(0).

\+ Four of 29 and 3 of 19 reconstituted mice in Group A and C, respectively, expressed serum donor allotype as detected by anti-$Igh^{a/d}$ sera, but did not type positive with either anti-Igh^a or anti-Igh^d specific sera. Consequently, these mice could not be assigned to the categories Igh^a only, Igh^d only, or Igh^a and Igh^d.

on the data of Figure 1. The results, shown in Table 1, indicate that the number of reconstituted scid recipients that expressed Igh^a only, Igh^d only or both Igh^a and Igh^d, agreed favorably with the number expected. Not shown is that both Igh^a and Igh^d were present in almost all scid recipients that received a non-limiting cell dose (5.0×10^5 fetal liver cells) containing ~5 Ig progenitor cells.

Extent of reconstitution

Given the preceding results, we could now compare the extent of lymphocyte reconstitution in scid recipients that were seeded on average with less than one vs. multiple (~5) Ig progenitor cells. Four criteria were used to assess reconstitution: 1) repopulation of lymphatic tissues as evident from histologic examination; 2) responsiveness of splenic cells to B and T cell mitogens; 3) serum Ig concentrations; and 4) Ig light (Igl) chain heterogeneity.

The histologic sections in Figure 2 illustrate the kind of lymphocyte repopulation typically seen in scid recipients that received 1-2 vs. 5 Ig progenitor cells. In the latter group of recipients, the thymus (A), lymph nodes (C) and spleen (E) appeared normal as evident from lymphocyte density and tissue architecture. For example, the reconstituted thymus displayed a prominent cortex and corticomedullary junction (Panel A) and the splenic follicles were completely repopulated with

Table 2. Mitogen responses of spleen cells from scid recipients seeded with different numbers of Ig progenitor cells.

scid recipient	Number of seeded Ig Prog. Cells	Serum Ig (mg/ml)	cpm $\times 10^{-3}$ Control	cpm $\times 10^{-3}$ CON A	SI	cpm $\times 10^{-3}$ LPS	SI
62-51	~5	≥5.00	1.91	381.68	200.00	39.83	21.00
63-42	~5	≥5.00	1.60	264.06	165.00	31.18	19.52
[illegible]	[illegible]	[illegible]	[illegible]	[illegible]	[illegible]	[illegible]	[illegible]
62-50	1-2	0.20	0.20	0.14	0.69	1.59	7.86
64-36	1-2	1.10	1.15	86.53	75.00	0.75	0.65
67-56	1-2	0.36	1.83	18.60	10.00	1.34	0.73
C.B17 scid		0.02	1.29	0.80	0.62	1.29	1.00
C.B17+/+		≥5.00	2.08	342.32	156.00	58.94	28.00

Spleen cells were plated at 5×10^{-5} cells/well (96-well plates) in RPMI 1640 with 10% FCS, 2×10^{-5} M 2-ME, 10 mM hepes and 2 mM L-glutamine. Medium was supplemented with either Con A (5μg/ml) or LPS (5 μg/ml). Cultures were incubated at 37°C in 5% CO_2 for 48 hours after which 1 μCi [^{3}H]-thymidine was added to each well for an additional 18 hours. Cells were harvested using a MASH filter harvester and assayed for [^{3}H]-thymidine incorporation by β-scintillation counting.

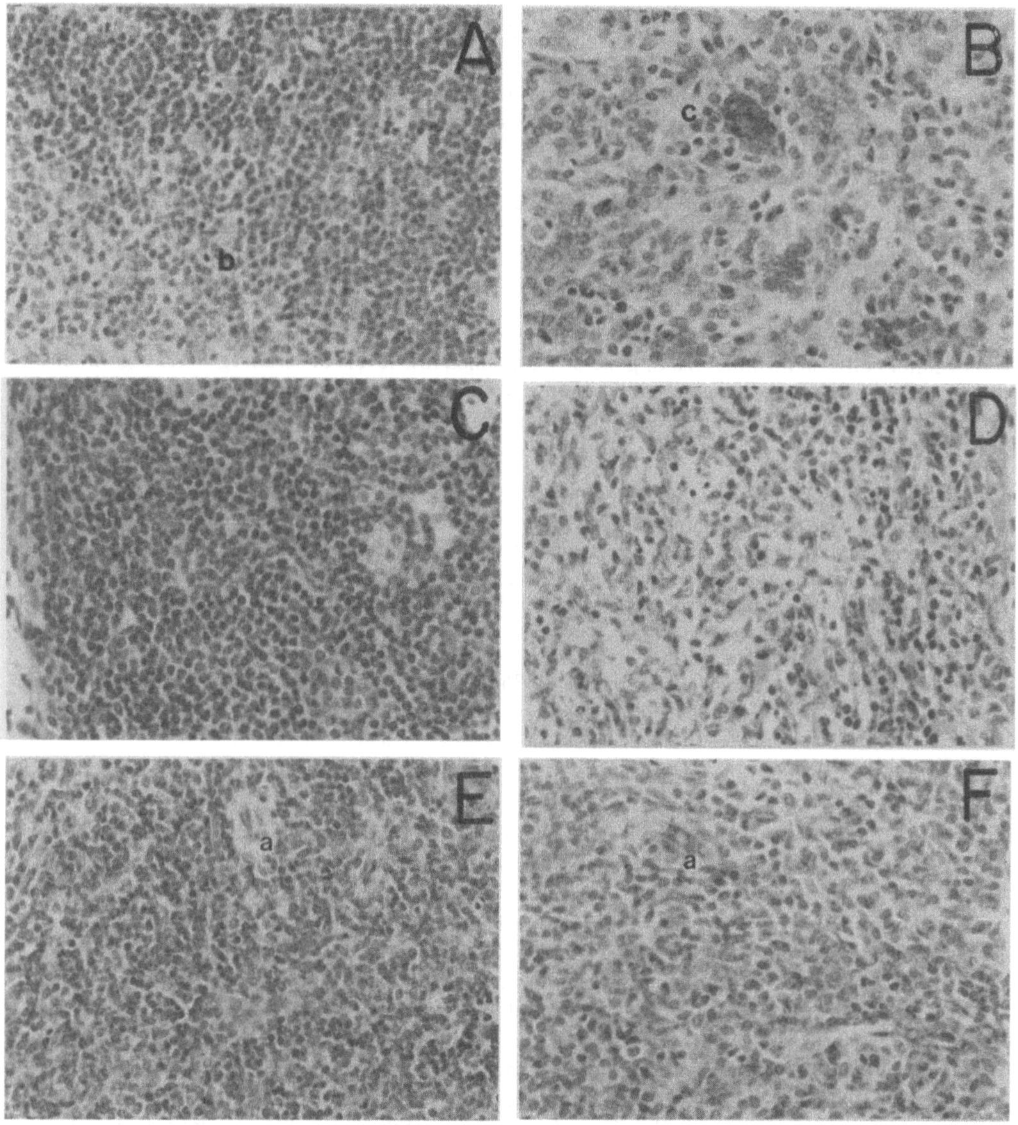

Figure 2. Histologic sections of lymphoid tissues from reconstituted scid mice. Scid neonates were injected with a mixture of BALB/c and C.AL-20 fetal liver cells and necropsied 14 weeks post engraftment. Panels A, C and E are from scid recipients that were seeded with ~5 Ig progenitor cells; panels B, D and F are from recipients that were seeded with 1-2 Ig progenitor cells. A. Fully reconstituted thymus with corticomedullary delineation (b); B. Poorly reconstituted thymus with abundant histiocytes and epithelial cells and very few lymphocytes; C. Fully reconstituted lymph node; D. Slightly reconstituted lymph node with many fibroblastic cells and few lymphocytes; E. Completely repopulated splenic follicle with lymphocytes concentrically arranged around arteriole (a); F. Sparsely repopulated spleen with many stromal cells and poorly differentiated lymphoid cells and few mature lymphocytes. (A-F, X250)

lymphocytes that were characteristically arranged in a concentric pattern around the central arteriole (Panel E). In contrast, mice seeded with 1-2 Ig progenitor cells showed relatively few lymphocytes in thymus (B), lymph nodes (D) and spleen (F). The thymus often contained numerous clusters of epithelial cells similar to the thymus of unreconstituted scid mice. In all of the tissues examined, histiocytes and fibroblasts were conspicuous as in unreconstituted scid mice.

Consistent with the histologic picture, both groups of reconstituted scid mice differed dramatically with respect to their serum Ig concentrations and mitogen responsiveness. As shown in Table 2, mice that were seeded with multiple Ig progenitor cells contained ≥5.0 mg/ml of serum Ig. Furthermore, their splenic cells gave normal proliferative responses to the T and B cell mitogens, concanavalin A (CON A) and lipopolysaccharide (LPS), respectively. Mouse 66-48 was exceptional in that it showed a low stimulation index to LPS. Scid mice seeded with only 1-2 Ig progenitor cells, on the other hand, generally contained low levels of serum Ig (<1.0 mg/ml) and their spleen cells showed relatively little, if any, proliferative response to CON A or LPS. None of the mice shown here (62-50, 64-36 and 67-56) had spleen cells that responded to both mitogens.

Interestingly, some of the mice seeded with 1-2 Ig progenitor cells showed normal to near-normal heterogeneity of serum Igl chains others showed very restricted heterogeneity. This is illustrated in Figures 3 and 4. In mice 64-39, 66-46 and 63-41, the isoelectric-focusing (IEF) patterns of Igl chains were complex and similar to those of normal BALB/c mice (C) (see Figure 3). In contrast, mice 66-45, 66-49 and 66-52 showed restricted IEF patterns characteristic of oligoclonal B cells (see Figure 4). The Igl chains of mice seeded with multiple Ig progenitor cells were generally as heterogeneous as those of normal BALB/c mice, including a mouse (66-47 in Figure 3) that produced Igh^d only; a few mice, like mouse 64-35 in Figure 4, displayed a restricted IEF pattern.

CONCLUSION

We have found that scid mice can be reconstituted with normal donor lymphocytes in a dose-dependent manner provided that reconstitution is initiated in newborn recipients. Injection of relatively low numbers (0.25 -1.0 x 10^5) of normal BALB/c or C.AL-20 fetal liver cells into scid neonates resulted in only partial lymphocyte reconstitution of some of the recipients within 12-14 weeks. The probability of repopulating a given scid recipient with donor allotype-producing cells followed a Poisson distribution. This relationship enabled us to ascertain the frequency of Ig progenitor cells in BALB/c and in C.AL-20 fetal liver cell inocula. Recipients were then injected with mixtures of BALB/c and C.AL-20 fetal liver cells using a total cell dose that either contained ~5 Ig progenitor cells or an average of less than 1 Ig progenitor cell. Those recipients that produced only one of two possible donor allotypes at the limiting cell doses were assumed to be seeded with 1-2 Ig progenitor cells. Though many or all of the defined progenitor cells may have given rise also to T cell progeny, this possibility was not directly addressed in the present study.

Scid recipients that received cell doses containing multiple Ig progenitor cells (≥5) generally showed complete reconstitution, indicating that scid mice fully support normal lymphocyte differentiation. In contrast, recipients that were seeded with 1-2 Ig progenitor cells were sparsely populated with lymphocytes. Their serum Ig levels were generally low and their splenic cells showed relatively little or no proliferation to CON A or LPS. This could reflect a limited differentiative and proliferative capacity of individual progenitor cells or the need for progeny of several progenitor cells to interact in order to reconstitute scid mice completely. Alternatively,

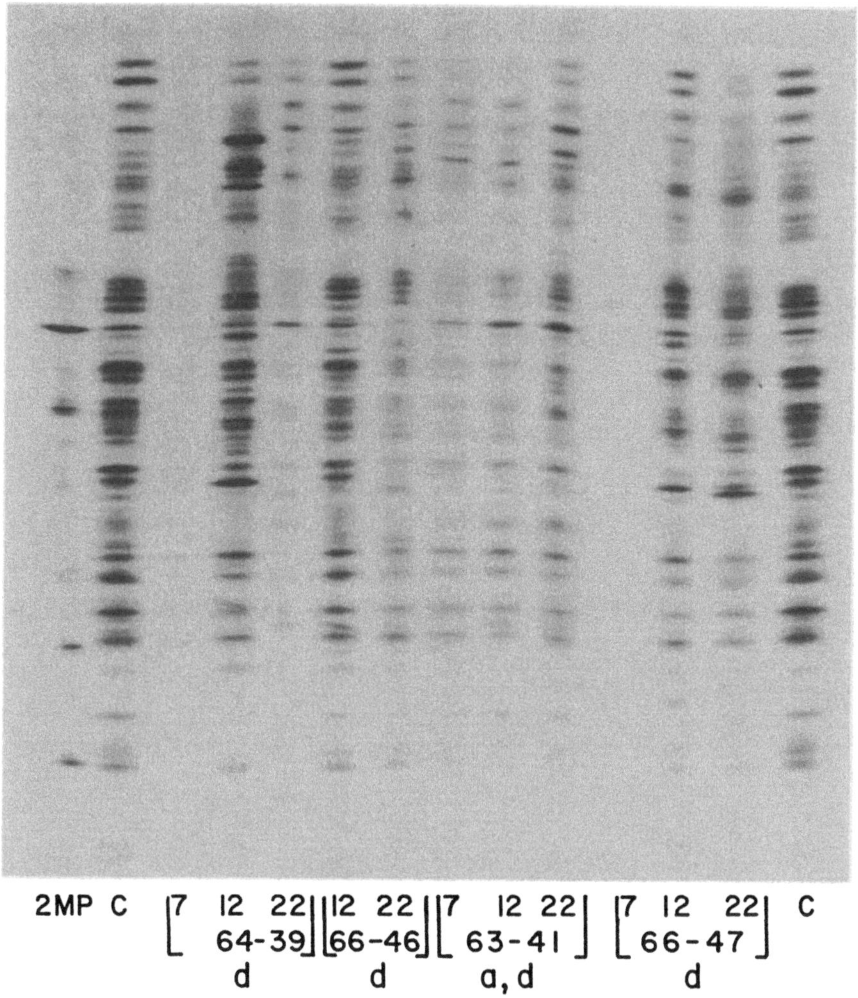

Figure 3. Illustration of diverse (polyclonal) IEF patterns of serum Igl chains from scid neonates that were injected with a mixture of BALB/c and C.AL-20 fetal liver cells. Serum samples were taken from reconstituted mice at 7, 12 and 22 weeks of age; the IgG was purified, reduced and alkylated with ^{14}C-iodoacetamide and the Igl chains were then isolated and isofocused on IEF gels as previously described (Gibson 1984; Bosma et al. 1988). The samples labeled 64-39, 66-46 and 63-41 were from mice seeded with 1-2 Ig progenitor cells; 66-47 was from a mouse seeded with ~5 Ig progenitor cells (the faint pattern seen in the 22 week sample of mouse 64-39 resulted from technical failure). A rerun of the 64-39 samples indicated equivalent heterogeneity in both the 12 and 22 week bleeds, but significant alteration in the representation of several clonotypes. For example, considerable increase in $\lambda 1$ light chains was noted in the 22 week bleed (see lane 1 for $\lambda 1$ position). The allotypes present in each sample are indicated with lower case letters: (a) for BALB/c and (d) for C.AL-20. Control samples included Igl chains of normal BALB/c mice (C) and a mixture of Igl chains from two mouse myeloma proteins (2 MP), HOPC-1 ($\lambda 1$, upper set of bands) and FLOPC-1 (lower set of bands) (Potter 1972).

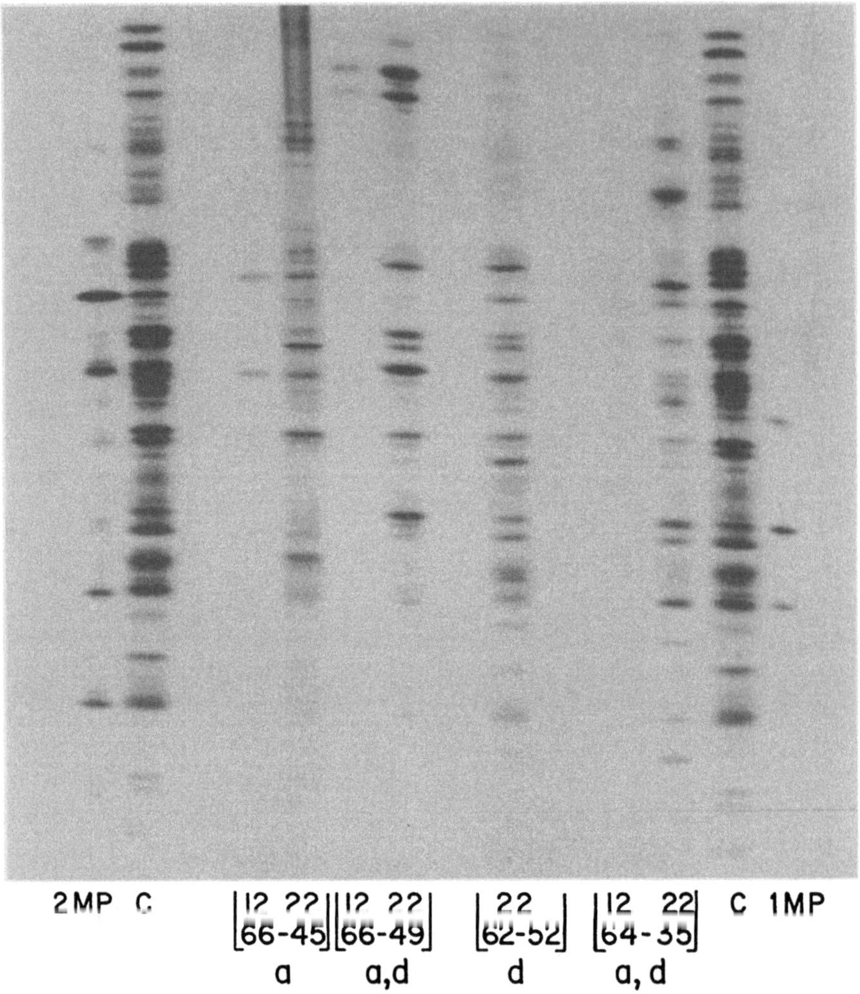

Figure 4. Illustration of restricted (oligoclonal) IEF patterns of serum Igl chains from scid neonates that were injected with a mixture of BALB/c and C.AL-2O fetal liver cells. The symbols on the face of the figure as well as the procedures used for isolating and isofocusing the Igl chains are described and referenced in the legend of Figure 3. The samples labeled 66-45, 66-49 and 62-52 were from mice seeded with 1-2 Ig progenitor cells; 64-35 was from a mouse seeded with ~5 Ig progenitor cells. The lane with one myeloma protein (1 MP) contained Igl chains from myeloma 467 (Potter 1977).

complete reconstitution of scid mice by the progeny of one progenitor cell may require 20-30 weeks or longer. Interestingly, some recipients that were seeded with 1-2 Ig progenitor cells produced Ig of near-normal diversity as judged from the IEF patterns of their serum Igl chains. In other recipients, also seeded with 1-2 Ig progenitor cells, the IEF patterns were very restricted and resembled those seen in leaky scid mice (Bosma et al. 1988; Gibson et al. this volume). Whether this reflects heterogeneity in the differentiative stage of Ig progenitor cells defined or variable clonal expansion of Ig-producing progeny in response to different naturally-occurring antigens is not clear.

In future experiments, lymphocyte progenitors in liver cells from both 13-14 and 10-11 day-old embryos will be evaluated for their potential to give rise to B and T cell progeny. With the use of 10-11 day-old embryos we hope to ensure that we define lymphocyte progenitor cells not yet committed to the B or T cell pathway.

ACKNOWLEDGEMENTS

We would like to thank Dr. S. Litwin for help with the statistical analysis, E. Cunningham for technical assistance, D. Cloutier and S. MacLean for help with the IEF analysis and M. Piatek for typing this manuscript. We also would like to pay tribute to the late Dr. R. P. Custer, a valued colleague and friend, who was responsible for all of the histopathological analyses of scid mice in this and earlier publications. This work was supported by NIH grants AI-13323, CA-04946, CA-06927 and RR-05539, by an appropriation from the Commonwealth of Pennsylvania, by the Medical Research Council and National Cancer Institute of Canada, and by the Fonds de la Recherche en Santa du Québec.

REFERENCES

Bosma GC, Custer RP, Bosma MJ (1983) A severe combined immunodeficiency mutation in the mouse. Nature (Lond) 301: 527-530

Bosma GC, Fried M, Custer RP, Carroll A, Gibson DM, Bosma MJ (1988) Evidence of functional lymphocytes in some (leaky) scid mice. J Exp Med 167: 1016-1033

Custer RP, Bosma GC, Bosma MJ (1985) Severe combined immunodeficiency in the mouse. Am J Pathol 120: 464-477

Fulop GM, Phillips RA (1986) Full reconstitution of the immune deficiency in scid mice with normal stem cells requires low-dose irradiation of the recipients. J Immunol 136: 4438-4443

Gibson DM (1984) Evidence for 65 electrophoretically distinct groups of light chains in BALB/c and NZB myelomas. Mol Immunol 21: 421-432

Litwin W (1988) Biostatistics in basic cancer research. Cancer Investigation 6(5): 533-540

Potter M (1972) Immunoglobulin producing tumors and myeloma proteins in mice. Physiol Rev 52: 631-719

Potter M (1977) Antigen binding proteins of mice. Adv in Immunol 25: 141-211

Reconstitution of Lymphocyte Subsets in Scid Mice by Transplantation of Fetal Primordia

Nanette W. Solvason and John F. Kearney

Department of Microbiology, Division of Developmental and Clinical Immunology, The Comprehensive Cancer Center, University of Alabama at Birmingham, Birmingham, AL 35294

INTRODUCTION

Since the original description of the Ly1+ B cell by Manohar et al. (1982), accumulated experimental data suggest that such cells constitute a separate subset of B cells. In the normal adult mouse Ly1+ B cells have an unusual distribution in that almost half of the B cells in the peritoneal cavity are Ly1+ while this B cell subset is essentially undetectable in the lymph nodes and peripheral blood and makes up less than 3% of B cells in the spleen (Hayakawa et al. 1983, 1984). In contrast, very early in development the majority of identifiable B cells in the fetal liver and spleen are Ly1+ (Hayawaka et al. 1983). The early B cell repertoire differs from the adult with respect to V_H gene usage and specificity of immunoglobulins produced. Because of their prevalence early in development, the Ly1+ B cell subset has been indirectly implicated as being responsible for the unique nature of the fetal and neonatal B cell repertoire. Many IgM antibodies secreted from hybridomas derived from fetal and perinatal liver and spleen bind to each other (Vakil and Kearney 1986; Vakil et al. 1986; Holmberg et al. 1984). This autoreactivity appears to be different from that described in autoimmune disease and it has been proposed that it is a normal and perhaps critical feature of the developing repertoire which offers the potential for early B cells to interact with each other via their surface immunoglobulin or secreted immunoglobulin products. We and others have proposed that the end result of this autospecificity is an expansion of certain target B cell clones which provide optimum protection against infection (Vakil and Kearney 1986; Vakil et al. 1986; Briles et al. 1987).

Ly1+ B cells are also correlated with autoimmune disease in certain strains of mice. In the autoimmune NZB mouse, the percentage of Ly1+ B cells increases as the animals age and it is this B cell population which hypersecretes IgM in vitro without the addition of mitogen or antigen (Hayakawa et al. 1983, 1984). Sorting experiments of normal human peripheral blood lymphocytes which divide B cell subpopulations into CD5+ (human Ly1 homologue) and CD5- populations have demonstrated that rheumatoid factor is secreted by the CD5+ subset (Hardy et al. 1987; Casali et al. 1987). CBA/n mice have an X-linked immunodeficiency (xid) which results in the inability of this strain to respond to certain T independent antigens and it is of interest that these mice have no Ly1+ B cells (Hayakawa et al. 1986). It has been demonstrated that the xid mutation in CBA/n mice when backcrossed into NZB does not affect the hyperactivity of B cells from females, however B cells from males lose this hyperactivity (Nakajima et al. 1979). These experiments indicate that the xid

Current Topics in Microbiology and Immunology, Vol. 152

mutation affects the Ly1+ B cell subset, although it is not clear what this relationship is.

Finally, the Ly1+ B cell subset does not appear to respond to T dependent antigens such as SRBC or TNP-KLH like conventional B cells (Hayakawa et al. 1984). However, in mice which receive allotype congenic peritoneal wash cells a large proportion of the normal serum IgM is donor derived at 20 weeks post transplant (Forster and Rajewsky 1987). Thus Ly1+ B cells are probably responsible for a large amount of the IgM found in the serum of normal unimmunized mice. It is not clear as to the function of this antibody but mice fed antigen free diets and kept in a germ free environment also have essentially normal levels of serum IgM. Because large amounts of this natural serum IgM are derived from Ly1+ B cells and because of the apparent affinity for self antigens described for this B cell subset, it has been proposed that this serum IgM plays a regulatory role in the maintenance of the adult B cell repertoire.

These results have given impetus to the notion that Ly1+ B cells are a separate lineage of B cells which are functionally distinct from conventional B cells and that they have a regulatory role by the nature of the inherent affinity of their immunoglobulin for self antigens. Several experiments suggest that this B cell subset does not appear to be reconstituted by adult bone marrow transplantation leading to the proposal that Ly1+ B cells are also developmentally a separate lineage of B cells with a defined site of origin separate from conventional B cells (Hayakawa et al. 1985). In a search for an alternative source of Ly1+ B cells we have examined the potential of fetal liver, bone, spleen and omentum to reconstitute Ly1+ B cells when transplanted into scid mice.

The omentum is included in this list of tissues because it is known that the omentum in the adult mouse and human contains aggregates of lymphoid cells called milk spots (Holub et al. 1979; Dux et al. 1977). Additionally, it has already been demonstrated that the fetal omentum when transplanted into the anterior eye chamber gives rise to donor derived Thy1+ and small numbers of Ly1+ cells (Kubai and Auerbach 1983). Some characteristics of the adult omentum and fetal omentum are listed in Table I and II, respectively.

<u>TABLE 1. CHARACTERISTICS OF ADULT OMENTUM:</u>

1. Is a fold of peritoneum extending from the stomach to adjacent organs in the abdominal cavity (human)

2. Contains follicular aggregations of lymphoid cells which are involved in antibody production (Dux et al. 1986)

3. Has the ability to self-regenerate lymphocytes in diffusion chambers following supralethal irradiation (Holub et al. 1971)

<u>TABLE 2. CHARACTERISTICS OF FETAL PRIMORDIAL OMENTUM:</u>

1. Becomes distinct from the splenic and pancreatic rudiments in mouse embryos at day 13 of gestation

2. Consists of a loose mesothelial sheet of tissue

TABLE 2. (continued)

3. In other species such as amphibia, lymphocytes are derived from a region delineated by the anterior limbs, foregut and mesonephros. This region defines the boundaries of the developing omentum

4. Transplanted omental anlage gives rise to Thyl+ and Lyl+ lymphoid cells

MATERIALS AND METHODS

Thirteen day gestation (C57Bl/6(Igh6b, $H2K^b$) x BALB/c(Igh6a, $H2K^d$)) F1 fetal omentum, liver, femur/tibia or spleen was transplanted under the kidney capsule of an unirradiated scid mouse. The anatomical position of the fetal omentum at 17 and 18 days of gestation in the mouse is indicated in Fig. 1.

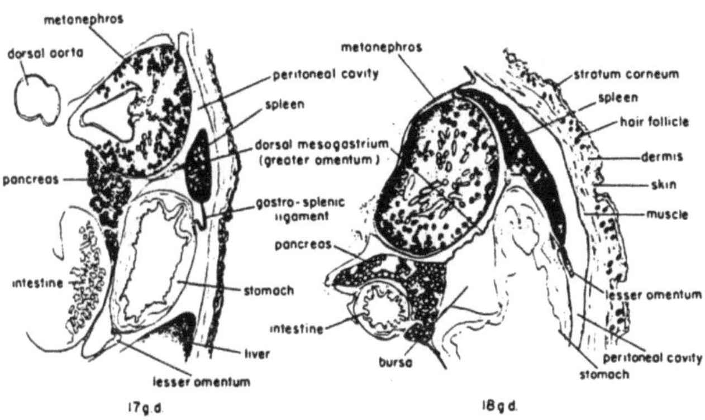

FIGURE 1. *Diagram indicating anatomical position of fetal omentum. Reprinted with permission from Burgess Press. (Rugh 1968)*

A photograph of the dissected omentum with spleen and pancreatic rudiments is shown in Fig. 2. Four months post transplantation, the scid mice were sacrificed and the lymphoid tissues (spleen, lymph nodes, and peritoneal wash cells) were examined for the presence of donor derived lymphocytes. Donor derived B and T cells were identified by using a combination of FITC anti-mouse IgM or FITC anti-T3 with PE anti-$H2K^b$, respectively (see Table 3 for a list of monoclonal antibodies used). Each mouse was also bled by cardiac puncture and the serum was collected for subsequent analysis by ELISA. Briefly, plates were coated with polyclonal anti-mouse IgM, blocked then incubated with serum which was serially diluted. After incubation, either biotinylated anti-Igh6a (RS3.1) or anti-Igh6b (AF6.78.25) was added followed by streptavidin conjugated alkaline phosphatase. The reaction was developed with phosphatase substrate and the OD_{450} was read on a Multiscan Titertek.

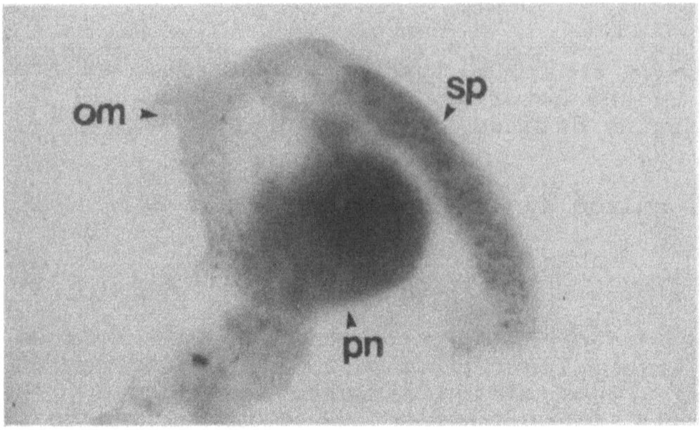

FIGURE 2. *Photograph showing the 13 day gestation fetal omentum (om), spleen (sp) and pancreas (pn).*

TABLE 3. ANTIBODIES USED IN SERUM AND FACS ANALYSIS

Name	Specificity	Isotype	Reference
AF6.78.25	Igh6b	IgG1	Stall and Loken 1984
RS3.1	Igh6a	IgG1	Schuppel et al. 1987
K10.56	$H2K^b$	IgG2b	Klein et al. 1983

RESULTS

All host lymphoid tissues tested from the mouse with liver transplant had normal proportions of donor derived B cells and T cells. Additionally >95% of the cells in the peritoneal cavity were donor in origin (data not shown) and 40% of these B cells were Ly1+. In contrast, in the lymphoid organs from mice with the omentum transplant only donor B cells were detected; no T cells were reconstituted. Again almost all cells in the peritoneal cavity were donor in origin and about 40% of the B cells were Ly1+ (Fig. 3). Neither fetal spleen nor bone gave any B or T cell reconstitution.

Quantitation of the serum immunoglobulin from each mouse demonstrated the presence of donor derived IgM in both liver and omentum transplanted mice, but no donor IgM was detected in the spleen or bone transplanted mouse (Fig. 4). Additionally there was a gradient of reconstitution of gamma isotypes in the liver transplanted mouse with the highest reconstitution being IgG1 isotype followed by IgG2a, IgG2b and finally IgG3 (data not shown). However, this IgG isotype distribution was completely reversed in the omentum transplanted mouse. IgG3 was present at the highest levels followed by IgG2b, and IgG2a, essentially no IgG1 was detected.

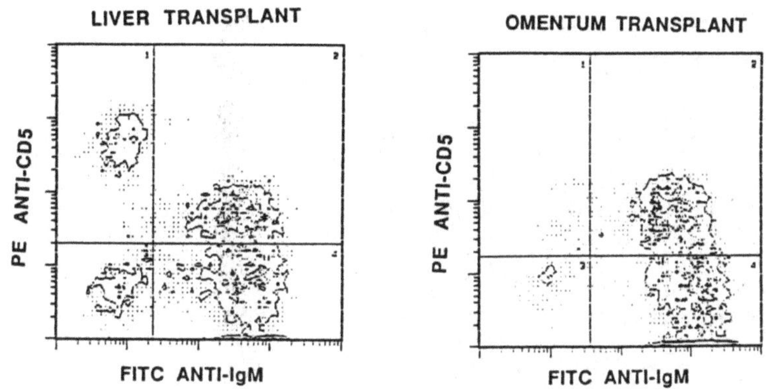

<u>FIGURE 3.</u> *Flow cytometry of peritoneal wash cells of mice with liver transplant (left) or omentum transplant (right). Cells were stained with a combination of PE anti-Ly1 and FITC anti-IgM. Ly1+ T cells are seen in quadrant I in the left panel, however no such population is seen on the right. Both liver and omentum transplants give rise to Ly1+ B cells as demonstrated in quadrant 2.*

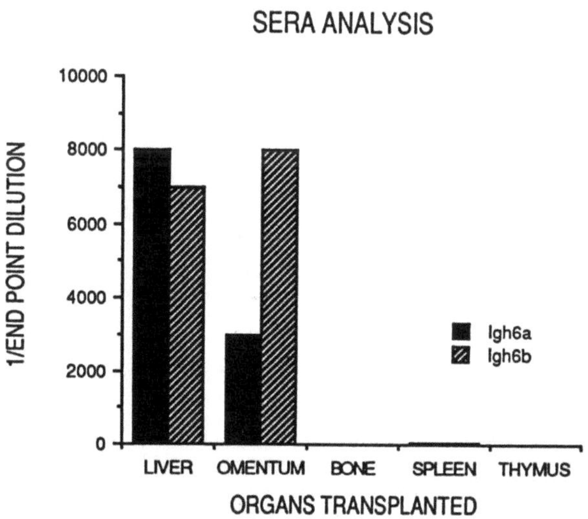

<u>FIGURE 4.</u> *1/end point dilution of Igh6a (RS3.1) and Igh6b (AF6.78.25) in serum of mice with fetal liver, omentum, spleen or bone transplants.*

DISCUSSION AND SUMMARY

As early as 1971, Holub et al. had demonstrated that after supralethal irradiation the mouse omentum could self-regenerate lymphocytes. Additionally, in 1983, Kubai and Auerbach demonstrated that 7 days post transplantation of 13 day gestation omentum, donor derived lymphocytes could be detected in the graft itself. Both these sets of experiments were limited in that it was not possible to identify B and/or T cells and these experiments did not examine the potential of omentally derived lymphocytes to migrate into peripheral lymphoid organs.

Results from experiments described in this paper clearly demonstrate that the 13 day gestation fetal omentum when transplanted into scid mice contains B cell precursors whose progeny can migrate to spleen, lymph nodes and peritoneal cavity. These B cell precursors can reconstitute Ly1+ B cells, however no donor T cells were identified in the lymphoid organs described. It should be stressed that Ly1- B cells were also found in the peritoneal wash cells of scid mouse with omentum transplant. Additionally, in the spleen of the mouse with omentum transplant essentially all of the B cells in the spleen were Ly1-. The fetal liver, in contrast, reconstitutes both T and B cells (Ly1+ and Ly1-) to normal proportions in all organs tested.

The isotype distribution which is seen in serum of mice with omentum transplant reflects the T dependency of each isotype switch. That is, IgG1 is the most T dependent switch and there is no detectable IgG1 in the mouse with omentum transplant probably because of the lack of T cell reconstitution. The switch to IgG3 however is the least T dependent switch and this IgG subclass is found at highest levels in mice with omentum transplants. The IgG subclass distribution in mice with liver transplants is comparable to the ratio seen in normal adult mice. IgG1 levels are the highest followed by IgG2a, IgG2b and finally IgG3 subclass.

In summary, the fetal omentum is a source of Ly1+ and Ly1- B cells but no omentally derived T cells have been identified to date. These B cells migrate into spleen, lymph nodes and peritoneal cavity and are functional as they secrete immunoglobulin into the serum of the transplanted mice.

ACKNOWLEDGEMENTS

The authors would like to thank Dr. John Eldridge for supplying many of the antibody preparations used in this study and Ann Marie Hamilton for critical reading of the text.

This work is supported by NIH grants CA 16673, CA 13148 and AI 14782.

REFERENCES

Briles DE, Carrol J (1987) Natural selection and the response to polysaccharides. In: Evolution and Vertebrate Immunity: The antigen receptor and MHC gene families. G. Kelsoe and D. Schulze, (eds) Texas University Press (Austin) pp 175-190

Casali P, Burastero SE, Nakamura M, Inghirami G, Notkins AL (1987) Human lymphocytes making rheumatoid factor and antibody to ss DNA belong to leu-1+ B-cell subset. Science 236:77-80

Dux K, Janik P, Szaniawska B (1977) Kinetics of proliferation, cell differentiation, and IgM secretion in the omental lymphoid organ of B10/Sn mice following intraperitoneal immunization with sheep erythrocytes. Cellular Immunol. 32:97-109

Dux K, Rouse R, Kyewski B (1986) Composition of the lymphoid cell populations from omental milky spots during the immune response in C57Bl/Ka mice. Eur. J. Immunol. 16:1029-1032

Forster I, Rajewsky K (1987) Expansion and functional activity of Ly-1+ B cells upon transfer of peritoneal cells into allotype-congenic, newborn mice. Eur. J. Immunol. 17:521-528

Hardy RR, Hayakawa K, Shimizu M, Yamasaki K, Kishimoto T (1987) Rheumatoid factor secretion from human Leu-1+ B cells. Science 236:81-84

Hayakawa K, Hardy RR, Parks DR, Herzenberg LA (1983) The "Ly-1 B" cell subpopulation in normal, immunodefective, and autoimmune mice. J. Exp. Med. 157:202-218

Hayakawa K, Hardy RR, Honda M, Herzenberg LA, Steinberg AD, Herzenberg LA (1984) Ly-1 B cells: Functionally distinct lymphocytes that secrete IgM autoantibodies. PNAS 81:2494-2498

Hayakawa K, Hardy RR, Herzenberg LA, Herzenberg LA (1985) Progenitors for Ly-1 B cells are distinct from progenitors for other B cells. J. Exp. Med. 161:1554-1568

Hayakawa K, Hardy RR, Herzenberg LA (1986) Peritoneal Ly-1 B cells: genetic control, autoantibody production, increased lambda light chain expression. Eur. J. Immunol. 16:450-456

Holub M, Hajdu I, Trebichavsky I, Jaroskova L (1971) Formation of lymphoid cells from precursors in irradiated mouse omenta. Eur. J. Immunol. 1:465-470

Holub M, Jaroskova L, Fischer H, Viklicky V (1979) Neoformation of lymphatics in the mouse omentum. In: Advances in Experimental Medicine and Biology, 114:427-432

Holmberg D, Forsgren S, Forni L, Ivars F, Coutinho A (1984) Reactions among IgM antibodies derived from normal neonatal mice. Eur. J. Immunol. 14:435-442

Klein J, Figuero F, David CS (1983) H-2 haplotypes, genes and antigens: second listing. II. The H-2 complex. Immunogenetics 17:553-596

Kubai L, Auerbach R (1983) A new source of embryonic lymphocytes in the mouse. Nature 301:154-156

Manohar V, Brown E, Leiserson M, Chused T (1982) Expression of Lyt-1 by a subset of B lymphocytes. J. Immunol. 129:532-538

Nakajima P, Datta SK, Schwartz RS, Huber BT (1979) Localization of spontaneously hyperactive B cells of NZB mice to a specific B cell subset. PNAS 76:4613-4616

Rugh R (1968) The mouse. Burgess Publishing Company, Minneapolis, Minnesota, p 266

Schuppel R, Wilke J, Weiler E (1987) Monoclonal anti-allotype antibody towards BALB/c IgM. Analysis of specificity and site of a V-C crossover in recombinant strain BALB/c-Igh-V^a/Igh-C^b. Eur. J. Immunol. 17:739-741.

Stall AM, Loken MR (1984) Allotypic specificities of murine IgD and IgM recognized by monoclonal antibodies. J. Immunol. 132:787-790.

Vakil M, Kearney J (1986) Functional characterization of monoclonal auto-anti-idiotype antibodies isolated from the early B cell repertoire of BALB/c mice. Eur. J. Immunol. 16:1151-1155

Vakil M, Sauter H, Paige C, Kearney J (1986) In vivo suppression of perinatal multispecific B cells results in a distortion of the adult B cell repertoire. Eur. J. Immunol. 16:1159-1162

Use of the Scid Mouse Transplantation System in Studies of Lymphocyte Differentiation

Kenneth Dorshkind

Division of Biomedical Sciences
University of California
Riverside, CA 92521-0121

INTRODUCTION

It is well established that lymphocytes can be reconstituted in mice with Severe Combined Immunodeficiency Disease (Bosma et al. 1983; Dorshkind et al. 1984; Custer et al. 1985; Fulop and Phillips 1986). Engraftment of donor lymphoid precursors is facilitated by the lack of endogenous lymphocytes and the presence of a normal hemopoietic microenvironment in the mice. These and other characteristics make the Severe Combined Immunodeficient (SCID) mouse an ideal recipient in which the lymphoid differentiative potential of selected donor populations can be measured. The purpose of this chapter is to briefly review studies that have demonstrated this.

RECONSTITUTION OF SCID MICE WITH CELLS FROM LONG-TERM LYMPHOID BONE MARROW CULTURES

Among the first experiments that demonstrated the versatility of the SCID mouse transplantation system were those in which cells from Whitlock-Witte long-term bone marrow cultures were transplanted into the mutants. These cultures support the growth of B lymphocytes and their precursors (Whitlock and Witte 1982) and have provided a system in which the regulation of B cell differentiation by components of the hemopoietic microenvironment can be studied. The cells in the cultures include newly produced B cells, pre-B cells, and lymphoid cells that do not yet express Immunoglobulin (Ig) or other B lineage associated antigens. At the time of their description, questions existed regarding the normal potential of the cultured cells, because they did not differentiate in vitro past the surface IgM stage or exhibit any functional response (Nagasawa et al. 1985).

The SCID mouse proved to be an ideal recipient in which to assess the full differentiative and functional potential of the cultured cells, because differentiation of donor cells occurred in a background free environment in which the repopulation of all lymphocytes in hemopoietic tissues could be measured. The developmental potential of cells from the long-term lymphoid cultures was measured by injecting varying numbers into SCID mice and measuring reconstitution ten weeks later. The SCID mice used in these experiments were not pretreated with irradiation or any other conditioning regimens.

Cultured cells reconstituted lymphocytes in the mice as measured by the presence of B cell colony forming units (CFU-B) and surface IgM expressing cells in the spleen. The levels attained were lower than observed with grafts of fresh bone marrow (Table 1) but were nevertheless significant in view of the total absence of such populations in unreconstituted SCID mice. The ability to detect such low level reconstitution is due to the lack of endogenous lymphoid cells and demonstrates the sensitivity of this transplantation model. Only background levels of CFU-B and surface IgM positive cells were observed in the bone marrow.

An advantage in the use of the SCID mouse is that multiple reconstitution parameters can be analyzed. For example, the mice lack Ig secreting B cells, making it possible to determine if donor cells can secrete Igs that can be detected in the serum. Both IgM and IgG of multiple isotypes were present in the serum of reconstituted mice, and the presence

Current Topics in Microbiology and Immunology, Vol. 152

Table 1. Reconstitution of Lymphocytes in SCID Mice

Mouse (N)	No. and Source of donor cells	Spleen Cell Profile 10 wk After Reconstitution	
		% IgM	CFU-B/10^5cells
SCID (8)	None	0	0
SCID (4)	5 x 10^6 Bone Marrow	24.8	1598
SCID	5-23 x 10^6 Lymphoid Bone Marrow Culture	3.2	215

Information in this table represents a compilation of data from several different experiments.

of the latter molecules indicated that the cultured cells, which express only IgM in vitro, can class switch. Class switching is dependent upon T cell help, and the presence of this function is consistent with the observed low level of T cell reconstitution in the long-term bone marrow culture reconstituted mice. The severe immunodeficiency of the mice also makes it possible to further assess the potential of the donor cells to mediate an immune response. All SCID mice that received cells from the long-term lymphoid cultures were able to respond to immunization with both T independent and T dependent antigens. The specificity of the response was demonstrated by the recovery of antigen specific IgM and IgG antibodies from the serum (Dorshkind et al. 1986).

The above findings indicate that the cultured cells have the potential for normal B cell differentiation and demonstrate the usefulness of the SCID mouse transplantation system. The ability to measure various immune responses following reconstitution further indicates that the mice will be important for analyzing the immune responsiveness of selected subpopulations of lymphoid cells. This was particularly well demonstrated in the course of characterizing cells from long-term fetal liver cultures. The predominant cells in those cultures, described by Denis et al. (1984), are B lineage cells that are more immature than those present in the lymphoid cultures established with marrow. Cultured fetal liver cells transplanted into SCID mice could secrete serum IgM and IgG_3 and confer immunity to the T independent antigen TNP-Ficoll. However, this was the extent of their autonomous functional activity, since they could not fully class switch and the mice reconstituted with the cultured fetal liver cells could not respond to T dependent antigens. Nevertheless, the cultured cells had the potential for normal differentiation, because these responses were demonstrable when a source of T cell help was also reconstituted (Denis et al. 1987).

USE OF SCID MICE TO STUDY ABNORMALITIES OF LYMPHOCYTE DEVELOPMENT

Recent investigations have raised the possibility that defects in the hemopoietic microenvironment can contribute to abnormalities of lymphocyte development, such as those that occur in the New Zealand Black mouse (NZB; Bray et al. 1984). B lymphopoiesis proceeds at an accelerated rate in NZB mice of four weeks of age and younger and then declines in adult mice of 15 weeks of age and older (Jyonouchi et al. 1982). By six months of age, the majority of NZB mice develop autoantibodies to a variety of self antigens. The relationship between accelerated B lymphopoiesis and autoimmune disease is unclear (Kincade et al. 1982). The ability to reconstitute lymphocytes in SCID mice led to the initiation of experiments to test whether NZB B cell defects could occur in the normal environment provided by the immunodeficient mice. Bone marrow cells from NZB mice were injected into SCID recipients and various reconstitution parameters measured eight weeks later.

Bone marrow cells from four week old NZB donors reconstituted B cells in the marrow and spleen of SCID mice, and the absolute number of both surface IgM expressing cells and CFU-B in the latter tissue were higher than in recipients of BALB/c or NZB.xid marrow cells. NZB.xid mice carry the xid gene that ameliorates the defects observed in NZB mice (Ohsugi et al. 1981). Thus, the hyperactive B cell activity of young NZB mice can be observed in the SCID mouse environment as well. The SCID mouse transplantation system was further used to investigate whether the cessation of B lymphopoiesis in old NZB mice was due to an absence of precursors in those animals. Bone marrow cells from 6 month old NZB donors reconstituted B cells in the marrow and spleen of SCID mice to levels comparable to levels present in recipients of BALB/c bone marrow. This suggests that a microenvironmental defect operative in the old mice inhibits B lymphopoiesis, but this process can proceed in the apparently normal SCID mouse environment. In addition, the serum of SCID mice that received cells from the 6 month old NZB mice contained both IgM and IgG at levels higher than observed in recipients of NZB.xid cells. However, only mice reconstituted with the old NZB marrow had anti-single stranded DNA antibodies in their serum, and these were primarily of the IgM class. Anti-single stranded DNA antibodies in intact NZB mice are also of the μ isotype, suggesting a similar mechanism for their synthesis in SCID hosts (Dorshkind et al. 1989).

Further work is needed to establish whether the NZB defects transferred into SCID mice are due to maintenance of mature cells in the donor inoculum or have occurred in populations that developed in the immunodeficient hosts. This will require additional experiments in which SCID mice receive grafts of NZB cells from which mature lymphocytes have been depleted. It will also be of interest to selectively reconstitute SCID mice with only B lineage cells to assess whether the B cell defects are due solely to intrinsic abnormalities in those cells or require other interactions. This must be considered in view of T cell reconstitution in the mice. The previously noted results, using cells from the long-term cultures, demonstrating that such selective reconstitution is possible makes this approach feasible.

REFERENCES

Bosma GC, Custer RP, Bosma MJ (1983) A severe combined immunodeficiency mutation in the mouse. Nature 301:527-530

Bray KR, Gershwin ME, Skelly RR, Ahmed A, Kincade PW (1984) Studies of congenitally immunologically mutant New Zealand mice. IX. Age-related microenvironmental defects on autoantibody production in NZB and NZB.xid mice studied by transplantation. J Immunol 132:2913-2918

Custer RP, Bosma GC, Bosma MJ (1985) Severe combined immunodeficiency (SCID) in the mouse: pathology, reconstitution, neoplasms. Am J Pathol 120:464-477

Denis KA, Treiman LJ, St Claire JI, Witte ON (1984) Long-term cultures of murine fetal liver retain very early B lymphoid phenotype. J Exp Med 160:1087-1011

Denis KA, Dorshkind K, Witte ON (1987) Regulated progression of B lymphocyte differentiation from cultured fetal liver. J Exp Med 166:391-403

Denis KA, Dorshkind K, Witte ON (1988) Analysis of B lymphocyte lineages and diversity using long-term lymphoid cultures. In: Witte ON, Klinman NR, Howard MC (eds), B Cell Development, p. 121-127. Alan R Liss, Inc, New York

Dorshkind K, Keller GM, Phillips RA, Miller RG, Bosma GC, O'Toole M, Bosma MJ (1984) Functional status of cells from lymphoid and myeloid tissues in mice with severe combined immunodeficiency disease. J Immunol 132:1804-1808

Dorshkind K, Denis KA, Witte ON (1986) Lymphoid bone marrow cultures can reconstitute heterogeneous B and T cell-dependent responses in severe combined immunodeficient mice. J Immunol 137:3456-3463

Dorshkind K, Yoshida S, Gershwin ME (1989) Bone marrow cells from young and old New Zealand Black mice can reconstitute B lymphocytes in severe combined immunodeficient recipients. J Autoimmunity 2:173-186.

Fulop GM, Phillips RA (1986) Full reconstitution of the immune deficiency in SCID mice with normal stem cells requires low-dose irradiation of the recipients. J Immunol 136:4438-4443

Jyonouchi H, Kincade PW, Landreth KS, Lee G, Good RA, Gershwin ME (1982) Age-dependent deficiency of B lymphocyte lineage precursors in NZB mice. J Exp Med 155:1665-1678

Kincade PW, Jyonouchi H, Landreth KS, Lee G (1982) B-lymphocyte precursors in immunodeficient, autoimmune and anemic mice. Immunol Rev 64:81-98

Nagasawa R, Kanagawa O, Tittle TV, Chiller JM (1985) In vivo maturation of pre-B cells derived from long-term cultured bone marrow. J Immunol 135:965

Ohsugi Y, Gershwin ME, Ahmed A (1981) Study of congenitally immunologically mutant New Zealand Mice V B-cell function of NZB.xid mice. J Immunogenetics 8:129-137

Whitlock CA, Witte ON (1982) Long-term culture of B lymphocytes and their precursors from murine bone marrow. Proc Natl Acad Sci USA 79:3608-3612.

The Scid Mouse as a Model to Identify and Quantify Myeloid and Lymphoid Stem Cells

G.M. Fulop, D.-D. Wu, and R.A. Phillips

Division of Hematology/Oncology, The Hospital for Sick Children and Departments of Medical Genetics and Immunology, University of Toronto, Toronto, Canada, M5G 1X8

INTRODUCTION

The process of B cell production and differentiation is achieved in a highly regulated manner within the bone marrow of adult mammals (Osmond 1980). Many studies indicate that hematopoiesis is organized in a system where a multipotent stem cell (Sp) gives rise to progeny in all the hematopoietic lineages (Wu et al. 1968; Abramson et al. 1977). Although Sp have finite proliferative potential which can be exhausted by serial transplantation, single stem cells when injected into lethally irradiated recipients can fully reconstitute both myeloid and lymphoid function. Recently retroviral vectors have been used to introduce selectable clonal markers into hematopoietic stem cells to follow their expansion and differentiation in transplantation experiments (Joyner et al. 1983; Williams et al. 1984; Dick et al. 1985; Keller et al. 1985).

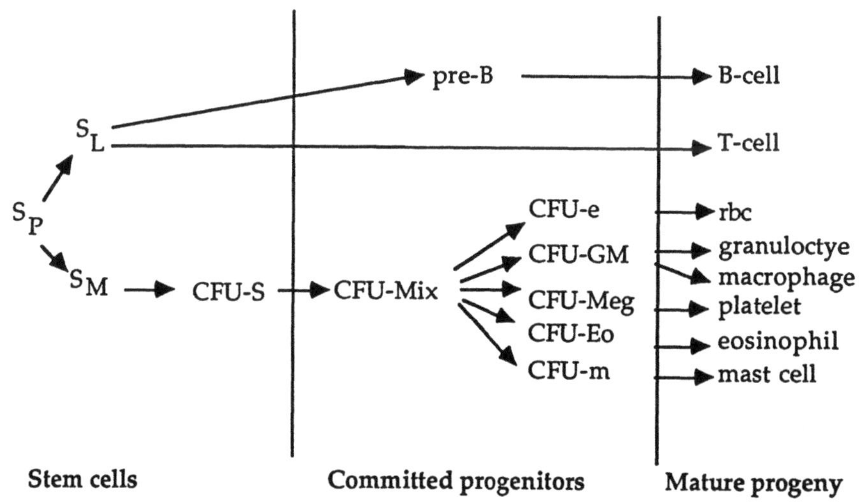

Figure 1. Working model for differentiation of hematopoietic stem cells.

Figure 1 outlines our working model for lineage relationships in the hematopoietic system. An important feature of the model is the hierarchy of cells. Stem cells are detected most reliably in in vivo transplantation assays, whereas the committed progenitors can be measured with quantitative, in vitro colony-forming assays. As cells progress along the myeloid and lymphoid lineages, the cells lose self-renewal and

Current Topics in Microbiology and Immunology, Vol. 152

proliferative potential and become restricted in their differentiative potential until they are restricted to producing a single cell type. S_P stem cells differentiate into stem cells restricted to the lymphoid (S_L) or myeloid (S_M) pathways (Abramson et al. 1977; Jones-Villeneuve and Phillips 1980). Myeloid-restricted stem cells, S_M, are well described (Abramson et al. 1977) and in turn give rise to another class of restricted stem cells, spleen colony-forming cells (CFU-S) (Till and McCulloch 1961; Wu et al. 1968). Although CFU-S have all the requirements of a stem cell, i.e., self-renewal, proliferative ability, capacity to differentiate and response to regulatory signals, they, in contrast to S_M, are unable to reconstitute the myeloid or lymphoid system of suitable recipient mice (Kitamura et al. 1981). The existence of a lymphoid-restricted stem cell (Phillips et al. 1988; Fulop and Phillips 1989) in bone marrow remains controversial with some investigators questioning its existence (Snodgrass and Keller 1987).

LONG-TERM BONE MARROW CULTURES (LTBMC)

Normal bone marrow cells are not only a source of stem cells capable of B lymphocyte production, but contain all cell members in the differentiation pathway of B lymphocyte development. Hence in a reconstitution assay, the large numbers of mature B and pre-B cells present in the inoculum may obscure contributions from lymphoid stem cells. However, bone marrow cells maintained under culture conditions described by Dexter et al. (1977) rapidly lose mature B lineage cells (Dorshkind and Phillips 1982, 1983; Phillips et al. 1984; Denis and Witte 1986) but retain S_L (Fulop and Phillips 1989). In medium containing 20% horse serum and 10^{-6} M hydrocortisone, bone marrow cells will over a period of 2 to 3 weeks establish a nearly confluent adherent stromal layer. At this time additional bone marrow is added to the culture and stem cells in this fresh inoculum adhere to the stromal layer where they self-renew, proliferate and differentiate releasing mature hematopoietic cells into the culture medium (Dexter et al. 1977). Predominant cell types found in this non-adherent layer during this phase of culture growth are cells of the granulocytic, monocytic and erythroid lineages, although mature erythrocytes are not seen in the cultures. On the stromal layer, pockets of active hematopoiesis are recognizable. In such long-term bone marrow culture (LTBMC) systems progenitor cells for the myeloid lineage (from both adherent and non-adherent layers) can be detected in <u>in vitro</u> colony-forming assays (Dexter et al. 1977). Despite the absence of detectable lymphoid cells in LTBMC, such cultures have the ability to reconstitute irradiated recipients for both myeloid and lymphoid function (Dorshkind and Phillips 1982, 1983; Phillips et al. 1984; Fulop and Phillips 1986). Furthermore, Dorshkind (1987) showed that when LTBMC are switched to a different growth medium, B lymphocytes appear in the cultures after a lag of several weeks.

The possibility of a lymphoid-restricted stem cell present in LTBMC was suggested initially by Jones-Villeneuve and Phillips (1980). LTBMC were established from CBA/N mice carrying the T6 chromosome marker and used to reconstitute lethally irradiated recipients. Two months post transplantation the T6 marker was observed only in the B and T lymphocyte population. Results from other experiments (Dorshkind and Phillips 1982, 1983; Fulop and Phillips 1986) showed that the adherent population of LTBMC contained most of the lymphoid reconstituting activity. Harrison et al. (1987) have also observed reduced myeloid reconstituting ability in LTBMC using a competitive repopulation assay to evaluate the proliferative and differentiative potential of stem cells in LTBMC. The implication of these results is that S_P and S_M do not survive well in LTBMC.

RECONSTITUTION ASSAYS FOR HEMATOPOIETIC STEM CELLS

Lethally irradiated mice

Stem cells are best assayed by transplantation (Phillips 1985). However, the assay must allow for the enumeration of the different classes of stem cells and their frequencies. Lethally irradiated mice have limited suitability as recipients in such stem cell assays. Radiation non-selectively destroys all hematopoietic cells in the recipient making survival dependent on a rapid reconstitution of all hematopoietic cells. This is achieved by transplanting large numbers of stem cells capable of both myeloid and lymphoid differentiation. The necessity for both lineages prohibits the detection and characterization of S_L and S_M stem cells with restrictive differentiative potential. In such lethally irradiated recipients, a S_L stem cell cannot reconstitute myeloid function and so the recipient will shortly die of anemia. The requirement of large numbers of cells to be transplanted immediately to help the irradiated recipient survive restricts studying the potential of single stem cells.

Mutant mice

The use of specific mutant mice that require no or little radiation prior to engraftment can overcome the problems of lethally irradiated recipients in transplantation assays. Such recipients allow detection and study of both pluripotent and restricted stem cells. Mice homozygous for mutations at the W locus (W/W^v) have been used extensively for stem cell assays of the myeloid system (Harrison and Astel 1976; Boggs et al. 1983; Dick et al. 1985). Several investigators have demonstrated that a single stem cell can cure a W/W^v mice of its anemia (Abramson et al. 1977; Boggs et al. 1982; Kitamura et al. 1981). Despite the presence of normal lymphoid differentiation in W/W^v mice, use of sublethally irradiated W/W^v recipients allows detection of stem cells with lymphoid potential. *Scid* mice have proved to be ideal recipients for studies on lymphoid differentiation. These mice, homozygous for the recessive *scid* mutation on chromosome 16, have an almost complete absence of normal B and T lymphocytes (Bosma et al. 1983). Their immune deficiency can be corrected by transplantation of normal hematopoietic stem cells (Bosma et al. 1983; Fulop and Phillips 1986, 1989).

RECONSTITUTION OF SCID MICE

As previously described (Fulop and Phillips 1986), full reconstitution of lymphoid function with either fresh bone marrow or LTBMC requires sublethal irradiation of the *scid* recipients. However, optimal engraftment of lymphoid function in the recipient is dependent upon the dose of irradiation given. As seen in Table 1, *scid* recipients analyzed 6 weeks post grafting with 10^6 bone marrow cells showed increased numbers of CFU-B/organ with increased radiation dosage. Optimal reconstitution for lymphoid function was achieved with 400 cGy, a whole body irradiation dose which does not compromise the survival of non-engrafted, irradiated *scid* recipients.

Characterization of lymphoid function in scid mice following grafting with stem cells from either fresh bone marrow or LTBMC was monitored with the colony-forming B cell assay (CFU-B) (Metcalf et al. 1975; Kincade et al. 1978). The colonies are scored after 6 days in culture. The CFU-B assay is sensitive, detecting 1 CFU-B/ 10^5 cells plated; normal scid mice have no detectable CFU-B (Dorshkind et al. 1984; Fulop and Phillips 1986).

Table 1. Requirement of whole body irradiation for the reconstitution of scid mice with stem cells

Dosage (cGy)	CFU-B/Organ ($\times 10^3$) Femur	Spleen
0	9	76
50	18	272
400	51	408
Normal BALB/cByJ mice	50	465

Long term reconstitution of lymphoid function requires seeding of stem cells into the bone marrow. The traditional method of transplanting stem cells has been by intravenous injection, but Mosier et al. (1988) found that the intraperitoneal route of injection was more efficient than intravenous injection in obtaining reconstitution of scid recipients with human peripheral blood leukocytes. We compared the reconstituting efficiency of bone marrow cells with these two routes of administration. Using a short term reconstitution assay (Dorshkind and Phillips 1983), intravenous injection gave good CFU-B reconstitution while scid mice grafted intraperitoneally had no detectable reconstitution (data not shown).

Table 2. Stem cell frequency in normal bone marrow and in LTBMC

Cells Injected	Curing units[a] per 10^6 cells	Cells Injected	Curing units per 10^6 cells
LTBMC	290	BM	93
	1300		400
	450		
	630		
Mean	572		193

[a] Cure was defined as any scid recipient having ≥ 500 CFU-B/femur.

Since the survival of scid recipients after sublethal irradiation does not depend on the success or failure of the graft, one can study the reconstituting ability of stem cells, even ones restricted in their repopulating ability. The ability of single stem cells to reconstitute irradiated scid recipients allowed us to determine the frequency of lymphoid reconstituting stem cells in fresh bone marrow and in LTBMC. Using a limiting dilution assay (Boggs et al. 1982), irradiated scid recipients were grafted with decreasing numbers of cells from fresh bone marrow or LTBMC and assayed for CFU-B activity 6 weeks post grafting. LTBMC routinely reconstituted scid mice for lymphoid function with lower numbers of cells than did fresh bone marrow (Table 2).

An essential characteristic of stem cells is their self-renewal ability. If the reconstituting cells in LTBMC and bone marrow are stem cells, they must be able to reconstitute lymphoid function in secondary recipients. Bone marrow from primary scid recipients reconstituted with either LTBMC or fresh bone marrow gave good reconstitution of B cell activity in secondary scid recipients. These results confirmed the self-renewal potential of lymphoid reconstituting cells in LTBMC indicating that these cells are stem cells.

DIFFERENTIATIVE POTENTIAL OF STEM CELLS TRANSPLANTED INTO SCID MICE

As shown in Fig. 1, there are three types of reconstituting stem cells, a pluripotent stem cell, S_P and two stem cells with restricted differentiative potential, S_M and S_L. The results described above may result from transplantation of either S_P or S_L; the experiments did not examine possible myeloid reconstitution. To identify the types of stem cells transplanted into different scid recipients, two types of experiments were done. In the first series of experiments, we used restriction fragment length polymorphisms (RFLP) to identify grafted donor cells. In a second series of experiments we used retroviruses to transduce unique genetic markers to allow identification of progeny from single stem cells.

BALB/cByJ mice and C.B-17 scid mice differ at the scid locus on chromosome 16 and at the immunoglobulin heavy chain locus on chromosome 12. BALB/cByJ mice have the Ig-h^a allele and C.B-17 mice carry the Ig-h^b locus originally derived from C57BL/6 mice. These differences allow easy distinction between the two mice because of an RFLP in the Cμ region of the Ig-h locus. On Southern blots of DNA digested with EcoR1 and probed with Cμ, the fragment containing Cμ is 12.5 kb in BALB/cByJ mice and 11 kb in C.B-17 mice. To examine the patterns of differentiation of stem cells from bone marrow and LTBMC, we grafted cells or cultures prepared from BALB/cByJ mice into scid recipients. Six to 10 weeks later, DNA was prepared from various tissues of reconstituted mice. As expected, all of the lymphoid cells, both B cells and T cells, contained the BALB/cByJ RFLP. However, in most of the scid mice reconstituted with LTBMC, the myeloid system was predominantly derived from the scid mouse. In contrast, the myeloid system in scid mice reconstituted with bone marrow was all derived from the BALB/cByJ donor.

We can make two conclusions from these results. First, donor stem cells can completely reconstitute the myeloid system in sublethally irradiated scid mice. Second, LTBMC are defective in myeloid repopulating ability. These data confirm previous results from our laboratory (Jones-Villeneuve and Phillips 1980) and others (Harrison et al. 1987) also indicating reduced myeloid reconstituting ability in LTBMC.

The failure of LTBMC to reconstitute the myeloid system of scid mice (above) and normal irradiated recipients (Jones-Villeneuve and Phillips 1980; Harrison et al. 1988) suggests that LTBMC are deficient in S_P and contain predominantly restricted stem cells, S_L, S_M and CFU-S. To directly test bone marrow preparations for the presence of restricted stem cells we infected bone marrow cells (Dick et al. 1985) with retroviruses containing the bacterial gene for neomycin resistance. Retroviruses integrate randomly but stably into stem cells so that all progeny of an infected stem cell have the same viral integration site. Characterization of the integration site in cells of different lineages identifies cells derived from a common stem cell. We have studied the pattern of reconstitution in scid mice grafted with infected stem cells obtained from mice pre-treated with 5-FU to enrich for stem cells. In a small series of experiments using mice engrafted with a limited number of stem cells, we have observed two distinct patterns of reconstitution. In some mice, all lymphoid cells contain a common viral integration site but myeloid cells remain scid in origin and have no integrated provirus. A simple interpretation of these data is that reconstitution occurred from a lymphoid-restricted stem cell. The other pattern of reconstitution is the opposite, reconstitution of the myeloid system by a single, infected donor stem cell and absence of lymphoid reconstitution. The best interpretation of this pattern is that the mice were reconstituted with a myeloid-restricted stem cell.

SUMMARY

Scid mice are excellent recipients for studying the characteristics of stem cells. Sublethal irradiation not only enhances engraftment of stem cells but enables one to graft limiting numbers of cells without compromising the survival of the recipient. This enables one to estimate the frequency of stem cells by limiting dilution analysis. Compared to fresh bone marrow, LTBMC are slightly enriched for stem cells capable of reconstituting lymphoid function in scid recipients. The stem cells have self-renewal ability since bone marrow from cured primary scid recipients can cure secondary recipients. Our results indicate that lymphoid reconstitution following engraftment with LTBMC occurs from a lymphoid-restricted stem cell; similar restricted stem cells also exist in normal bone marrow.

ACKNOWLEDGEMENTS

The research was supported by grants from the Medical Research Council and by The National Cancer Institute of Canada.

REFERENCES

Abramson S, Miller RG, Phillips RA (1977) The identification in adult bone marrow of pluripotent and restricted stem cells of the myeloid and lymphoid systems. J Exp Med 145: 1567-1579

Boggs DR, Boggs SS, Saxe DF, Gress LA, Canfield DR (1982) Hematopoietic stem cells with high proliferative potential. Assay of their concentration in marrow by the frequency and duration of cure in W/W^v mice. J Clin Invest 70: 242-253

Bosma GC, Custer RP, Bosma MJ (1983) A severe combined immunodeficiency mutation in the mouse. Nature 301: 527-530
Denis KA, Witte ON (1986) In vitro development of B lymphocyte from long term cultured precursor cells. Proc Natl Acad Sci USA 83: 441-445
Dexter TM, Allen TD, Lajtha LG (1977) Stimulation of differentiation and proliferation of hematopoietic cells in vitro. J Cell Physiol 82: 461-473
Dick JE, Magli MC, Huszar D, Phillips RA, Bernstein A (1985) Introduction of a selectable gene into primitive stem cells capable of long-term reconstitution of hematopoietic system of W/W^{v} mice. Cell 42: 71-79
Dorshkind K (1987) In vitro differentiation of B lymphocytes from primitive hematopoietic precursors present in long-term bone marrow cultures. J Immunol 136: 422-429
Dorshkind K, Phillips RA (1982) Maturational state of lymphoid cells in long-term bone marrow cultures. J Immunol 129: 2444-2450
Dorshkind K, Phillips RA (1983) Characterization of early B lymphocyte precursors present in long-term bone marrow cultures. J Immunol 131: 2240-2245
Dorshkind K, Keller GM, Phillips RA, Miller RG, Bosma GC, O'Toole M, Bosma MJ (1984) Functional status of cells from lymphoid and myeloid tissues in mice with severe combine immunodeficiency disease. J Immunol 132: 1804-1808
Dorshkind K, Pollack SB, Bosma MJ, Phillips RA (1985) Natural killer (NK) cells are present in mice with severe combined immunodeficiency. J Immunol 134: 3798-3801
Fulop GM, Phillips RA (1986) Full reconstitution of the immune deficiency in scid mice with normal stem cells requires low-dose irradiation of the recipients. J Immunol 136: 4438-4443
Fulop GM, Phillips RA (1989) Use of scid mice to identify and quantitate lymphoid-restricted stem cells in long term bone marrow cultures. Blood (in press)
Fulop GM, Bosma GC, Bosma MJ, Phillips RA (1988) Early B-cell precursors in scid mice: Normal numbers of cells transformable with Abelson murine leukemia virus (A-MuLV). Cell Immunol 113: 192-201
Harrison DE, Astle CM (1976) Population of lymphoid tissues in cured W-anemic mice by donor cells. Transplant 22: 42-46
Harrison DE, Lerner CP, Spooncer E (1987) Erythropoietic repopulating ability of stem cells from long-term marrow culture. Blood 69: 1021-1025
Jones-Villeneuve EV, Phillips RA (1980) Potentials for lymphoid differentiation by cells from long term cultures of bone marrow. Exp Hematol 8: 65-76
Joyner A, Keller G, Phillips RA, Bernstein A (1983) Retrovirus transfer of a bacterial gene into mouse hematopoietic progenitor cells. Nature 305: 556-558
Keller G, Phillips RA (1984) Maintenance of hemopoiesis in long-term bone marrow cultures from Sl/Sl^{d} and W/W^{v} mice. Exp Hematol 12: 822-824
Keller GM, Paige C, Gilboa E, Wagner E (1985) Expression of a foreign gene in myeloid and lymphoid cells derived from multipotent hematopoietic precursors. Nature 318 : 149-154
Kincade PW, Paige CJ, Parkhouse RME, Lee G (1978) Characterization of murine colony-forming B cells. 1. Distribution, resistance to anti-immunoglobulin antibodies, and expression of Ia antigens. J Immunol 120: 1289-1296
Kitamura Y, Yokoyama M, Matsuda H, Ohno T, Mori KJ (1981) Spleen colony-forming cell as a common precursor for tissue mast cells and granulocytes. Nature 291: 159-160

Lauzon RJ, Siminovitch KA, Fulop GM, Phillips RA, Roder JC (1986) An expanded population of natural killer cells in mice with severe combine immunodeficiency (scid) lack rearrangement and expression of T cell receptor genes. J Exp Med 164: 1797-1802.

Malynn BA, Blackwell TK, Fulop GM, Rathbun GA, Furley AJW, Ferrier P, Heinke LB, Phillips RA, Yancopoulos GD, Alt FW (1988) The scid defect affects the final step of the immunoglobulin VDJ recombinase mechanism. Cell 54: 453-460

Metcalf D, Nossal GJV, Warner NL, Miller JFAP, Mandel TE, Layton JE, Gutman GA (1975) Growth of B lymphocyte colonies in vitro. J Exp Med 142: 1534-1549

Mosier DE, Gulizia RJ, Baird SM, Wilson DB (1988) Transfer of a functional human immune system to mice with severe combined immunodeficiency. Nature 335: 256-259

Osmond DG (1980) Production and differentiation of B lymphocytes in the bone marrow. In Battisto JR, Knight KL (eds) Immunoglobulin genes and B cell differentiation. Elsevier/North Holland, Amsterdam, pp 135-158

Phillips RA (1980) Enhanced lymphoid and decreased myeloid reconstituting ability of stem cells from long-term cultures of mouse bone marrow. J Supramolecular Structure 14: 77-83

Phillips RA (1985) Comparison of different assays for multipotent hematopoietic stem cells. In Ford RJ, Maizel AL (eds) Mediators in cell growth and differentiation. Raven Press, New York, pp 135-145

Phillips RA, Bosma M, Dorshkind K (1984) Reconstitution of immune-deficient mice with cells from long-term bone marrow cultures. In Greenberger JS, Nemo GJ, Wright DG (eds) Long term bone marrow cultures. Alan R Liss Inc., New York, pp 309-321

Phillips RA, Wu D-D, Fulop GM (1988) Lymphoid-restricted stem cells. Curr Top Microbiol Immunol 141: 11-15

Schuler W, Weiler IJ, Schuler A, Phillips RA, Rosenberg N, Mak TW, Kearney JF, Perry RP, Bosma MJ (1986) Rearrangement of antigen receptor genes is defective in mice with severe combined immune deficiency. Cell 46: 963-972

Snodgrass R, Keller G (1987) Clonal fluctuation within the haematopoietic system of mice reconstituted with retrovirus-infected stem cells. EMBO J 6: 3255-3260

Till JE, McCulloch EA (1961) A direct measurement of the radiation sensitivity of normal mouse bone marrow. Radiat Res 14: 213-222

Whitlock CA, Robertson D, Witte ON (1984) Murine B cell lymphopoiesis in long-term culture. J Immunol Meth 67: 353-369

Williams DA, Lemischka IR, Nathan DG, Mulligan RC (1984) Introduction of new genetic material into pluripotent hematopoietic stem cells of the mouse. Nature 310: 476-480

Wu AM, Till JE, Siminovitch L, McCulloch EA (1968) Cytological evidence for a relationship between normal hematopoietic colony-forming cells and cells of the lymphoid system. J Exp Med 127: 455-463

VI. Potential Uses: Engraftment with Human Tissues and Cells

The Scid-hu Mouse: Current Status and Potential Applications

J. M. McCune, H. Kaneshima, M. Lieberman[1], I.L.Weissman, and R. Namikawa

SyStemix, Inc.
3400 West Bayshore Road
Palo Alto, CA 94303 USA

The detailed analysis of physiology and pathophysiology in man is a difficult and frequently impossible task. In many instances, volunteers will not serve as experimental subjects; multiple variables cannot be simultaneously explored; and placebo-controlled trials are confounded by intervening social dynamics. Animals, used instead as surrogates for man, then only provide approximate models of uncertain relevance. For most diseases affecting man, even animal models are unavailable, precisely because these diseases affect only man.

Such is the case with the acquired immunodeficiency syndrome. The etiologic agent, human immunodeficiency virus (HIV), produces disease only in man. The human CD4 receptor, present on T and myelomonocytic cells, appears to be necessary for infection, but certainly not sufficient: other, human-specific factors are also required for HIV viral entry, replication, and/or gene expression (Maddon et al. 1986). Although there are other lentiviruses which produce disease in other animals, none is identical to HIV. Thus, the pathophysiology of HIV infection can only be studied in tissue culture or during the process of clinical trials in man. The former situation is potentially too far removed from events in vivo to be relevant; the latter is too closely tied to the real-life exigencies of hospital wards to be easily studied. Neither is optimal as a route towards understanding the biology of HIV infection. Neither, likely, will lead to an efficient and meaningful search for therapeutic agents acting against HIV.

At least on paper, the best route towards such a search may be described. First, it would be useful to have a small animal model for HIV infection. Given the lore of information that has previously been acquired about the murine immune system, that small animal might best be a mouse. Secondly, the virus used for infection should be HIV, not some surrogate lentivirus from another species (e.g., a yet-to-be found small animal lentivirus). Thirdly, there should be no known restriction to the replication and spread of HIV infection within the small animal model; the cells that are initially infected must be human and there should be uninfected human

[1]HK and RN are from SyStemix, Inc. ML is from the Cancer Biology Research Laboratory and ILW from the Laboratory of Experimental Oncology, Department of Pathology, Stanford University School of Medicine, Stanford, CA 94305

cells around to support further cycles of infection. Finally, the cells which are human should be those that are likely to be infected in vivo in man, namely, the cells of the hematolymphoid organs. The first cells to actually "see" HIV in vivo may well be those within lymph nodes draining the site of infection (or, alternatively, those cells found within the thymus of the developing fetus). These organs contain $CD4^+$ T cells and $CD4^+$ myelomonocytic cells (including antigen-presenting dendritic cells), all of which are spatially grouped within closely apposed functional units. Such cells are not easily grown in vitro; it is the interaction of these cells with HIV which we would hope to closely observe, after infection in vivo.

In sum, we would hope to be able to create a mouse with a human immune system, permissive for infection with HIV in vivo. Here, we describe our approach towards this goal, our preliminary results, and future directions.

The SCID-hu mouse

The mammalian hematolymphoid system is an anatomically diffuse collection of cells, some mobile and others not. These cells interact co-ordinately with one another, usually within the well-defined microenvironmental space of lymphoid organs. The system, at its minimal construction, includes pluripotent hematopoietic stem cells and the micro-environments in which these cells self-renew and differentiate (fetal liver/bone marrow), the thymus (in which T cells differentiate into functionally mature subpopulations), and peripheral lymphoid organs (e.g., lymph node, spleen, mucosal tissue, and skin) in which immune responses are orchestrated between T cells, B cells, and antigen-presenting cells. At its maximal extent, with all elements intact and functioning, the system is comprised of multiple T and B cell clones, each able to interact with distinct epitopes. Recognition of foreign major histocompatibility complex (MHC) antigens (or exogenous antigen in general) by these receptors drives an immune response, most often in a process which itself is constrained by self-MHC/peptide presentation. In an associative manner, self-MHC and autologous antigens are tolerated.

Although the exact mechanisms of tolerance induction remain unclear, classic experiments by Medawar and his colleagues in the 1950's indicated that, during fetal life, lineage-restricted lymphoid precursors could be rendered tolerant to non-self (Billingham et al. 1953). Later, in the 1970's, the thymus was formally demonstrated to be the site in which developing T cell acquired MHC restriction and tolerance (Zinkernagel et al. 1978a,b). After T cells leave the thymus, they circulate throughout the periphery as mature, functionally capable cells, able to discriminate self from non-self. Therefore, a precondition to placing a human immune system into a mouse is that the cells be "taught" to recognize mouse as self; otherwise graft versus host disease might ensue.

To accomplish that end, we reasoned that the cells might best be allowed to mature within the mouse. Thus, if one could implant into a mouse a source of human hematopoietic stem cells deplete of mature T cells (fetal liver/fetal bone marrow) as well as human fetal thymus, the former might migrate to the latter, differentiate through, and later emerge as mature, functional human T cells with "self-recognition" defined by both mouse and human MHC. These mature T cells might home to and function within peripheral human lymphoid organs, also implanted within the mouse. Not coincidentally, such organs (inclusive of the implanted thymus) and the cells differentiating through them are precisely the targets that one would hope to infect with HIV.

If the mature human T cells must be tolerant of the mouse, so too must the mouse accept them. To meet this condition, we chose an immunodeficient strain of mouse, one that could not mount an immune response of its own. Since we were interested in providing the mouse with a human immune system which might complement an inherent deficiency, we sought a mouse strain which appeared to mimic the deficiency found after HIV infection in man. In the most satisfactory of biologic assays, we might then replete the murine immunodeficiency with a human immune system, create immunodeficiency again by infecting with HIV, and then ask the questions: which drugs (or vaccines) protect the human immune system of the mouse from HIV? in which doses? by which routes? And most importantly, what is the mechanism of protection in vivo?

Of the various strains we surveyed, the C.B17 scid/scid stock seemed most suitable (Bosma et al. 1983; Dorshkind et al. 1984; Custer et al. 1985; Ware et al. 1985). As described in detail in this volume, the T and B lymphoid lineages of the scid/scid homozygote (hereafter designated SCID) are numerically and functionally defunct: allografts and xenografts are accepted with ease; antibody responses cannot be elicited after primary immunization. The defect appears to reside in a qualitative and/or quantitative defect in a recombinase-associated function; T and B cell antigen receptors do not rearrange as they normally would during differentiation (Schuler et al. 1986; Malynn et al. 1988; Hendrickson et al. 1988). Otherwise, the mice have a full complement of erythrocytes, platelets, granulocytes, and monocytes; they do not die of anemia, bleeding, or acute infection (Czitrom et al. 1985; Dorshkind et al. 1985). Given the severity of the immunodeficiency, the animals have been observed to often succumb to opportunistic infection. It struck us, most especially, that the infection of which they die is the very same which afflicts so many patients with AIDS: *Pneumocystis carinii* pneumonia (see Shultz, this volume). We reasoned that these mice would accept engrafted hematolymphoid organs from man and that human progenitor cells might thereafter give rise to mature human T and B cells circulating in the peripheral circulation. We wondered whether such cells would functionally complement the defect of the SCID, and thereby protect the mice from the acquisition of opportunistic infection.

Our first experiments involved a group of 21 SCID mice. Ten were maintained as controls in a clean animal room, within bonneted microisolator cages. After four months, all had assumed the appearance of the mouse shown in the lower part of Fig. 1: with ruffled fur, a hunched-over posture, a rapid respiratory rate, and weight loss. All, within this time period, died. Given a concomittant outbreak of *Pneumocystis carinii* pneumonia in the breeding colony from which these mice were derived (see Shultz, this volume), we presume that these mice also died of that disease.

The remaining mice were removed from the barrier facility, engrafted with human fetal thymus and human fetal liver cells, and maintained in an experimental colony elsewhere. Although these animals intially became ill, all improved until they assumed the appearance of a normal BALB/c mouse. The top mouse in Fig. 1 is a member of this group. Apparently, the engraftment of human fetal tissue into these mice had protected them, directly or indirectly, from the acquisition of opportunistic infection.

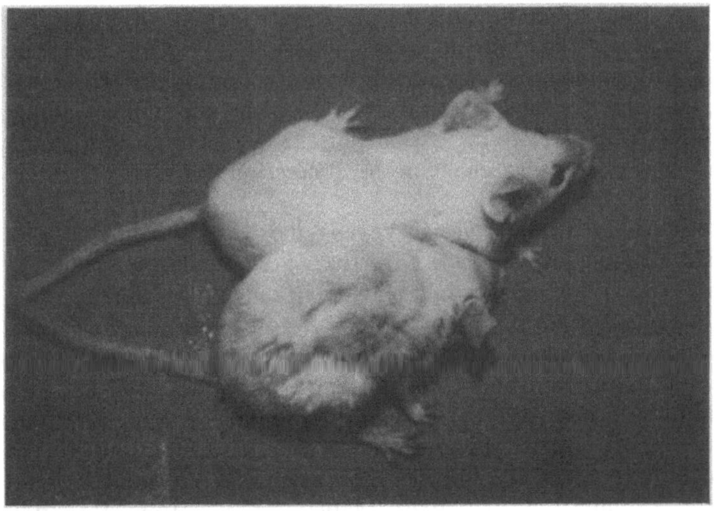

Fig. 1. The SCID-hu mouse. These mice were originally from the same litter. The SCID mouse on the bottom was maintained in a barrier facility as an untreated control. The SCID-hu mouse on the top had been surgically implanted with a human fetal thymus and given human fetal liver cells, approximately 2 months before this picture was taken.

In the course of evaluating this phenomenon more closely, a number of observations were made. Described in detail elsewhere (Mc Cune et al. 1988), the following represent salient points upon which present investigations are based.

First, the thymic implants of the SCID-hu mice were found to be vascularized and to grow well beneath the renal capsule (our experience to date suggests that the "take rate" for such grafts approaches 70-80%). When examined microscopically, the architecture of the grafts was virtually indistinguishable from that of normal, age-matched human fetal thymus with respect to distribution of antigens found on cortical epithelial cells (CDR2), medullary epithelial cells (MD1), and thymocytes representative of the cell surface markers CD1, CD4, and CD8. The only difference noted to date is that the thymic implants of the SCID-hu mouse are populated with dendritic cells of murine origin, expressing murine Class II antigens. These cells are of bone-marrow origin and likely migrated into the human thymus after engraftment. Accordingly, when human thymic progenitors pass through the SCID-hu thymus (see below), "self" is seen not only in the context of human but also mouse MHC antigens. Potentially such progenitors may be tolerized to the MHC antigens of both.

When a source of human stem cells is given to the SCID-hu mouse intravenously (i.v.), these cells are found to home to and differentiate through the engrafted human fetal thymus. Thereafter, differentiated progeny are found in the peripheral circulation of the mouse as phenotypically mature human T cells, single positive for CD4 or CD8 in a ratio of 3-4/1 (a normal ratio for age-matched controls) (Yachie et al. 1981). Further, human stem cells given i.v. are found to give rise to differentiated myelomonocytic cells, found within the engrafted human fetal thymus as interdigitating (dendritic) cells. Thus, the SCID-hu mouse system may be used as a "read-out" for those undifferentiated cells in the human fetal liver which would otherwise differentiate along the myeloid or lymphoid lines. To the extent that these populations might be subfractionated from the total mixture of cells present in the fetal liver (or, alternatively, from the fetal or adult bone marrow), this read-out might represent an assay system for their isolation.

When fetal liver progenitors are given i.v., the read-out of mature T cells in the SCID-hu is time-dependent (not seen before 3-4 weeks) and time-limited (not lasting after 10-12 weeks). At its peak, the transient wave of T cells represents 2-40% of the total mononuclear cells in the peripheral blood. As detected by serologic techniques, these cells stain with a variety of antibodies against human, and not murine, cell surface antigens; no staining above background is observed with isotype-matched controls. Further, total cellular DNA from these cells has been shown to hybridize with DNA probes specific for human, and not murine, DNA. Subtle signs of graft-versus-host disease are not observed: the CD4/CD8 ratio remains normal and human T cells are not found to infiltrate into murine hematolymphoid organs. We reasoned that the human T cells disappeared from the peripheral circulation as a consequence of (a) destruction (e.g., by nonspecific effector mechanisms in the mouse, potentially NK cells), (b) segregation (e.g., into the engrafted human lymph nodes or into mouse skin, spleen, liver, or lung), and/or
(c) absence of self-renewal or replenishment. The first possibility is being addressed by backcrossing the SCID mouse stock onto other stocks deficient in NK cell activity (Shultz, work in progress). The second possibility has been partially addressed: significant numbers of human cells are not observed in the mouse lymph nodes, spleen, or thymus; the status of cell

migration into skin, liver, and lung is being evaluated now (Streeter et al., work in progress).

In consideration of the last possibility, we reasoned that the absence of self renewal of stem cells might be secondary to the fact that they were introduced intravenously, in the absence of the microenvironmental stromal cells which normally promote self-renewal and differentiation in the bone marrow and/or fetal liver. Accordingly, we have constructed SCID-hu mice that contain fetal thymus and entire fragments of fetal liver (containing both the hematopoietic precursors as well as the stromal microenvironment). In these mice, preliminary observations suggest that human T cell repopulation in the periphery lasts for much longer periods of time. Again, the CD4/CD8 ratio remains constant between 3-4/1. By inference, stem cell self renewal has been maintained. Also by inference, the stromal cells which maintain such self-renewal are present in the SCID-hu mouse. If so, the growth factors which mediate these processes may be assayed in vivo. In sum, short-term reconstitution by i.v. injection of fetal liver hematopoietic cells leads to an assay for stem cells. Long term reconstitution in the presence of fetal liver stromal cells leads to an assay for growth factors acting on human pluripotent progenitors.

When, in addition to fetal liver and fetal thymus, human fetal lymph node is implanted into the SCID-hu mouse, human plasma cells with IgG and/or IgM in the cytoplasm are observed and human IgG is found to be circulating in the mouse peripheral blood. The specificity of the IgG is unknown and preliminary attempts to immunize SCID-hu mice by intentional immunization have thus far proved unsuccessful. The presence of differentiated human plasma cells suggests, however, that functional interactions might occur between antigen-presenting cells, T cells, and B cells. The interacting cells may have been resident within the human lymph nodes upon implantation into the SCID mouse. Alternatively, T cells circulating within the SCID-hu peripheral blood might home to such nodes and contribute to the response. The observation of the phenomenon suggests that the appropriate stratum for primary immunization is there. Ongoing experiments will thoroughly test this possibility.

Potential uses of the SCID-hu mouse

The foregoing discussion suggests a number of potential uses of the SCID-hu mouse.

1) Analysis of viral infections affecting man

The human T and myelomonocytic cells found in the engrafted human thymus and lymph node, as well as the mature T cells in the peripheral blood, contain subpopulations which could be permissive for infection by HIV. Progenitor cells in the engrafted fetal liver may also be infected by this virus. Once infected in the SCID-hu mouse, these cells will exist in the microenvironments within which infection is likely to occur in man, that is, within hematolymphoid organs. Thus, the interactions that take place between infected cells and adjacent uninfected cells can be studied. At a more basic level, the viral gene products that are active within different subpopulations of human cells might be identified.

In preliminary experiments, it is clear that this is true within the engrafted human thymus and lymph node (Namikawa et al. 1988). In a time- and dose-dependent manner, both organs within the SCID-hu mouse are found to be permissive for infection with molecularly-cloned isolates of HIV-1. With time, signs of infection spread throughout the organs, indicating that uninfected cells are being infected or, alternatively, that progenitor cells might give rise to migrating progeny. When analyzed simultaneously for the presence of viral mRNA transcripts (by in situ hybridization) and for viral structural proteins (by immunohistochemistry), many cells in the thymus show evidence of both, but some show only viral mRNA products. This raises the possibility that some subsets of infected cells may harbor a pattern of HIV gene expression in which active viral production is suppressed. Indeed, such a pattern of infection has been observed before in other lentivirus infections, most notably that associated with visna (Narayan et al. 1983; Gendelman et al. 1986; Davis et al. 1987).

If further analysis of the SCID-hu mouse provides insights about the pathophysiology of acute HIV infection in vivo, within human lymphoid organs, it might also lead to a useful model system in which to test anti-viral agents before they are tested in the clinical setting. The following questions can now be addressed: what is the volume of distribution of the given agent (i.e., does it actually enter the infected thymus and lymph node)? if it enters the organs, with which subpopulations of cells does the agent interact? how is the agent metabolized, within those cells? and finally, once it has interacted with and/or been metabolized by such cells, does it actually suppress ongoing viral infection, i.e., is the agent efficacious? As a case in point, can azidothymidine be demonstrated to enter the organs, to be phosphorylated within both T cells and myelomonocytic cells, and to suppress ongoing or acute HIV infection in both? Could it, for instance, be used as an agent to give prophylactically in the event of an accidental needle stick in the clinical situation? Such experiments, performed within the SCID-hu mouse, might provide information readily transferable to the clinical setting.

2) Analysis of human hematopoietic stem cells and lineage-restricted precursors

The SCID-hu mouse represents a read-out system for the identification of those subpopulations of human fetal liver cells which have progenitor activity for the human T and myelomonocytic lineages. As such, it is unique. The mouse pluripotent hematopoietic precursor, for instance, was purified by asking the question: which subpopulation of cells will protect and reconstitute lethally irradiated mice (Spangrude et al. 1988)? In contrast, isolated populations of human bone marrow cells can only be introduced into humans who would have otherwise received whole bone marrow for life-threatening hematologic disorders. These studies are not easily done nor easily controlled. In the setting of the SCID-hu mouse, subpopulations of human fetal liver cells (or, alternatively, human bone marrow cells) could be isolated by a variety of techniques (magnetic bead separation, FACS separation, etc.) and then provided a controlled and quantitative fashion to SCID-hu mice bearing human fetal thymus implants. Those cells which are found to lead to T and myelomonocytic differentiation are likely to be inclusive of both lineage-restricted progenitors and pluripotent stem cells. Further subfractionation of these subpopulations, as well as further refinement of the SCID-hu (e.g., in such a way that it can be used to identify erythroid, granulocytic, and megakaryocytic lineages) might then lead to more precise purification of the respective progenitor cells.

Once available, these cells could be used in therapeutically useful ways in man, for instance, in the settings of bone marrow transplantation as well as in those instances in which selected cell populations are deficient (anemias, thrombocytopenia, granulocytopenia, lymphopenia). To the extent that the stem cells can be altered, they might also be administered to humans in such a fashion as to convey new activities. For instance, the introduction of genes that convey resistance to various chemotherapeutic agents into stem cells might permit chemotherapeutic regimens against various tumors that are now not possible, simply because they are too toxic to the hematopoietic system. Similarly, anti-viral activities might be introduced into stem cells in such a fashion that "intracellular immunity" to particular viruses might be achieved and conveyed to all progeny (Baltimore 1988; Friedman et al. 1988).

3) Analysis of the growth factors which regulate hematopoiesis

When fetal liver cells are administered to the SCID-hu mouse i.v., in the absence of microenvironmental stromal cells and the growth factors that they produce, the production of mature cells is time-limited. If human stem cells can be isolated, they will in turn serve as assays, either in vivo or in vitro, for the identification of those factors which regulate self-renewal and/or differentiation. This line of investigation has in the past been hampered, precisely for lack of a suitable assay. The biologic processes which connote such cellular development are likely to be complicated. In much the same way that enzymes may be competitively or noncompetitively inhibited or allosterically activated, the signals that "turn on" or "turn off" a given cell within a lineage are likely to be multiple, antagonistic, and complex. Indeed, the very designation of such factors as "growth factors" tends to blur this concept and to belie such complexity; as likely, for every growth factor there is a "no growth" or "no differentiation" factor, perhaps made by different cells at different times in different locations. The SCID-hu represents not only a means by which to purify stem cells and their progeny but also an assay system in which to monitor growth and differentiation. Once isolated, these cells in that context represent a means by which to decipher this complexity.

If combinations of growth factors can be identified, methods available by recombinant technology will permit their production in bulk. A number of avenues will then open. On the one hand, it may be possible to use them in conjunction with limiting numbers of purified cells to grow such cells in vitro. Once grown, it will predictably be easier to prepare these cells for use in vivo, in man. Therein, their activities might be more definitively illustrated. Finally, both the stem cells, the lineage-restricted precursors, and their appropriate factors might be given simultaneously in vivo to patients, in an effort to synergize those activities.

4) Analysis of the functions of mature human T and B cells

At this stage in experimentation, the human T cells found in the peripheral circulation or lymphoid organs of the SCID-hu are phenotypically mature and apparently tolerant of mouse MHC antigens (i.e., they have a normal CD4/CD8 ratio). It is possible that they have acquired tolerance by transit through a chimeric human fetal thymus which expresses both mouse and human MHC antigens. Alternatively, they may be completely nonfunctional. In an intermediate sense, they may be representative of such a limited clonal diversity that mouse MHC antigens are not "seen," simply for lack of the appropriate xenoreactive human T cell receptors. Each of these possibilities suggests a number of obvious experiments. If tolerance is found to be enforced, then the mechanisms by which such processes occur might be systematically studied within the SCID-hu. If, simultaneously, the T cells are found to be functionally reactive against newly-introduced antigens (e.g., HIV), then clones of such cells might be isolated to study their activities (e.g., inhibition of HIV infection, immune selection of HIV variants, etc.). If HIV is introduced into the thymus implant of the SCID-hu mouse, then tolerance to HIV might be studied as a potential pathophysiologic correlate of infection of the infected human fetus. If the cells are found, instead, to be nonfunctional, then the biologic correlates (growth factors, immunologic stimulae, etc.) which render them functional could be systematically explored. Each of these studies would be difficult to support in man; each will likely contribute to our understanding of human T cell biology and, possibly, to the derivation of T cell functions which might be therapeutically useful in human disease states.

The status of the human B cells within the SCID-hu mouse is not well defined, but even more directly engaging. Human IgG is made by human plasma cells found within the engrafted lymph nodes. Likely, this is the result of interaction between antigen-presenting cells, T cells, and B cells; this inference demands further verification. At present time, there is no technique for the initiation of primary human humoral responses, other than that commonly employed in vaccinations in vivo. In vitro, using antigens that would be dangerous to administer in man, the results are unsatisfactory and do not regularly lead to the generation of hybridoma cell lines which produce high levels of high affinity human monoclonals, of defined class, subclass, and allotype. If the processes which lead to the production of IgG in the SCID-hu can be understood, or at the very least empirically reproduced, then it may be possible to use this model as a means by which to produce such antibodies. As therapeutic, prophylactic, or diagnostic agents for use in man, such reagents would potentially serve a number of useful purposes.

These and other applications of the SCID-hu mouse will be explored with time. Their relative merits will be aligned in order. At the very least, observations will be made in the setting of the model as an "in vivo tissue culture flask" (Weiss 1988). Some of these observations may be interesting; some may be of practical importance to the treatment of diseases in man. All will serve to further underline the general importance of the SCID mouse as a research tool for the study of the mammalian immune system.

Acknowledgements: We thank L Shultz, ISY Chen, and T Hamm for interesting discussions and for sharing reagents. This work was supported by NCI grant CA03352, NIAMS grant AR5P40RR03624-029, and an equipment grant from the Weingart Foundation

References

Baltimore D (1988) Intracellular immunization. Nature 335: 395-396

Billingham RE, Brent L, Medawar PB (1953) Actively acquired tolerance to foreign cells. Nature 172: 603-605

Bosma GC, Custer RP, Bosma MJ (1983) A severe combined immunodeficiency mutation in the mouse. Nature 301: 527-530

Custer RP, Bosma GC, Bosma MJ (1985) Severe combined immunodeficiency (SCID) in the mouse. Am J Pathol 120: 464-477

Czitrom AA, Edwards S, Phillips RA, Bosma MJ, Marrack P, Kappler JW (1985) The function of antigen-presenting cells in mice with severe combined immunodeficiency. J Immunol 134: 2276-2280

Davis JL, Molineaux S, Clements JE (1987) Visna virus exhibits a complex transcriptional pattern: one aspect of gene expression shared with the acquired immunodeficiency syndrome retrovirus. J Virol 61:1325-1331

Dorshkind K, Keller GM, Phillips RA, Miller RG, Bosma GC, O'Toole, Bosma MJ (1984) Functional status of cells from lymphoid and myeloid tissues in mice with severe combined immunodeficiency disease. J Immunol 132: 1804-1808

Dorshkind K, Pollack SB, Bosma MJ, Phillips RA (1985) Natural killer cells are present in mice with severe combined immunodeficiency (SCID). J Immunol 134: 3798-3801

Friedman AD, Triezenberg SJ, McKnight SL (1988) Expression of a truncated viral transactivator selectively impedes lytic infection by its cognate virus. Nature 335: 452-454

Gendelman HE, Narayan O, Kennedy-Stoskopf S, Kennedy PGE, Ghotbi Z, Clements JR, Stanley J, Pezeshkpour G (1986) Tropism of sheep lentiviruses for monocytes: susceptibility to infection and virus gene expression increase during maturation of monocytes to macrophages. J Virol 58: 67-74

Hendrickson EA, Schatz DG, Weaver DT (1988) The scid gene encodes a trans-acting factor that mediates the rejoining event of Ig gene rearrangement. Genes and Dev 2: 817-829

Maddon PJ, Dalgleish AG, McDougal JS, Clapham PR, Weiss RA, Axel R (1986) The T4 gene encodes the AIDS virus receptor and is expressed in the immune system and the brain. Cell 47: 333-348

Malynn BA, Blackwell TK, Fulop GM, Rathburn GA, Furley AJW, Ferrier P, Heinke LB, Phillips RA, Yancopoulos GD, Alt FW (1988) The scid defect affects the final step of the immunoglobulin VDJ recombinase mechanism. Cell 54: 453-460

Mc Cune JM, Namikawa R, Kaneshima H, Shultz LD, Lieberman M, Weissman IL (1988) The SCID-hu mouse: murine model for the analysis of human hematolymphoid differentiation and function. Science 241: 1632-1639

Namikawa R, Kaneshima H, Lieberman M, Weissman IL, Mc Cune JM (1988) Infection of the SCID-hu mouse by HIV-1. Science 242: 1684-1686

Narayan O, Kennedy-Stoskopf S, Sheffer D, Griffin DE, Clements JE (1983) Activation of caprine arthritis-encephalitis virus expression during maturation of monocytes to macrophages. Infect Immun 41: 67-73

Schuler W, Weiler IJ, Schuler A, Phillips RA, Rosenberg N, Mak T, Kearney JF, Perry RP, Bosma MJ (1986) Rearrangement of antigen receptor genes is defective in mice with severe combined immunodeficiency. Cell 46: 963-972

Spangrude GJ, Heimfeld S, Weissman IL (1988) Purification and characterization of mouse hematopoietic stem cells. Science 241: 58-62

Ware CF, Donato NJ, Dorshkind K (1985) Human, rat or mouse hybridomas secrete high levels of monoclonal antibodies following transplantation into mice with severe combined immunodeficiency (SCID). J Immunol Methods 85: 353-361

Weiss R (1988) HIV and Kaposi's sarcoma in mice. Nature 337:112-113

Yachie A, Miyawaki T, Nagaoki R, Yokoi T, Mukai M, Uwadana N, Taniguchi N (1981) Regulation of B cell differentiation by T cell subsets defined with monoclonal OKT4 and OKT8 antibodies in human cord blood. J Immunol 127: 1314-1317

Zinkernagel RM, Callahan GN, Althage A, Cooper S, Klein PA, Klein J (1978a) On the thymus in the differentiation of "H-2 self-recognition" by T cells: evidence for dual recognition? J Exp Med 147: 882-896

Zinkernagel RM, Callahan GN, Althage A, Cooper S, Streilein JW, Klein J (1978b) The lymphoreticular system in triggering virus plus self-specific cytotoxic T cells: evidence for T help. J Exp Med 147: 897-911

Studies of HIV Infection and the Development of Epstein-Barr Virus-Related B Cell Lymphomas Following Transfer of Human Lymphocytes to Mice With Severe Combined Immunodeficiency

Donald E. Mosier[1], Richard J. Gulizia[1], Stephen M. Baird[2,3], Stephen Spector[3], Deborah Spector[3], Thomas J. Kipps[4], Robert I. Fox[4], Dennis A. Carson[4], Neil Cooper[4], Douglas D. Richman[2,3], and Darcy B. Wilson[1]

[1]Medical Biology Institute, 11077 North Torrey Pines Road, La Jolla, CA 92037, USA
[2]Veterans Administration Medical Center, La Jolla, CA 92161, USA
[3]University of California - San Diego, La Jolla, CA 92103, USA
[4]Research Institute of Scripps Clinic, La Jolla, CA 92037, USA

Abstract: Mice with severe combined immunodeficiency (C.B-17 *scid*, hereafter SCID) accept xenografts of adult human peripheral blood leukocytes (PBL). The transplanted human PBL expand in number and survive for at least thirteen months and have been shown to reconstitute human immune function at both the T and B cell levels. Human immunoglobulin production is restored, and secondary antibody responses to antigens such as tetanus toxoid can be induced. All SCID mice reconstituted with 50 x 10^6 or more PBL from donors with evidence of exposure to Epstein-Barr virus (EBV) have developed human B cell lymphomas at 8-16 weeks after PBL engraftment, whereas mice reconstituted with PBL from EBV-seronegative donors fail to develop tumors. These tumors involve both lymphatic and non-lymphatic organs, and histologically they resemble large cell or immunoblastic lymphomas. The tumors are associated with high levels of human immunoglobulin secretion and serum electrophoresis reveals oligoclonal immunoglobulin banding patterns. Analysis of tumor DNA shows the presence of EBV genomes and oligoclonal patterns of immunoglobulin J_H gene rearrangement. Taken together, these observations suggest an EBV-related proliferation of B lymphocytes leading to the rapid appearance of oligoclonal B cell malignancies following transfer of B lymphocytes from "normal" donors to SCID mice.

SCID mice reconstituted with PBL from EBV-seronegative donors have been infected with the LAV-1 strain of human immunodeficiency virus (HIV-1). Virus has been recovered from most infected animals by co-culture of mouse tissue with human T lymphoblasts. Some mice with high virus titers have developed an acute wasting syndrome and depletion of human T cells. These preliminary results suggest the use of HIV-infected human PBL-reconstituted SCID mice as a small animal model for AIDS research.

INTRODUCTION

We (Mosier *et al.*, 1988) and others (McCune *et al.*, 1988; Kamel-Reid and Dick, 1988) have recently shown that mutations that render mouse strains immunodeficient, such as *scid* (Bosma *et al.*, 1983) or *nu.xid.bg* (Andriole *et al.*, 1985), permit the xenotransplantation of human hematopoietic cells to such mice. Our group has concentrated on the transplantation of mature human lymphoid and monocytic cells from peripheral blood of adult donors after making the surprising observation that obvious xenogeneic graft-versus-host disease did not occur using this protocol. Our initial success in engrafting functional human lymphocytes in SCID mice has been reported (Mosier *et al.*, 1988). This presentation will summarize our recent findings in two areas: (1) the genesis of Epstein-Barr virus (EBV)-related B cell lymphomas in

Current Topics in Microbiology and Immunology, Vol. 152

PBL-reconstituted SCID mice; and (2) the appearance of an acute AIDS-like syndrome in SCID mice infected with HIV-1.

METHODS

Mononuclear PBL were isolated from the blood of EBV-seropositive or seronegative donors by Ficoll-Hypaque gradient separation. From 10-100 x 10^6 PBL were injected intraperitoneally into C.B-17 *scid* mice. The immunologic phenotypes of fresh human mononuclear cells or cells recovered from SCID mice were determined by staining with monoclonal antibodies specific for CD3, CD4, CD5, CD8, CD14 (Mac-3), CD19 (B4), CD20 (B1), and CD25 (IL-2R). In addition, the presence of murine lymphocytes (which occur in 5-10% of SCID mice in low numbers) was evaluated by staining with monoclonal antibodies to murine Thy-1 and Ly-5 (B220). Single cell suspensions were prepared from the spleens, lymph nodes, peripheral blood, peritoneal cavity washings, and tumor nodules of SCID mice repopulated with human PBL. Following reaction with the panel of monoclonal antibodies, staining was evaluated by flow cytometry.

Serum was obtained from PBL-reconstituted SCID mice at weekly intervals after intraperitoneal injection of human cells. Human immunoglobulin M, G, and A were detected by an ELISA technique as previously described (Mosier *et al.*, 1988). This assay was shown not to detect mouse immunoglobulin, and SCID mice used in these experiments were shown to have less than 5 µg/ml mouse immunoglobulin in a mouse IgM-specific ELISA. Serum samples were also subjected to immunoelectrophoresis in cellulose acetate.

Tissues from PBL-reconstituted SCID mice were fixed in either 10% buffered formalin or cold acetone. Sections of representative tissues were examined after staining with hematoxyln and eosin or immunoperoxidase labelling of monoclonal anti-human lymphocyte antibodies.

DNA from normal or tumor tissue was extracted, subjected to electrophoresis, transferred to membranes by Southern blotting, and hybridized with cDNA probes specific for human *Alu* sequences, the immunoglobulin H chain J region, and the internal repetitive sequence of EBV. In addition, EBV or HIV sequences were detected by the polymerase chain reaction (PCR).

HIV replication was detected by co-cultivation of fresh human T lymphoblasts and cells from PBL-reconstituted SCID mice. Viral replication was assessed by the appearance of p24 core protein in the culture supernatant using the Abbot p24 capture assay. Virus presence was confirmed by PCR and *in situ* hybridization.

RESULTS AND CONCLUSIONS

EBV-associated lymphomagenesis: In our initial studies of the reconstitution of SCID mice with human cells, we injected from 10 to 100 x 10^6 Ficoll-Hypaque separated PBL from healthy adult male donors either intraperitoneally (i.p.) or intravenously (i.v.) into SCID mice recipients. We observed engraftment of human cells in all mice injected i.p., but no evidence of engraftment of PBL given i.v. Eight mice that received more than 40 x 10^6 PBL i.p. began to show evidence of clinical illness at 6-8 weeks post-transplantation, while four recipients of 10 x 10^6 PBL are alive and well over one year after injection. Necropsy of affected animals revealed the presence of numerous tumor masses in the peritoneal cavity, liver, lungs, and kidneys, and diffuse enlargement of the spleen and peripheral lymph nodes. Histological examination of these tumor masses revealed a lymphoblastic proliferation, sometimes including plasma cells, and tumor DNA hybridized both with a human *Alu* sequence probe and a probe specific for the internal repeat elements of EBV. Analysis of tumor DNA also showed oligoclonal patterns of immunoglobulin J_H gene rearrangement. Tumor cells were stained by antibodies to human immunoglobulin and to the B lymphocyte-specific markers CD19 and CD20. Both donors involved in these experiments were found to be seropositive for antibodies to EBV. These preliminary results led us to postulate that prior EBV infection might cause either a polyclonal or monoclonal B cell lymphoproliferative disease following transfer of sufficient numbers of PBL's to SCID mice.

To evaluate this hypothesis, we have compared the appearance of tumors in SCID mice receiving equivalent numbers of PBL's from either EBV-positive or EBV-negative donors (the latter kindly provided by Dr. Neil Cooper). The results of these studies are presented below in Figure 1. It can be seen that the transfer of 50 x 10^6 PBL from an EBV-positive donor led to the death of all SCID recipients by 9 weeks post-transplant, whereas all recipients of the same number of PBL from an EBV-seronegative donor are alive at greater than 24 weeks post-transplant. Similar results were obtained in three separate experiments involving three different donor pairs.

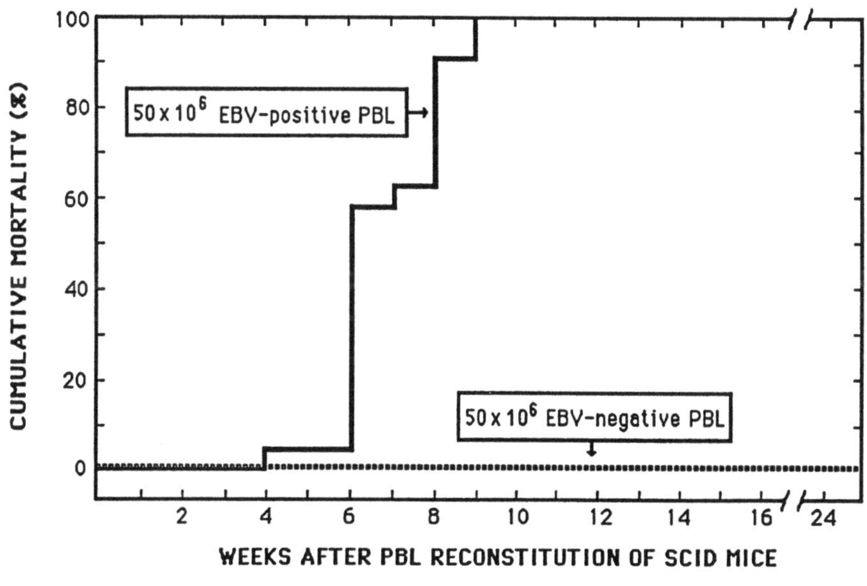

Figure 1: Mortality of SCID mice transplanted with PBL from EBV-seropositive or seronegative donors. Exposure to EBV or an associated agent is required for the development of B cell lymphomas.

These results suggest that EBV exposure is related to the development of these spontaneous lymphomas. Further data show that these tumors are often bi- or triclonal, and involve both IgM- and IgG-producing B cells, although IgM-positive tumors are more common. We conclude that the development of EBV-associated B cell lymphomas in SCID mice provides a new model for studying early steps in lymphomagenesis.

HIV Infection of PBL-reconstituted SCID mice: One of the original goals in transplanting human cells to SCID mice was to create a small animal model for AIDS research. Recent experiments have suggested that we are close to meeting this goal. SCID mice reconstituted with 10 x 10^6 PBL from an EBV-seronegative donor have been infected 8 weeks later with the LAV-1 strain of HIV-1, given as a mixture of free virus and *in vitro* infected syngeneic T lymphoblasts. As is shown in Figure 2, we have recovered HIV by virus culture from almost all of these mice. Moreover, four animals with very high levels of virus have developed an AIDS-like syndrome characterized by wasting, weight loss, and depletion of human T lymphocytes. Spleens of these animals show a low of lymphoid cells from the periarteriolar sheath and follicular areas that become repopulated with human cells after PBL engraftment of SCID mice. While preliminary, these results are encouraging for the prospect of developing a useful animal model for understanding the pathogenesis of AIDS. A recent report (Namikawa *et al.*, 1988) has shown evidence of HIV infection of human thymus cells in SCID mice.

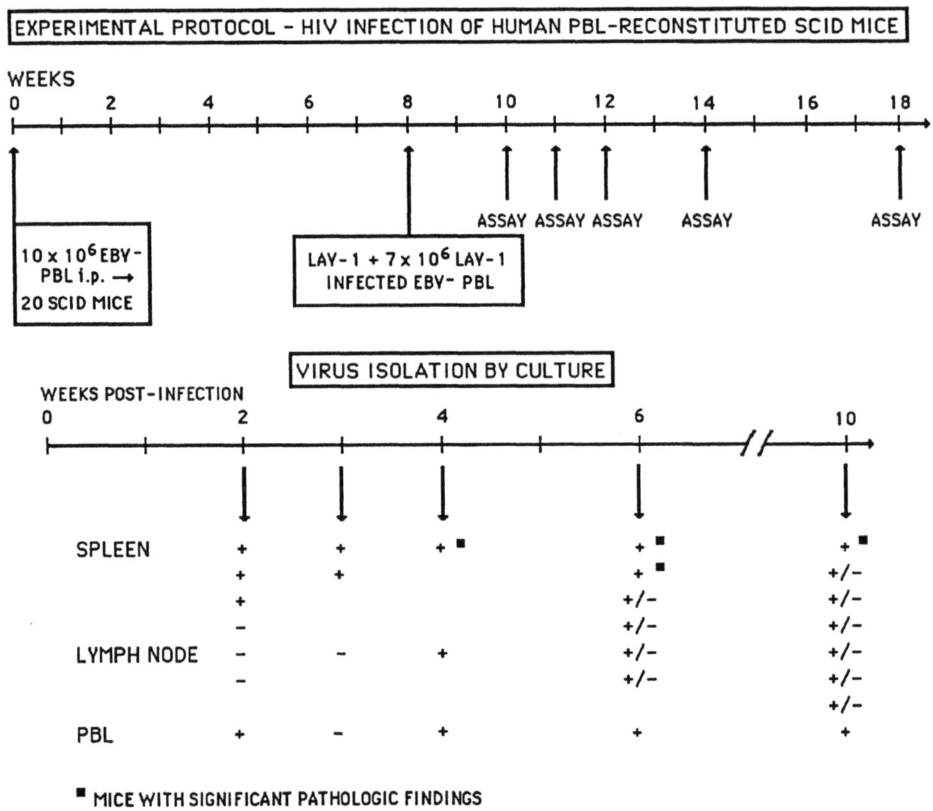

Figure 2: HIV isolation from PBL-reconstituted SCID mice at 2-10 weeks post-infection. Cultures were sampled twice weekly for p24 and maintained for 4 weeks. +/- indicates that final culture results are not yet available.

This research is supported by the following awards: NIAID UO-1 AI-23607; NIAID RO-1 AI-27703; NIAID RO-1 AI-22871; and NIAID PO-1 AI-24526.

REFERENCES

Andriole GL, Mulé JJ, Hansen CT, Linehan WM, Rosenberg SA (1985) Evidence that lymphokine-activated killer cells and natural killer cells are distinct based on an analysis of congenitally immunodeficient mice. *J. Immunol.*, **135**:2911-2913

Bosma GC, Custer RP, Bosma MJ (1983) Severe combined immunodeficiency mutation in the mouse. *Nature* **301**:527-530

Kamel-Reid S, Dick JE (1988) Engraftment of immune-deficient mice with human hematopoietic stem cells. *Science* **242**:1706-1709

McCune JM, Namikawa R, Kaneshima H, Shultz LD, Lieberman M, Weissman IL (1988) The SCID-hu mouse: Murine model for the analysis of human hematolymphoid differentiation and function. *Science* **241**:16321639

Mosier DE, Gulizia RJ, Baird SM, Wilson DB (1988) Transfer of a functional human immune system to mice with severe combined immunodeficiency. *Nature* **335**:256259

Namikawa R, Kaneshima H, Lieberman M, Weissman IL, McCune JM (1988) Infection of the SCID-hu mouse by HIV-1. *Science* **242**:1684-1686

Human Lung Tumors, Patients' Peripheral Blood Lymphocytes and Tumor Infiltrating Lymphocytes Propagated in Scid Mice

R.B. Bankert, T. Umemoto, Y. Sugiyama, F.A. Chen, E. Repasky and S. Yokota
Dept. of Molecular Immunology, Roswell Park Memorial Institute, 666 Elm St., Buffalo, NY 14263 USA

INTRODUCTION

Studies of the pathogenesis of human tumors and the evaluation of new therapeutic modalities have been limited by the lack of a suitable experimental animal model. Currently very little is known regarding patients' immune response to their tumors and the consequences of this response to tumor growth. Further, the evaluation of novel tumor therapies has been largely restricted to animal tumor models which may not accurately reflect the efficacy of such therapies for human tumors (Bankert et al. 1989). While athymic homozygous nu/nu mice have been used extensively for the heterotransplantation of human tumors (Fogh and Trempe 1975), their use in evaluating therapies for human cancer is limited by their ability to respond to thymic independent antigens, by the presence of immunoglobulin in their circulation and tissues and by phenotypic and karyotypic changes in the tumor that have been observed during tumor growth in nude mice. We have previously reported (Reddy et al. 1987) that xenografts of human lung tumors can be successfully propagated in the B cell and T cell deficient mouse mutant strain, CB-17 scid (described by Bosma et al. 1983). Several characteristics of the scid mouse make it particularly attractive as a potential model to evaluate new therapeutic approaches to the treatment of human cancers. In this report we summarize our findings that scid mice support the growth of human tumors and we establish here that this mouse is a viable model with which to evaluate the antibody directed delivery of cytotoxic agents to human lung tumors.

In this paper we also report that peripheral blood lymphocytes as well as tumor infiltrating lymphocytes are able to reconstitute scid mice thereby confirming and extending the report by Mosier et al. 1988. While exploring the possibility that scid mice co-transplanted with patients' lymphocytes and tumor cells could act as a model to study the effect of host immunity upon tumor growth, several observations were made and are reported here regarding the scid-Hu model.

METHODS

Animals--C.B-17 mice, heteroygous for the scid gene were provided by Dr. Mel Bosma, ICR, Philadelphia, PA. These mice were set up as breeding pairs to obtain mice that are homozygous for the scid gene, i.e. <1 μg of immunoglobulin/ml of serum. The homozygous scid mice were then set up as monogamous breeding pairs and the line propagated as brother X sister matings. All caging, food, water and beddings were autoclaved before use and cages were maintained within a high-efficiency particulate air filtered laminar flow hood.

Current Topics in Microbiology and Immunology, Vol. 152

Cells--The A549 tumor cell line was derived from a human alveolar lung tumor (Lieber et al. 1976), and the cells maintained in culture in RPMI-1640 medium supplemented with 10% fetal calf serum.

Monoclonal Antibodies--The murine monoclonal anti-tumor antibody 5E8 and the control monoclonal antibodies 2C3 and 4B1 were described previously (Saji et al. 1984; Bankert et al. 1987).

Radioiodination of Monoclonal Antibodies--Affinity purified antibodies were radiolabeled as described by Sugiyama et al. (1988).

Preparation of Liposomes--Large unilamellar liposomes were prepared and loaded with cytosine arabinoside according to a modified procedure of Martin and Papahadjopoulos (1982) and monoclonal anti-tumor antibodies were covalently coupled to the liposomes as reported previously (Bankert et al. 1989).

Preparation of Human Lymphocytes and Leukocytes--Lymphocyte enriched populations were obtained by centrifugation of diluted whole blood (derived from normal donors or lung cancer patients) with Ficoll-Paque (Pharmacia, Piscataway, NJ). Peripheral blood leukocytes were obtained from volunteer blood donors.

Quantification of Human Ig--A standard inhibition ELISA plate assay was used to measure human Ig in the scid-Hu mice. Plates are coated with 5 μg/ml of normal human Ig and unbound areas of the plate saturated with bovine serum albumin. The inhibition of binding of HRP-labeled anti-human antibody to the plate by known concentrations of human Ig was determined and a standard curve generated that was used to quantify the human Ig in scid mouse serum.

Antibody Assays and Immunofluorescence--Assays for anti-dextran antibody (Abu-Hadid et al. 1987) and for antibodies to A549 cells by immunocytoadherence and immunofluorescence assays were as previously reported (Saji et al. 1984; Zylstra 1986).

RESULTS AND DISCUSSION

Growth of Human Lung Tumors in SCID Mice

SCID mice were inoculated either subcutaneously (s.c.) or intravenously (i.v.) with single cell suspensions of a human alveolar lung tumor cell line termed A549. This tumor is one of several human lung tumor cell lines that expresses the tumor associated antigen gp160/130 which has also been shown to be expressed on a large proportion of non-small cell lung biopsies (Saji et al. 1984; Zylstra et al. 1986). Twenty-seven scid mice were inoculated subcutaneously with A549 cells and all of these mice developed a tumor nodule 18 days after inoculation. The tumor grew progressively reaching a diameter of 3cm within twelve weeks, at which time the mice were sacrificed. Postmortem examination revealed no gross or microscopic evidence of tumor growth outside of the s.c. site of inoculation. One half (7 of 14) of the scid mice inoculated with A549 cells i.v. developed tumor within 29 weeks. Tumor growth was observed exclusively in the lung where there were multiple tumor nodules.

It was previously established that serial transplantation did not alter the phenotype of the lung tumor (Reddy et al. 1987). Tumor morphology, number and type (all human) of chromosomes and the expression of the tumor associated antigen gp160/130 were maintained for at least five passages of the tumor in vivo.

While the A549 tumor cell line also grows in athymic nude mice, the scid mouse is a better model for evaluating the efficacy of immunospecific drug delivery protocols for the treatment of human cancer. One obvious advantage of the scid mouse is that these mice, as opposed to nude mice, have little or no immunoglobulin in their serum or interstitial fluid. This simplifies the task of precisely localizing anti-tumor antibody following administration of the antibody-cytotoxic conjugate. Another advantage is that scid mice lack functional B- and T-cells and do not mount an immune response to T-independent antigens which may be expressed either on the tumor cell itself or on the cytotoxic complex (for example the dextran used to conjugate drug and antibody). And finally, the human lung tumor does produce tumor nodules in the lungs of scid but not nude mice.

Use of SCID Mice to Evaluate Immunospecific Targeting Protocols for the Treatment of Human Lung Cancer

SCID mice were used successfully to evaluate two immunospecific targeting protocols, i.e. the intravenous administration of ^{125}I-labeled monoclonal anti-tumor antibody to tumor bearing mice and the i.v. treatment of tumor bearing mice with drug containing immunospecific liposomes. The target of both delivery protocols is a glycoprotein which is a non-covalently associated heterodimer with a 160 kDa α chain and a 130 kDa β chain (mentioned above). A monoclonal antibody (5E8) that is specific for the α chain was used as the immunospecific ligand for these two protocols.

The first question to be addressed was whether or not the 5E8 anti-tumor antibody was able to extravasate and to penetrate the solid tumor nodule. Figure 1B establishes that 4 hours after an intravenous injection of 5E8 anti-tumor antibody (labeled with horseradish peroxidase) the antibody is easily detectable in the tumor nodule (primarily at the periphery). Twenty hours

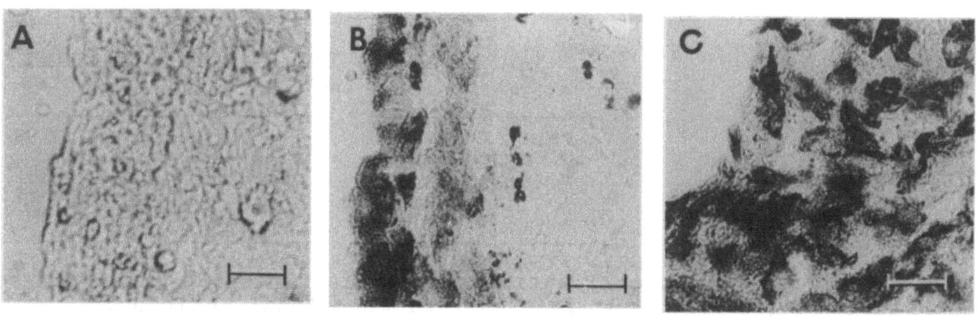

Fig. 1. Localization of HRP-labeled 5E8 anti-tumor antibody to lung tumor growing in scid mouse. SCID mice bearing subcutaneously growing A549 tumor nodules were injected i.v. with 0.5 ml (1.5 mg/ml) of either 5E8-HRP or a control antibody 2C3-HRP. Tumors, spleen, liver and lung removed at 4 hours or 20 hours, and frozen sections developed to reveal presence of enzyme. Shown here are tumors from an untreated mouse (A), tumor 4 hours after intravenous injection of 5E8-HRP (B), and tumor 20 hours after an intravenous injection of 5E8-HRP (C). Mag bar = 30 μm.

after administration (Fig. 1C) the antibody is present throughout the entire tumor nodule. The only normal tissue in which the HRP-5E8 dependent staining was above background levels was in the spleen. Here the staining was diffuse compared to the heavily stained focal areas in the tumor (data not presented). Very little staining was observed in either the tumor or normal tissues (except for spleen) when a control anti-hapten antibody (2C3-HRP) was injected into tumor bearing scid mice. These findings are confirmed by the biodistribution data of ^{125}I-5E8 and ^{125}I-4B1 presented in Table 1 which indicate that by 48 hours ^{125}I-labeled 5E8 (but not the control radiolabeled monoclonal antibody 4B1) preferentially localized to the tumor.

Table 1
Localization of ^{125}I-Labeled Monoclonal Anti-Tumor Antibody to Human Lung Tumors in SCID Mice

	Percentage of Injected cpm/gram of Tissue[a]		
	^{125}I-Labeled 5E8		^{125}I-Labeled 4B1
Organ	24 hr.	48 hr.	48 hr.
Tumor	17.3 ± 1.4	12.7 ± 0.3	4.5 ± 0.02
Lung	6.3 ± 0.3	5.0 ± 1.2	3.4 ± 0.3
Liver	6.0 ± 0.4	3.4 ± 0.3	3.4 ± 0.1
Spleen	12.1 ± 0.9	4.4 ± 1.5	4.0 ± 0.2

[a] SCID mice bearing human alveolar lung tumor A549 which expresses gp160/130 were given an intravenous injection of either ^{125}I-5E8 (8.04 μCi), a monoclonal antibody specific for gp160/130, or ^{125}I-4B1 (12.25 μCi), a control antibody, which has the same light and heavy chain isotype as 5E8, i.e. $gamma_1$,kappa.

The data presented in Table 2 indicate that ^{125}I-labeled 5E8 anti-tumor antibody but not a radiolabeled 4B1 control antibody delays or prevents the growth of the lung tumor in scid mice. Only one of the nine mice that were given the radiolabeled anti-tumor

Table 2
Effect of ^{125}I-Labeled Monoclonal Antibody on the Growth of Human Lung Tumors in SCID Mice

Antibody[a]	Dose	No. of Mice with Tumor[b] / No. of Mice Inoculated with Tumor
^{125}I-5E8	10 μg(58.3μCi)	1/9[c]
^{125}I-4B1	9.4μg(115.5μCi)	4/4[d]
---	---	25/25[d]

[a] 5E8 is a monoclonal antibody specific for gp160/130 expressed on A549 lung tumor. 4B1 is a control monoclonal antibody.
[b] Treated and control groups of mice were inoculated with 1 X 10^6 A549 alveolar lung tumor cells one week prior to administration of antibody.
[c] One mouse developed tumor twelve weeks after tumor inoculation.
[d] All mice developed tumors three to eight weeks after tumor inoculation.

antibody (one week after tumor inoculation) developed a tumor 12 weeks after tumor inoculation. All of the untreated control mice and mice given the ^{125}I-4B1 developed tumor 3 to 8 weeks after tumor inoculation.

Tumor bearing scid mice were also subjected to treatment with immunospecific liposomes. Large unilamellar liposomes containing cytosine arabinoside (Ara-C) were prepared and 5E8 anti-tumor antibody was covalently coupled to the liposomes (Bankert et al. 1989). The data in Figure 2 indicate that treatment of scid mice with these immunospecific liposomes delayed the growth of the lung tumor when compared to untreated tumor bearing mice or to mice given equal doses of Ara-C either as a free drug or as drug encapsulated within large unilamellar liposomes which were not modified by the attachment of anti-tumor antibody.

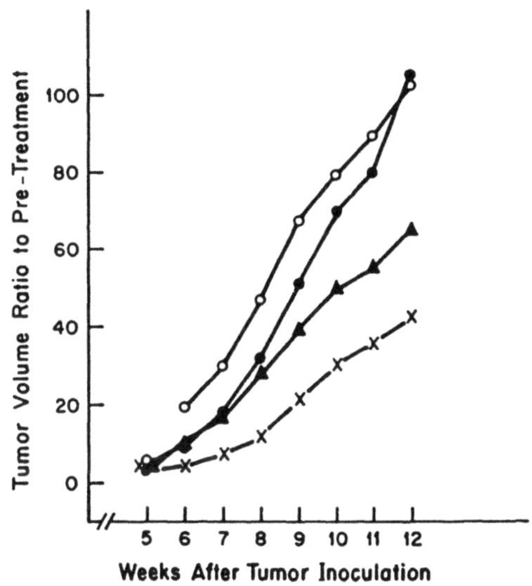

Fig. 2. Effect of immunospecific liposomes on lung tumor growth in scid mice. SCID mice bearing subcutaneously growing A549 cells treated with i.v. injections of 2.7 mg Ara-C/kg body weight of either free drug (closed circles), drug encapsulated into control, i.e., no antibody, liposomes (triangles), drug encapsulated into immunospecific liposomes, (X's), or not treated with drug (open circles).

These findings indicate that the scid mouse can be used to evaluate new therapeutic approaches to the treatment of human cancer. In future studies targeting of drugs and radiation will be tested in scid mice bearing the human lung tumor as multiple nodules in the lung, a model which more closely approximates the clinical situation in lung cancer patients.

Engraftment of SCID Mice with Peripheral Blood Lymphocytes from Patients with Primary Lung Tumors

Based upon a previous report indicating that human peripheral blood lymphocytes (PBL) can be successfully transplanted to scid mice (Mosier et al. 1988) an attempt was made to reconstitute scid mice with PBL obtained from patients bearing a primary adenocarcinoma of the lung. The ultimate goal of this endeavor was to utilize the scid mice reconstituted with patients' PBL (scid-HuP) to study patients' immune response to their tumor and to determine the effect of this response on tumor progression. During the course of transplanting either patients' PBL or patients' tumors to scid mice several observations were made and are summarized below.

Three groups of scid mice were given 1 X 10^7 PBL via an intraperitoneal injection. Each group received cells from a different patient bearing a primary lung adenocarcinoma. Human immunoglobulin (Ig) levels in the sera of these mice were determined weekly in order to monitor the presence of human B-cells. Since there were no significant differences between groups the results from the three groups of mice were combined and reported in Fig. 3. One week after transplantation the concentration of human Ig in the scid mice ranged from 20-170 μg/ml (Fig. 3, closed circles). Control scid mice that were not given human PBL have no detectable levels of human Ig (<1 μg/ml). The concentration of human Ig in the serum of the mice given human PBL increased steadily during the next five weeks reaching a concentration of approximately 1000 μg/ml. This rate of increase in human Ig in the serum of mice reconstituted with patients PBL is similar to the increase observed following scid mice reconstitution with normal PBL (Fig. 3, open circles) and is close to the values reported by Mosier et al. (1988) for mice reconstituted with 5 X 10^7 human PBL.

Our attempts to demonstrate that the human B- and T-cells were functional were unsuccessful. SCID-HuP mice were immunized either with the thymic independent antigen fraction S B1355 dextran or with irradiated A549 lung tumor cells. No anti-dextran antibodies were detected in the sera of scid-Hu mice immunized once with dextran and no antibodies to the A549 cells were detected after one or two immunizations with the irradiated tumor cells. One possible explanation for this failure to mount an immune response would be that the human B- and T-cells are preoccupied with an ongoing response to the tremendous array of scid mouse antigens which should be viewed as foreign by the donor human lymphocytes. This may be at least one viable explanation since we observed that erythrocytes of the scid-HuP mice had human IgM bound to their plasma membranes. This was determined by immunofluorescence, by a mixed hemagglutination and by the agglutination of the erythrocytes with goat anti-human IgM antibody (data not presented).

Nine mice were monitored daily for up to 17 weeks post transplantation. Three of these mice were sacrificed between 6 and 8 weeks after transplantation. The six surviving mice appeared normal for up to 14 weeks. At this time all of the mice began to show clinical signs of GVHD, i.e. hunched posture, rough hair coats and loss of weight. Mice were sacrificed and organs examined histologically. Changes in the skin and liver were consistent with GVHD in all six mice. Three of the six mice had lymphocytic or plasmacytic tumors with infiltration of multiple organ sites. Two of the tumors were determined by immunofluorescent staining to express human IgM on their cell surface and were tentatively classed as B-cell lymphomas. The remaining tumor was tentatively identified as a plasma cell myeloma on the basis of morphology and staining.

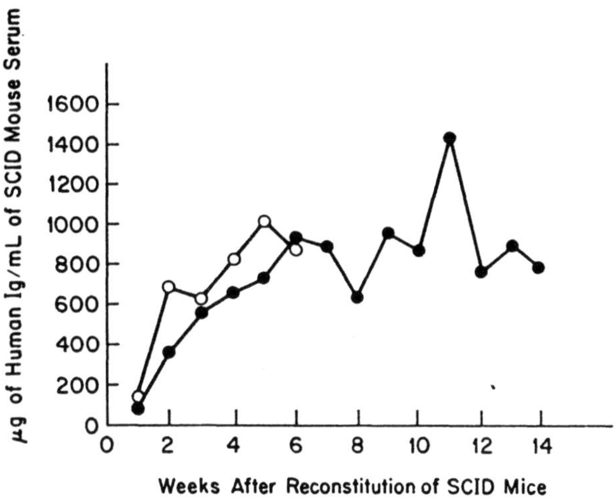

Fig. 3. Human immunoglobulin concentration in sera of scid mice following i.p. injection of normal or patients' peripheral blood lymphocytes. Mice were injected i.p. with 1 X 10^7 PBL obtained from patients with primary lung adenocarcinoma (closed circles) or from normal blood donors (open circles). Each point represents the average of nine mice.

Engraftment of SCID Mice with Tumor Infiltrating Lymphocytes

Attempts were made to transplant patient's lung tumors by injecting small pieces of lung tumor tissue obtained by surgical biopsy. In contrast to the relative ease and reproducibility of successfully xenografting human lung tumor lines, most of our attempts to establish engraftment of fresh tumor tissue were unsuccessful. Since many human tumors are infiltrated with chronic inflammatory cells, including lymphocytes, the possibility was considered that tumor infiltrating lymphocytes could be propagated in the scid mice following a tumor inoculation. In order to explore this possibility the sera of mice were assayed for human Ig 6-8 weeks after tumor transplantation. The results presented in Table 3 indicate that lymphocytes derived from the tumor can survive in the scid mice since one of these mice had over 900 µg of human Ig per ml of serum. The data in Table 3 are also consistent with the notion that the tumor infiltrating lymphocytes that reconstitute the scid mice block the successful engraftment of the tumor. In mice with the higher levels of human Ig the tumor failed to grow and in a mouse with a relatively low level of human Ig (20 µg/ml) tumor growth was present but circumscribed. When the tumor from this scid mouse was transplanted to a second scid mouse (Table 3, RPMI-18.2S) the tumor grew progressively. No human Ig (i.e., <1 µg/ml) was detected in this mouse. This tumor was subsequently subpassaged again into two more scid mice. Once again the tumors grew and neither mouse had any detectable level of human Ig in the serum. If tumor infiltrating lymphocytes are responsible for the failure of some human tumor engraftment then elimination of the lymphocytes (by treatment of the tumor tissue with anti-T cell antibodies plus complement prior to transplantation) should facilitate the growth of the tumor in the scid mice. These experiments are now in progress.

Table 3
Growth Inhibition of Human Lung Tumors in SCID Mice Associated with Tumor-Infiltrating Lymphocytes

Patient #	Tumor Type	Tumor Growth in scid[a]	Human Ig $\mu g/ml$[a]
RPMI-17.1	Lung Adeno	-	995.2
RPMI-18.1	Lung Adeno	-	69.0
RPMI-18.2	Lung Adeno	±	20.8
RPMI-18.2(S)[b]	Lung Adeno	+	< 1.0
RPMI-18.2(S,S)	Lung Adeno	+	< 1.0
RPMI-18.2(S,S)	Lung Adeno	+	< 1.0
Cell Line A549	Lung Alveolar	+[c]	< 1.0

[a] 6-8 weeks after tumor transplant.
[b] First subpassage (S), second subpassage (S,S).
[c] Tumor growth in 24 out of 24 mice inoculated.

Engraftment of SCID Mice with Peripheral Blood Leukocytes

In contrast to the original report on scid mouse reconstitution with normal human peripheral blood (Mosier et al. 1988), we were unable to demonstrate normal lymphocyte function in scid mice injected with PBL from lung cancer patients as assessed by immunization with two antigens. Moreover, unlike the report of Mosier et al. (1988) we observed signs of GVHD in all of the mice fourteen weeks after the injection of only 1 X 10^7 cells and lymphoid tumors in three of six mice examined. These differences between Mosier's results and the ones reported here could have been due to the fact that we used lung cancer patients and not normal donors as a source of PBL.

To further explore this animal model scid mice were injected either i.p. or i.v. with 1 X 10^7 - 1 X 10^8 peripheral blood leukocytes obtained from healthy volunteers. These preparations (from which platelets are removed) contained approximately 50% lymphocytes along with monocytes and granulocytes.

The concentration of human Ig in the sera of these mice at intervals of time following the injection of the leukocytes is shown in Fig. 4. In all of the mice injected with leukocytes i.p. the level of human Ig in the serum rose rapidly reaching levels of 1 mg/ml of serum or greater just two weeks after the transplantation of human cells. In contrast to the mice given human cells via intraperitoneal injection only one of three mice given human cells i.v. had any evidence of human Ig in the serum, and this was not detectable until four weeks after the human cell transplantation.

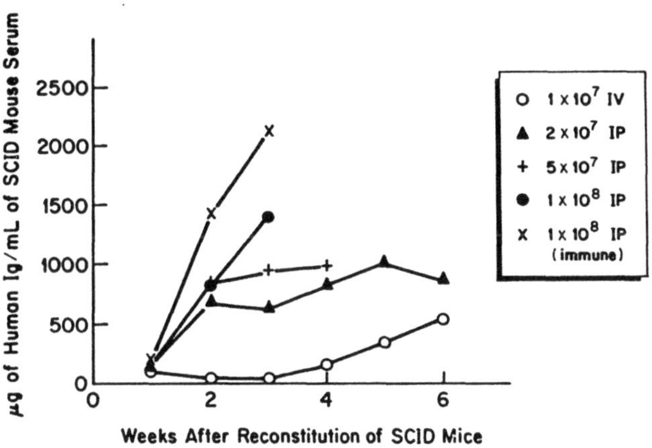

Fig. 4. Human immunoglobulin concentration in sera of scid mice following i.p. or i.v. injections of peripheral blood leukocytes. Each point represents the average of at least three mice.

As with the mice given the patients' PBL, none of the scid mice reconstituted with normal human cells exhibited any detectable levels of antibody following a single immunization with dextran or multiple immunizations with irradiated A549 cells.

All of the 7 mice given 1 X 10^8 human leukocytes exhibited clinical signs of GVHD within 3-4 weeks after transplantation, and all 7 mice were dead six weeks after cell transfer. Mice given 5 X 10^7 began to show clinical signs of GVHD at six weeks post-transplantation. Mice given 2 X 10^7 normal leukocytes (containing approximately 1 X 10^7 lymphocytes) resembled mice given 1 X 10^7 lymphocytes obtained from the lung cancer patients, i.e. there was no evidence of tumors, GVHD or death for at least ten weeks after the cell transplantation.

These results indicate that scid mice given either peripheral blood lymphocytes from cancer patients or leukocytes from normal donors do develop signs of GVHD, tumors and ultimately die following cell transfer and that these sequelae occur earlier when higher numbers, i.e. 5 X 10^7 - 1 X 10^8 of human cells are transplanted. Moreover it is apparent that scid mice reconstituted with human lymphocytes are unable to respond to at least some antigens. It is clear that we still have much more work to do in order to fully understand the scid-Hu model to study the human immune response and its effect on the pathogenesis of cancer.

REFERENCES

Abu-Hadid MM, Bankert RB, Mayers GL (1987) Antigen-specific drug targeting used to manipulate an immune response <u>in vivo</u>. Proc Natl Acad Sci USA 84: 7232-7236

Bankert RB, Repasky EA, Mazzaferro PK, Kubo RT (1987) Generation and use of an antigen-specific hybrid to study B-cell function. J Mol Cell Immunol 3: 279-291

Bankert RB, Yokota S, Ghosh SK, Mayhew E, Jou Y-H (1989)

Immunospecific targeting of cytosine arabinonucleoside-containing liposomes to the idiotype on the surface of a murine B-cell tumor _in vitro_ and _in vivo_. Cancer Res 49: 301-308
Bosma GC, Custer RP, Bosma MJ (1983) A severe combined immunodeficiency mutation in the mouse. Nature 301: 527-530
Fogh J, Trempe G (1975) New human tumor cell lines. In: Fogh J (ed) Human Tumor Cells _in Vitro_, pp 115-159. Plenum Publishing Corp, New York
Lieber M, Smith B, Szakal A, Nelson-Rees W, Todaro G (1976) Continuous tumor-cell line from a human lung carcinoma with properties of type II avelar epithelial cells. Int J Cancer 17: 62-70
Martin FJ, Papahadjopoulos D (1982) Irreversible coupling of immunoglobulin fragments to preformed vesicles. J Biol Chem 257: 286-288
Mosier DE, Gulizia RJ, Baird SM, Wilson DB (1988) Transfer of a functional human immune system to mice with severe combined immunodeficiency. Nature 335: 256-259
Reddy S, Piccione D, Takita H, Bankert RB (1987) Human lung tumor growth established in the lung and subcutaneous tissue of mice with severe combined immunodeficiency. Cancer Res 47: 2456-2460
Saji S, Zylstra S, Schepart BS, Ghosh SK, Jou Y-H, Takita H, Bankert RB (1984) Monoclonal antibodies specific for two different histological types of human lung carcinoma. Hybridoma 3: 119-129
Sugiyama Y, Chen F-A, Takita H, Bankert RB (1988) Selective growth inhibition of human lung cancer cell lines bearing a surface glycoprotein gp160 by ^{125}I-labeled anti-gp160 monoclonal antibody. Cancer Res 48: 2768-2773
Zylstra S, Chen F-A, Ghosh SK, Repasky EA, Rao U, Takita H, Bankert RB (1986) Membrane-associated glycoprotein (gp160) identified on human lung tumors by a monoclonal antibody. Cancer Res 46: 6446-6451

Acknowledgements--The authors wish to thank Dr. Maurice Barcos, Department of Pathology, Roswell Park Memorial Institute, Buffalo, NY for his evaluation of the histopathology of scid mice. This work was supported in part by grants from the National Institutes of Health, CA33462, CA25253 and CA22786 and a grant from the American Cancer Society CH-421A.

Adoptive Transfer of Human Peripheral Blood Lymphocytes (PBL) in Scid Mice

Klaus Pfeffer, Klaus Heeg, Ruth Bubeck, Peter Conradt and Hermann Wagner

Department of Medical Microbiology and Immunology, University of Ulm, Albert-Einstein-Allee 11, D-7900 Ulm, FRG

Summary

Two protocols were examined for the ability to transfer a human T cell system into SCID mice. Upon intraperitoneal injection (i.p.) of human peripheral blood lymphocytes (PBL) into SCID mice the injected cells could be recovered over weeks from the peritoneal cavity, yet human T cells did not seed into secondary lymphoid organs such as the spleen, lymph nodes or bone marrow. In contrast, SCID mice grafted with human embryonal thymus tissue contained high numbers of $CD4^+CD8^-$ and $CD8^+CD4^-$ human T cells in their lymph nodes and spleen when they had been injected i.p. with human PBL.

Introduction

The transfer of a functional human immune system to mice with severe combined immunodeficiency (SCID) has recently been reported (Mosier et al. 1988; McCune et al. 1988). Two strategies have been used. In one (Mosier et al. 1988), 10-100 x 10^6 human peripheral blood lymphocytes (PBL) were injected i.p. into SCID mice. Afterwards, human T and B cells were recovered in expanded cell numbers from the spleen, lymph nodes (LN) and blood of the recipients. The grafted mice did not develop lethal graft-versus-host (GvH) disease, produced human immunoglobulins and reacted to tetanus toxoid with human-type antibody production. In the other (McCune et al. 1988), SCID mice were grafted with human embryonal thymus and LN tissue. This allowed differentiation of human T cells following i.v. injection of human bone marrow cells.

These observations suggested to us that SCID mice reconstituted with human lymphoid cells may be used as a model system mimicking an operating human T cell system. We therefore wished to reproduce these results. Because of ethical constraints in using embryonal human tissues, we primarily tested the efficacy of i.p. injection of human PBL to reconstitute the peripheral lymphoid pool of SCID mice.

Materials and Methods

C.B.17-SCID H-2^d mice initially supplied by Dr. Mel Bosma were bred under pathogen-free conditions in the animal facilities of Ulm university by Dr. Schnappauf. They were used at an age of 8-

Current Topics in Microbiology and Immunology, Vol. 152

20 weeks. Three SCID mice were grafted below the kidney capsule with fragments of human embryonal (11 weeks) thymic tissue.

Human PBL from EBV-seropositive donors were prepared by Ficoll Hypaque separation. Sterile PBL suspensions were injected i.p. into 8-20 week-old SCID mice. Injected mice were routinely scored for signs of illness. Fifteen normal SCID mice and 3 thymus-grafted SCID mice were injected with 50 x 10^6 human PBL originating from a bank of frozen cells. Individual mice from this first group were killed for analysis over 20 weeks. A second group of normal SCID mice (n = 8) received i.p. 100 x 10^6 human PBL from a frozen bank. A third group of SCID mice (n = 12) received 50 x 10^6 freshly prepared human PBL.

Upon cervical dislocation, the peritoneal cavity of killed mice was thoroughly washed out and the peritoneal wash-out cells were prepared. Thereafter the spleen, bone marrow and, if possible, LN and peripheral blood cells were taken and single cell suspensions were prepared. Lymphocytes recovered after Percoll density centrifugation (1.077 g/cm^3) were stained with a mixture of murine monoclonal antibodies to human CD2 (Leu5b), CD3 (Leu4), CD4 (OKT4), CD8 (OKT8), CD5 (Leu1) and CD14 (Mac-3) and to human monomorphic class I (W6/32) and class II MHC determinants (L243). If available, monoclonal antibodies against the mouse counterparts were also included. Staining was analysed on a Coulter EPIC's 753 flow cytometer. Tissue sections of spleens of SCID mice were processed using standard techniques and scored microscopically.

Immune reactivity in terms of cell proliferation and acquisition of cytolytic activity was determined using the limiting dilution (LD) approach. Irradiated (4000 rad) hybridoma cells producing anti-T3 monoclonal antibodies (Mab), were used as stimulator cells under conditions described previously (Brucker et al. 1987). The frequency of potentially proliferating and cytolytic T cells was determined as described (Heeg et al. 1987).

Results

Normal SCID mice (n = 15) injected i.p. with 50 x 10^6 human PBL from a frozen bank were screened for the presence of human T cells over a period of 20 weeks. Individual mice were sacrificed and the peritoneal cells (PEC), splenocytes, bone marrow cells, and LN cells were stained with a variety of Mab. The following pattern was observed. (1) Within the peritoneal cavity (site of injection) human $CD2^+CD3^+$, $CD4^+CD8^-$, $CD8^+CD4^-$ human class I and class II MHC positive cells could be recovered even 20 weeks post injection (Fig. 1). Human cells represented between 40%-70% of wash-out cells (Table 1). The recovered human T cells were immunologically reactive in that about 1 out of 300 cells proliferated in response to anti-T3 hybridomas; the frequency of proliferating and cytolytic cells was in the same order of magnitude (Table 2). Unlike mock injected SCID mice which contained about 10 x 10^6 splenocytes, animals injected i.p. with human PBL exhibited a splenomegaly, the cellularity ranging between 30 and

300 x 10^6 cells. Microscopically the splenic white pulp appeared as empty as in normal SCID mice, the hypercellularity being clearly restricted to the red pulp. Within the red pulp megakaryocytes were the dominant cell type. With one exception none of the individually tested mice contained splenic lymphocytes expressing classical human T cell markers such as CD2, CD3, CD4, and CD8. About 2 weeks after i.p. injection of human PBL, however, the spleen of SCID mice contained about 4-8% splenocytes expressing human class I but not class II MHC antigens. These human class I positive cells were also CD5 (Leu1) positive, but CD2, CD3, CD4 and CD8 negative. Attempts to activate positively selected $CD5^+$ class I MHC^+ human cells in order to assay for their potential function have failed so far.

Table 1: Lymphocyte subset distribution in PBL-injected and control SCID mice (representative results of 27 mice tested individually)

	SCID-hu PBL			SCID control		
Mab	PEC	spleen	bone marrow	PEC	spleen	bone marrow
human:						
CD2 + CD3	55%	-*	-	-	-	-
CD4	n.d.	-	n.d.	n.d.	-	n.d.
CD8	n.d.	-	n.d.	n.d.	-	n.d.
HLA-DR	40%	3%	-	-	-	-
HLA-I + CD5	8%	-	-	-	-	
mouse:						
$H2D^d$	32%	80%	85%	35%	80%	83%
Thy1	n.d.	n.d.	n.d.	n.d.	5%	n.d.
L3T4	n.d.	n.d.	n.d.	n.d.	<1%	n.d.

n.d. not determined
* = not detectable

Table 2: Frequency of proliferative and cytolytic human T cells parked for 4 weeks in the peritoneal cavity of SCID mice

Polyclonal stimulus	Frequency of PTL-p	CTL-p
anti-T3 hybridoma	1/46 (1/33 - 1/78)	1/229 1/165 - 1/376)
PHA	1/236 (1/150 - 1/556)	n.d.

Graded concentrations of PEC were cultured in the presence of 15000/well irradiated αT3 hybridoma cells (or PHA) plus 20 U recombinant IL-2 for 8 days. Afterwards the cultures were split

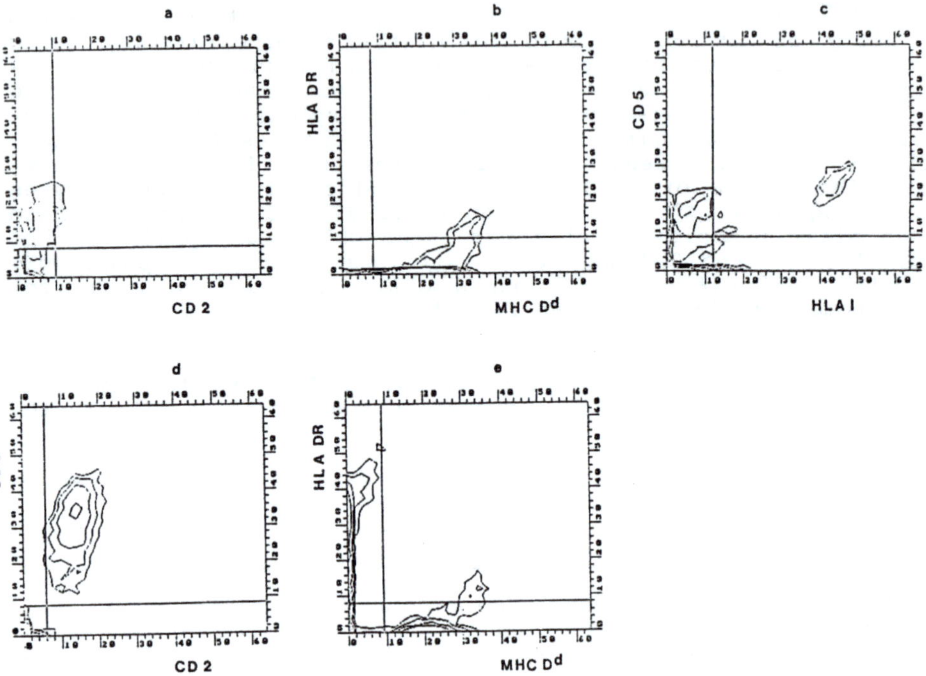

Fig. 1: Lymphocyte subset distribution in PBL-injected SCID mice. Two-colour fluorescence histogram of spleen cells (a-c) or peritoneal exudate cells (d, e) of SCID-hu-PBL mice. Stained with phycoerythrin-labelled Mab against CD8, HLA-DR, CD5 (y-axis, red fluorescence) or fluorescein isothiocyanate-labelled Mab against CD4, HLA-I or MHC-D^d (x-axis, green fluorescence). Three decade fluorescence plots of representative results are shown; contour intervals used were 2, 4, 8, 16 cells per pixel.

and assayed for proliferative or cytolytic responses. In the cytotoxicity assay ^{51}Cr-marked αT3 hybridoma cells were used.

As mentioned above, one SCID mouse within this group was atypical. When tested 4 weeks post injection, a tumour-like large nodule was visible retroperitoneally in the right perirenal region. In this nodule up to 50% of individuals wells were $CD2^+CD3^+$; again the splenocytes of this mouse were negative for human T cell markers. We cannot exclude the possibility that during i.p. injection part of the human cell inoculum was deposited directly at the retroperitoneal site and afterwards developed into the nodule observed.

Following the protocol described by Cune et al. (1988) we also constructed SCID-human mice (n = 3) by grafting human embryonal fetal thymus below the kidney capsule. Two weeks after grafting the SCID-human thymus, mice were injected i.p. with 50 x 10^6 HLA disparate human PBL. It is only under these conditions that significant numbers of human T cells were found in secondary lymphoid organs such as LN and spleen (Fig. 2, Table 3). Note that human $CD4^+$ cells were preferentially found in LNs, while the spleen contained relatively more human CD8 cells. Clearly more mice of this type have now to be evaluated.

Table 3: Lymphocyte subset distribution in SCID-hu-thymus mice

Surface marker	% cells staining in LN	Spleen
$CD2^+/CD3^+$ (double positive)	73	21
$CD4^+8^-$	55	9
$CD8^+4^-$	8	21
murine H-2D^d	10.5	41
human HLA-DR	51	15
CD5 human HLA-I (double positive)	7.5	22

Discussion

In agreement with a previous report (Mosier et al. 1988) the data presented here indicate that SCID mice transplanted i.p. with human (xenogeneic) PBL rich in T cells do not succumb to lethal GvH disease. However, such mice develop splenomegaly; note that the splenic hypercellularity is confined to the red pulp of the spleen. In contrast to published data (Mosier et al. 1988) we have never found significant numbers of human T cells in the secondary lymphoid organs of i.p. transplanted SCID mice. So far 27 individual mice have been tested. The only evidence that cells expressing human markers have the potential to migrate from the peritoneal cavity (site of injection) to secondary lymphoid organs such as spleen refers to a lymphoid cell expressing human class I MHC antigens and the CD5 marker, but lacking CD2, CD3, CD4, CD8 and human class II MHC antigens. We do not yet know whether this cell type classes as a $CD5^+$ B cell or a T cell precursor.

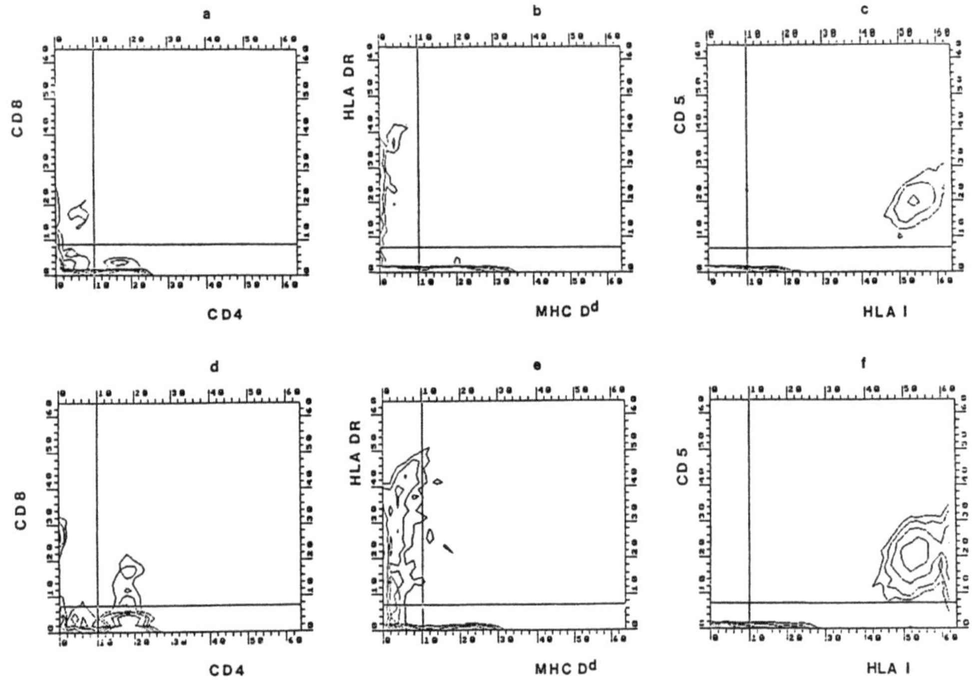

Fig. 2: Lymphocyte subset distribution in thymus-grafted and PBL-injected SCID mice. Two-colour fluorescence histogram of spleen cells (a-c) or lymph node cells (d-f) of SCID-hu-thymus mice. Stained with Mab against CD8, HLA-DR, CD5 labelled with phycoerythrin (y-axis, red fluorescence) or against CD4, MHC-D^d, HLA-I labelled with fluorescein isothiocyanate (x-axis, green fluorescence). Three decade fluorescence plots are shown; contour intervals used were 2, 4, 8, 16 cells per pixel

Although SCID mice have high levels of NK activity (unpublished data) xenogeneic human PBL injected i.p. survive for months in the peritoneal cavity; the T cells recovered express the classical T cell markers. If one accepts expression of human DR molecules as a marker of activation, presumably some of the human T cells were activated in response to xenogeneic mouse antigens. In this way one could explain the observed splenomegaly, which reflects a mouse-type splenic hypercellularity, as part of a subclinical GvH response.

Although we failed to transfer a human T cell system into the lympoid organs of SCID mice by mere i.p. inoculation of human PBL, our results in SCID-human thymus chimeric mice are promising. Upon i.p. injection of human PBL, human type T cells of the $CD4^{+}CD8^{-}$ and $CD8^{+}CD4^{-}$ subsets are abundant in the enlarged LN as well as in the spleen. We have not yet tested the functional reactivity of these SCID mouse-derived human T cells, nor do we know whether these cells have differentiated within the grafted human thymus, or have merely migrated out from the peritoneal cavity. However, we believe that SCID-human thymus chimeric mice (McCune et al. 1988) may represent a more reproducible model system in which it may be possible to transfer or to generate a functional human T cell system.

References

Brucker C, Reimann J, Wagner H, Kabelitz D (1987) Clonal activation of cytotoxic lymphocyte precursors by monoclonal anti-CD3 antibody: Analysis of feeder cell requirements. Immunol Letters 14: 121-127

Heeg K, Steeg C, Schmitt J, Wagner H (1987) Frequency analysis of class I MHC reactive $Lyt2^{+}$ and class II MHC reactive $L3T4^{+}$ IL-2 secreting T lymphocytes. J Immunol 138: 1421-1429

McCune JM, Namikawa R, Kaneshima H, Schultz LD, Liebermann M, Weissmann IL (1988) The SCID-hu mouse: murine model for the analysis of human hematolymphoid differentiation and function. Science 241: 1632-1639

Mosier DE, Gulizia RJ, Baird SM, Wilson DB (1988) Transfer of a functional human immune system to mice with severe combined immunodeficiency. Nature 335: 256-259

Establishment of Assays for Human Hematopoietic Cells in Immune Deficient Mice

John E. Dick
Department of Genetics, Research Institute, Hospital for Sick Children, and Departments of Medical Genetics and Medical Biophysics, University of Toronto Toronto, Canada

Introduction

Understanding the process of differentiation and development remains as a major challenge in biology. The hematopoietic system offers unique advantages for such studies. The blood-forming system of mice and humans consists of a heterogeneous array of cells, ranging from large numbers of differentiated cells with defined function to rare pluripotent stem cells with considerable developmental and proliferative potential (Till and McCulloch 1980). The reason this complex developmental system has proved to be so malleable for experimentation rests in the availability of quantitative colony assays for committed and multilineage progenitor cell types. Although no colony assays exist for the earliest cells in the murine stem cell hierarchy, they can nevertheless be indirectly assayed by reconstitution of lethally irradiated or genetically deficient mice. A large body of knowledge has accumulated over the past 25 years about the various classes of progenitor and stem cells, as well as the factors and cell-cell interactions that regulate their developmental program. Although in vitro colony assays exist for committed and multilineage human progenitor cells, our understanding of the biology of the human hematopoietic system has suffered relative to that in the mouse because of the lack of an in vivo assay system for human pluripotent stem cells. Several groups have recently reported the engraftment of immunedeficient mice with human lymphoid (Mosier et al 1988; McCune et al 1988) and myeloid hematopoietic cells (Kamel-Reid and Dick 1988). These animals offer an enormous opportunity to examine the human hematopoietic system and may lay the foundation for a human stem cell assay. This paper will briefly review our efforts to establish normal human hematopoietic cells in immunedeficient animals and examine the potential application of this approach to study human leukemic transformation.

Transplantation of Human Hematopoietic Cells into Immune-Deficient Mice

Conventional in vitro colony assays for human progenitors and even the newly developed techniques for assaying progenitors with high replating capacity (Leary and Ogawa 1987; Rowley et al 1987) are limited by the short length (2-3 weeks) of time the cultures can be maintained and the lack of evidence these cells have repopulating potential. The long term bone marrow culture (LTBMC) system has been developed to provide an environment where more primitive progenitors (stem cells?) can be assayed (Gartner and Kaplan 1980). A LTBMC is established by growing bone marrow cells in liquid culture under conditions that permit the stromal cells to form an adherent layer which provides a microenvironment for stem cells. These stem cells differentiate to produce a nonadherent layer of mature cells and committed progenitor cells, BFU-E, CFU-GM, and CFU-GEMM. It is possible that stem cells more primitive than CFU-GEMM reconstitute hematopoiesis in the LTBMC but the relationship of these cells to the repopulating stem cell is not clear (Andrews et al 1986). LTBMC can be maintained for 8-12 weeks but the production of progenitors drops off very rapidly soon after establishment. Although erythroid progenitors can be detected in LTBMC the conditions greatly favor the granulocyte/macrophage lineages. The steroids present in the LTBMC do not permit the growth of lymphoid cells, making detection of a pluripotential stem cell impossible.

One of the major limitations of in vitro colony assays and long term bone marrow cultures is the limited amount of cell proliferation in these assays and the fact that it is difficult to assay the earliest pluripotent stem cells using in vitro cultures. One needs only to look at the amount of information available about murine stem cells that was gained through the use of lethally irradiated or genetically anemic animals to realize the value of a similar human model system (Till and McCulloch 1980). The CFU-S or spleen colony assay for example has contributed in large measure to fundamental concepts of murine stem cell function such as self renewal, cycling status, and differentiation capacity. Long term reconstitution experiments have extended these studies to stem cells more primitive than CFU-S.

Previous attempts to grow human bone marrow by direct transplantation into lethally irradiated animals have been unsuccessful (Louwagie and Verwilghen 1970). Implantation into mice of human bone marrow in fusion chambers has also been generally unsuccessful and has not provided evidence for anything more than maintenance of mature human cells for short periods of time (Barr et al 1975).

Current Topics in Microbiology and Immunology, Vol. 152

We initially considered that two major barriers may prevent growth of transplanted human bone marrow in irradiated recipient mice: the rapid recovery of immune function and the absence of human hematopoietic growth factors. Even immune-deficient mice which lack functional T and B lymphocytes such as scid and nude have high levels of NK activity which may mediate a host response against the donor cells (Dorshkind et al 1985; Fodstad et al 1984). In order to overcome this problem we used triple recessive immune-deficient mice called bg/nu/xid (bnx) (a gift from Dr. Carl Hansen, NIH). The nude (*nu*) mutation makes the animals athymic, while the beige (*bg*) mutation reduces the number of NK cells, and the xid mutation affects B cell development and reduces lymphokine activated killer cells (LAK), an important activity in the host response to foreign cells (Andriole et al 1985).

A number of studies on hematopoietic growth factors have indicated that several factors acting on primitive cell types are species specific; for example, murine IL-3 does not stimulate the growth of human multipotent progenitor cells and vice versa. Therefore, in our initial experiments, human IL-3 and GM-CSF were continuously infused into the bnx animals using an osmotic minipump (Alza). Human bone marrow cells (10^7) were injected intravenously into sublethally irradiated CBA/J and bnx recipients; recipients were irradiated because many studies in mouse and man had indicated that conditioning was required for stem cell engraftment. After various lengths of time (5-14 days) the bone marrow and spleen were removed and the DNA from these tissue samples was analyzed by Southern blot to quantitate the proportion of human cells in the hematopoietic tissues of the recipient. The probe was a human chromosome 17 specific α-satellite probe that does not hybridize with mouse DNA (Waye and Willard 1987). Human DNA was detected in the bone marrow and spleen from transplanted bnx animals (Kamel-Reid and Dick 1988). Quantitative analysis comparing band intensities at different exposures indicated that human cells accounted for approximately 0.1-1.0% of the spleen and bone marrow. No evidence of engraftment was observed in the CBA/J mice confirming the importance of bnx mice as recipients.

Although the DNA analysis provided evidence that human cells had engrafted the animals, it was important to determine whether progenitor cells had also seeded to the bone marrow or spleen of the animal. The availability of quantitative colony assays for human myeloid progenitors enabled us to determine whether any progenitors had engrafted the animals. Bone marrow from engrafted animals was plated in methylcellulose colony assays under conditions that were selective for the growth of human granulocyte/macrophage progenitors (CFU-GM). Significant numbers of human progenitors were detected in the animals 14 days after transplant.

Table 1. Human progenitor colonies from transplanted normal and immunedeficient mice.

Mouse strain	Time post transplant (days)	total hCFU-M per mouse
CBA/J	14	0
bnx	5	5×10^4
bnx	14	2.0×10^5
bnx	35	1.6×10^5
bnx	100	8×10^4
scid	14	360

Mice were engrafted with normal human bone marrow as described in the text. After the length of time indicated the animals were sacrificed, bone marrow was obtained by aspirating both femurs and tibia, counted and plated in methylcellulose cultures. The total number of progenitors per mouse was calculated from the number of colonies that grew on the plates times the total amount of bone marrow in a mouse. The input number of CFU-M injected into the mice was approximately $5\text{-}10 \times 10^3$.

PCR technique was used to detect the presence of human chromosome 17 α-satellite DNA sequences in the individual hematopoietic colonies that grew in these cultures. Taken together, their dependence on human growth factors and the molecular identification of human DNA confirms that the in vitro colonies arising from the engrafted animals are indeed human. As expected from the DNA analysis no progenitors could be detected in CBAJ mice transplanted under the same conditions. A limited number of scid animals transplanted according to the same protocol have also been analyzed; these animals were engrafted with very low numbers of progenitor cells. While the reason for this difference in the level of engraftment is unknown it may be due to the higher NK or LAK activity in scid animals relative to the bnx mice. The studies described above clearly indicate the importance of the recipient animal for the successful engraftment.

It was however interesting that the exogenous supply of human IL-3 and GMCSF was not required because animals receiving factor were engrafted to the same level as animals that received either no pump or a pump filled with saline. Several possibilities could explain this result. The human cells could be responding to human factor secreted by stromal cells or other hematopoietic cells that had also engrafted the mouse. Alternatively they could be responding to some murine factor that is not species specific or to the murine bone marrow microenvironment. This latter possibility is quite intriguing because it is known that the earliest cells in the hematopoietic system of both mouse and man are regulated by cell-cell interactions with stromal cells or other microenvironmental cell populations.

Examination of the kinetics of engraftment of the human progenitors in animals sacrificed at various times up to three months after transplant suggested that an earlier cell type had engrafted these animals. The total human progenitors in the bone marrow of the animals increased significantly, reaching maximum levels by day 14 posttransplant when the number rose >40 fold over the input number. Since in vitro progenitors have little self-renewal potential (Till and McCulloch 1980) this large increase implies an earlier stem cell has seeded the bone marrow and spleen and is responsible for the continuous production of human progenitors. The continued production of progenitors for greater than 3 months post transplant lends further support to the suggestion that an earlier cell type had engrafted the hematopoietic tissues of the recipient animal.

While our work indicates that myeloid progenitor cells can engraft immune-deficient mice, two other groups have recently reported engraftment of human lymphoid cells into immune-deficient mice carrying a mutation at the scid locus. scid animals are T and B cell deficient and make very low or undectable quantities of immunoglobulin. Mosier et al (1988) transplanted human peripheral blood leukocytes (PBL) intraperitoneally (IP) into the scid mice. After several weeks significant levels of human T and B cells and a small number of macrophages could be detected in the lymphoid organs and the peripheral circulation of the animals. The engrafted cells were maintained for longer than 20 weeks although there were distortions in the normal ratios of T cell subsets. High levels of human serum immunoglobulin were also measured in the animals. A large proportion of the animals developed EBV positive tumors. Among other things these animals open the way for an important model system to study EBV mediated leukemogenesis.

Using a completely different approach McCune et al (1988) surgically implanted human fetal thymus and lymph node into scid animals; following this the animals were transplanted with human fetal liver as a source of stem cells. The fetal thymus and lymph node provided the microenvironment to permit differentiation of significant numbers of mature functional T and B cells which then migrated into the peripheral blood of the animal. Both $CD4^+$ and $CD8^+$ cells were found in the blood; in addition the B cells secreted large amounts of human serum immunoglobulin. Evidence was also presented indicating the migration of cells from the fetal liver to the fetal thymus and through the thymus to the blood. The donor T and B cells seemed to disappear after about 10 weeks post transplant. While animals with human lymphoid systems have many uses, their efficacy as a potential small animal model for the early events associated with AIDS infection has now also been demonstrated. The same group has recently been able to infect the human lymphoid cells from engrafted mice with high titers HIV (Namikawa et al 1988). Not only were the lymphoid cells infected but evidence was also provided suggesting the virus was spreading in the infected animals.

Genetic Manipulation

The ability to transfer new genetic information into hematopoietic cells provides a new and promising approach to addressing questions concerning stem cell commitment and proliferation (Dick et al 1986a; Dick 1987). For example, it should be possible to introduce key growth regulatory genes such as oncogenes into various hematopoietic cell populations and determine how the developmental program of the human stem cells has been perturbed. Furthermore, insertion of new genetic sequences at unique chromosomal positions provides a means of marking the progeny of an individual stem cell clone to assess the developmental potential of that particular stem cell. The introduction of clinically relevant genes into human hematopoietic stem cells offers a novel approach for correcting a variety of human genetic diseases (Anderson 1984; Belmont and Caskey 1986). The problem with all these approaches in the human has been the lack of a suitable assay to study the earliest cells into which genes can be introduced. We and others have developed conditions for high efficiency gene transfer into progenitors that can be assayed by in vitro colony and long term culture assays using retrovirus vector mediated gene transfer (Dick et al 1986b; Hock and Miller 1986; Hogge and Humphries 1987; Laneuville et al 1988). We have now combined this genetic approach with the in vivo assay for human hematopoietic cells by transplanting animals with bone marrow which had previously been infected with a retrovirus vector carrying the dominant selectable neo gene (unpublished). After infection, the cells were preselected to enrich the population for infected progenitor cells and transplanted into <u>bnx</u> animals. After 35 days of engraftment, the bone marrow from the transplanted animal was plated under human conditions, colonies grew up and these human colonies were then assayed for the presence of neo sequences by PCR on individual colonies. As shown in Table 2, a significant number of these colonies contained the neo gene.

Table 2. Gene transfer into progenitor cells able to engraft immune-deficient <u>bnx</u> mice

Number of CFU-GM Analyzed	Number of CFU-GM positive for neo gene	Gene Transfer Efficiency
10	4	40%

Mice were engrafted with bone marrow that had been preincubated in hematopoietic growth factors for 24 hours, cocultivated with a Neo vector producing cell, and preselected in G418. After 35 days of engraftment the bone marrow was plated in methylcellulose assays under conditions that supported human progenitors. Individual colonies were picked and analyzed by PCR for the presence of the neo gene.

While we cannot yet confirm that these progenitors arose from a common earlier stem cell and are not merely the progenitors which were infected directly, it should be possible to establish the clonal relationship of individual hematopoietic colonies by employing the PCR technique to clone out the flanking sequences around each retrovirus integration site (Frohman et al 1988). One of the long term applications of this technology once a complete hematopoietic system has been established in animals with both lymphoid and myeloid cells would be to perform a similar lineage analysis on infected human stem cells as has been successfully employed with murine hematopoietic stem cells (Dick et al 1985; Lemischka et al 1986; Snodgrass and Keller 1987). Such an analysis should provide insight into the nature of the stem cells engrafting the animal and identification of the cells that comprise the human stem cell hierarchy.

Models of Human Leukemia

One of the major problems in studying human leukemias is the difficulty in expanding leukemic cell populations from patients. Often if a leukemic cell line is derived after a long period of time in culture extensive selection for growth and culture has occurred and may result in a population of cells which does not accurately reflect the primary leukemia from which it arose. Engraftment of primary leukemias in immunedeficient animals may provide a more suitable environment for the growth of human leukemic cells which is less selective than in vitro culture conditions. Furthermore, it should also be possible to genetically manipulate these leukemic cell populations using the gene transfer approaches outlined above to examine the role of oncogenes in the leukemogenic process. In order to examine this possibility we have engrafted a number of human leukemic cell lines into both <u>bnx</u> and <u>scid</u> animals in order to determine how an autonomously growing acute leukemia would behave in the immunedeficient animals. We have recently obtained evidence that a lymphoid leukemic cell line derived from a pediatric patient in relapse from ALL can engraft <u>scid</u> mice. For the first 4 weeks post transplant significant growth occurred only in the bone marrow, but between 4 and 8 weeks the cells began to proliferate in the spleen as well. After 10 weeks leukemic cells were seen in the blood and massive infiltrates occurred in many tissues; the animals die after 12 weeks. These experiments suggest it will be possible to establish animal models of human leukemia. Interestingly, this cell line grows much better in scid than bnx recipients for reasons that are not known. Preliminary experiments have also been performed on transplantation of primary leukemic cell populations into both scid and bnx animals. Animals examined several weeks after transplant with primary human CML, AML, and ALL all appear to have been engrafted based on DNA analysis.

SUMMARY

Our knowledge of the organization and regulation of the murine hematopoietic stem cell hierarchy has, in large measure, been the result of the development of in vivo reconstitution assays for cells of various levels of differentiation. For example, many basic stem cell concepts were established using CFU-S as a paradigm. Although there are no clonal assays for the earliest pluripotential cells, they can nevertheless be quantitated based on their ability to reconstitute animals with a deficient hematopoietic system. A similar understanding of the human hematopoietic system has lagged because of the lack of a suitable assay for human stem cells. Using several different approaches it has become possible to transplant human hematopoietic cells into immune-deficient mice. Since both lymphoid and myeloid cells have been engrafted, these experiments may lay the foundation for an assay for human pluripotent stem cells. Moreover it should now become possible to establish experimental animal models of a large number of human hemopathies including leukemia.

REFERENCES

Anderson WF (1984) Prospects for human gene therapy. Science 226: 401-409

Andrews RG, Singer JW, Bernstein ID (1986) Monoclonal antibody 12-8 recognizes a 115 kd molecule present on both unipotent and multipotent hematopoietic colony-forming cells and their precursors Blood 67:842-845

Andriole GL, Mule JJ, Hansen CT, Linehan WM, Rosenberg SA (1985) Evidence that lymphokine-activated killer cells and natural killer cells are distinct based on an analysis of congenitally immunodeficient mice. J Immunol 135: 2911-2913

Barr RD, Whang-Peng J, Perry S (1975) Hemopoietic stem cells in human peripheral blood. Science 190: 284-285

Belmont JW, Caskey CT (1986) Developments leading to human gene therapy. In : Kucherlapati R (ed) Gene Transfer pp 411-441, Plenum, New York

Bernstein A, Dick J, Huszar D, Robson I, Magli M-C, Estrov Z, Freedman M, Phillips R (1986) Genetic engineering of mouse and human stem cells. Cold Spring Harbor Symp Quant Biol 51: 1083-1091

Dick J (1987) Retrovirus mediated gene transfer into hematopoietic stem cells. Annals New York Acad Sci 507: 242-251.

Dick J, Magli M-C, Huszar D, Phillips R, Bernstein A (1985) Introduction of a selectable gene into primitive stem cells capable of long term reconstitution of the hemopoietic system of W/W^v mice. Cell 42: 71-77

Dick J, Magli M-C, Phillips R, Bernstein A (1986a) Genetic manipulation of hematopoietic stem cells. Trends in Genetics 2: 165-170

Dick J, Magli M-C, Estrov Z, Freedman M, Phillips R, Bernstein A (1986b) Retrovirus-mediated gene transfer and expression in murine and human stem cells. In: Gale RP, Champlin R (eds) Progress in Bone Marrow Transplantation UCLA Symposium on Molecular and Cellular Biology, New Series, Vol 53 Liss, New York, pp 951-962

Dorshkind K, Pollock SB, Bosma MJ, Phillips RA (1985) Natural killer (NK) cells are present in mice with severe combined immunodeficiency (scid). J Immunol 134: 3798

Fodstad O, Hansen CT, Cannon GB, Statham CN, Lichtenstein GR, Boyd, MR (1984) Lack of correlation between natural killer activity and tumor growth control in nude mice with different immune defects. Cancer Res 44: 4403-4408

Frohman MA, Dush MK, Martin GR (1988) Rapid production of full-length cDNAs from rare transcripts: Amplification using a single gene-specific oligonucleotide primer. Proc Natl Acad Sci USA 85: 8998-9002

Gartner S, Kaplan H (1980) Longterm culture of human bone marrow. Proc Natl Acad Sci USA 77:4756

Hock RA, Miller AD (1986) Retrovirus-mediated transfer and expression of drug resistance genes in human haematopoietic progenitor cells. Nature 320: 275-277
Hogge DE, Humphries RK (1987) Gene transfer to primary normal and malignant human hemopoietic progenitors using recombinant retroviruses. Blood 69: 611-617
Kamel-Reid S, Dick J (1988) Engraftment of immune-deficient mice with human hematopoietic stem cells. Science 242: 1706-1709
Laneuville P, Chang W, Reid S, Fauser A, Dick J (1988) High efficiency gene transfer and expression in normal human hematopoietic stem cells. Blood 71: 811-814
Leary AG, Ogawa M (1987) Blast cell colony assay for umbilical cord blood and adult bone marrow progenitors. Blood 69:953-956
Lemischka I, Raulet D, Mulligan R (1986) Developmental potential and dynamic behavior of hematopoietic stem cells. Cell 45: 917-927
Louwagie A, Verwilghen R (1970) Growth of haemopoietic spleen colonies after grafting of human bone marrow in mice. Nature 225: 383
McCune JM, Namikawa R, Kaneshima H, Shultz LD, Lieberman M, Weissman IL (1988) The SCID-hu mouse: Murine model for the analysis of human hematolymphoid differentiation and function. Science 241: 1632-1639
Mosier DE, Gulizia RJ, Baird SM, Wilson DB (1988) Transfer of a functional human immune system to mice with severe combined immunodeficiency. Nature 335: 256-259
Namikawa R, Kaneshima H, Lieberman I, Wiessman I, McCune J (1988) Infection of the SCID-hu mouse by HIV-1. Science 242: 1684-1686
Rowley SD, Sharkis SJ, Hattenburg C, Sensenbrenner LL (1987) Culture from human bone marrow of blast progenitor cells with an extensive proliferative capacity. Blood 69:804-808
Snodgrass R, Keller G (1987) Clonal fluctuation within the hematopoietic system of mice reconstituted with retrovirus-infected cells. EMBO J 6: 3955-3960
Till JE, McCulloch EA (1980) Hemopoietic stem cell differentiation. Biochem Biophys Acta 605: 431-459
Waye JS, Willard, HF (1987) Nucleotide sequence heterogeneity of alpha satellite repetitive DNA: a survey of alphoid sequences from different human chromosomes. Nuc Acids Res 15: 7549

VII. Potential Uses: Model For Studies of Immune Function

Scid Mice in Reproductive Biology

D.A. Clark, D.K. Banwatt, J. Manuel, G. Fulop[a], B.A. Croy[b]

Molecular Virology-Immunology Program, McMaster University, 1200 Main St. West, Hamilton, Ontario, Canada L8N 3Z5

Introduction

Immunodeficient mice of genotype scid/scid have been used to test two major theories relating immunology to reproduction. The first idea is that fetuses express antigens foreign to the mother and may be aborted by a process analogous to classical allograft rejection, particularly when there is subnormal maternal suppressor cell activity at the implantation sites (Clark 1984; Clark et al. 1986a, 1989a). The second idea is that maternal T cells recognize foreign antigens on the placenta and respond by secreting growth factors such as GM-CSF and IL-3 that stimulate placental trophoblast growth and function thereby preventing abortion (the immunotrophic hypothesis)(Wegmann 1984, 1987, 1988; Athanassakis et al. 1987).

Relevant background

There are several models of failure in mice, and two have been extensively investigated for an immunological component. Transfer of Mus caroli X Mus caroli blastocysts into the laboratory mouse Mus musculus (the C-M Model) leads to implantation followed by infiltration by maternal non-specific killer T cells and resorption; more rapid failure occurs if the female recipients have been pre-immunized against Mus caroli (Croy et al. 1982, 1985). However, pregnancy failure also occurs when Mus caroli blastocysts are transferred to recipients of genotype scid/scid (on the C.B-17 background) where normal T and B cell-dependent immune responses would not have occurred (Croy et al. 1987; Crepeau et al. 1988). A further challenge to the concept that abortion may be due to classical graft rejection has been provided by studies of the second model, CBA/J females mated to DBA/2 males. This mating combination suffers a high rate of abortion, also associated with infiltration of the conceptus by NK and CTL type cells (Chaouat et al. 1989; de Fourgolles and Baines 1987; Gendron and Baines 1988). However, this is prevented by injection of antibody to the asialo-GM1 marker (associated with lymphokine activated killer cells, some NK cells and monocytes) but not antibody against the IL-2 receptor that blocks CTL generation and classical allograft rejection (Clark et al., in preparation) or monoclonal antibodies to Lyt 2 and L3T4 (Chaouat et al. 1988). In addition, it was noted that fetal death occurs prior to the ontogenetic development of sensitivity to lysis by CTL (Croy and Rossant 1987). Taken together, these data have led to rejection of the concept of the "fetal allograft" and the emergence of the thesis that abortion may be due to rejection by non-specific natural effector cells (Clark et al. 1987a). In this context it is interesting to note scid/scid mice like nu/nu mice retain activity of natural effectors such as NK cells and unirradiated scid/scid mice may resist (reject ?) foreign bone marrow grafts (Rolstad and Benestad 1984; Carlson and Marshall 1985; Lauzon et al. 1986).

---

a) Division of Hematology-Oncology, Hospital for Sick Children, Toronto, Ontario.
b) Department of Biomedical Sciences, Ontario Veterinary College, Guelph, Ontario.

Current Topics in Microbiology and Immunology, Vol. 152

Investigation of the C-M and CBA-DBA/2 models of pregnancy failure led to the discovery of local active suppression in the maternal uterus during pregnancy. Two types of suppressor cells were defined. During the period prior to implantation and for the first 4 days after implantation, a population of large-sized (modal sedimentation velocity 6-9 mm/hr) Lyt 2^+ suppressor cells capable of non-specifically down-regulating the CTL response was found (Brierley and Clark 1987). These type A suppressor cells appeared to be activated or recruited by the hormonal changes of pregnancy (Brierley and Clark 1987). In vivo injection of monoclonal anti-Lyt 2 antibody increased the rate of abortion in the CBA-DBA/2 system (Clark et al. 1989 and submitted). With formation of the distinct fetus and placenta 5 days after implantation, a second type B suppressor cell appeared. This small-sized (modal sedimentation velocity 3 mm/hr) cell appeared to be a granulated lymphocyte, lacked T cell markers, bore Fc receptors, and released a potent suppressor factor blocking activation of effector cells by cytokines such as IL-2 (Slapsys et al. 1986; Clark et al. 1986a). The suppressor factor has recently been shown to be closely related to transforming growth factor-beta 2 and to be able to inhibit the activation of both specific CTL and natural effector cells (Clark et al. 1988, 1989a,b). A deficiency of these suppressor cells and of soluble suppressor factor activity has been associated with subsequent abortion (Clark et al. 1986a, 1989a). Since type B suppressor activity is activated in response to soluble signals from fetal trophoblast cells that form the feto-maternal interface, a defect in trophoblast function in pregnancies destined to abort has been postulated (Slapsys et al. 1988).

Some fetal trophoblast cells at the maternal-fetal interface express Class 1-like MHC antigens, and in the CBA-DBA/2 model, generation of an immune response against paternal $H-2^d$ MHC antigens (by vaccination or a pregnancy by the more immunogenic Balb/c strain male) reduced the abortion rate (Chaouat et al. 1985, Chaouat et al. 1988). Associated with successful immunisation was an increase in type B suppressor activity (Clark et al. 1987b). This has led to the idea that antigen-specific immune responses might benefit pregnancy. One specific model for the effect is the immunotrophic model wherein T cells are proposed to react to Class 1 trophoblast antigens and secrete lymphokines that act as growth factors and stimulants for trophoblast (Wegmann 1987, 1988; Athanassakis et al. 1987; Chaouat et al. 1989). The adoptive transfer of protection against abortion using anti-$H-2^d$ specific serum factors (antibody ?) did not exclude a role for maternal T cells. The extent to which antigen-specific immune responses were essential for successful (non-abortive) pregnancies in general was unclear, however, particularly in view of the success of syngeneic pregnancies in inbred strains of mice and rats and of allogeneic pregnancies in immunodeficient mice (Slapsys et al. 1986; Croy et al. 1987).

<u>Pregnancy failure in SCID mice</u>

Table 1 summarizes data on resorption rates in naturally mated <u>scid/scid</u> females on both the C.B-17 (Balb/c) background and when crossed to C57Bl/6 <u>bg/bg</u> stock. <u>Scid/scid</u> X <u>scid/scid</u> matings, carried out in a defined flora barrier facility (<u>Lactobacillus</u> comprises the flora), had a 24 % resorption rate, almost as high as the CBA-DBA/2 system. However, the high resorption rate was not attributable to the <u>scid/scid</u> genotype; an even higher resorption rate occurred in the C.B-17 <u>+/+</u> X <u>+/+</u> matings where normal T cell function was expected. The Balb/c background is known to be associated with a small litter size (Croy et al. 1989) and when the scid/scid genotype was crossed with the C57Bl/6 stock carrying the bg/bg low NK genotype, a low resorption rate was seen. +/scid.bg/bg females also had a low resorption rate (2/29=7%) and the single scid/scid.bg/+ examined had 1/11 (9%) resorptions. These data imply it is the background genotype and not the scid or bg genes that determine resorption rates in the immunodeficient host.

Table 1

Mating	Number of mice	Resorptions	Expected litter size
scid/scid X scid/scid (C.B-17)	12	24/100 (24%)	6.3[a]
+/+ X +/+ (C.B-17)	9	24/62 (39%)*	4.2
scid/scid.bg/bg X scid/scid. bg/bg (C57Bl/6)	8	7/70 (10%)	7.9

a) Calculated from (100-24)/12
*) Significant increase ($P< 0.05$)

Further analysis of the high resorption rate in the SCID-C.B-17 system showed resorption rates of 25-27% mating scid/scid males with +/+ females and vice versa, but when an F1 cross (scid/+ x scid/+) was tested, a lower litter size typical of +/+ x +/+ occurred. The latter result could not be explained by a higher rate of resorption of +/+ embryos. The data taken together suggested embryos of +/+ genotype within a +/+ or +/scid female were associated with an increase in the resorption rate of embryos of all genotypes.

To further investigate possible mechanisms of resorption in immunodeficient scid/scid females, we decided to test three parameters that might be related to pregnancy failure:

(1) Deficiency of growth factor production in decidua
(2) Deficiency of suppressor cell activity
(3) Presence of natural effector cells capable of attacking trophoblast.

Placental cell growth factors from decidua of pregnant SCID mice

Decidual cell supernatants have been prepared from pregnant uteri by removing the fetoplacental unit (with a small amount of adherent decidua), and scraping the remaining decidua from the underlying muscle. The fragments are then passed through a 60 mesh screen in culture medium, washed, and incubated 48 hours at 37° C in a CO_2 incubator. Incubation volume is 1 ml per donor uterus which corresponds to approximately 6-8 implants (viable and resorbing) per ml. The supernatants are harvested and then tested for suppressive activity (to be described below) and growth stimulating activity (Athanassakis et al. 1987). Briefly, placentae from day 12.5 of pregnancy are disaggregated and the cells plated in culture wells together with medium supplemented with GM-CSF or with decidual cell supernatants. After 3 days, proliferation of the placental cells is assessed by ^{3}H-thymidine uptake. We used a supernatant conditioned by rat spleen cells stimulated with concanavalin A as a source of GM-CSF and related growth factors. As illustrated by Table 2, we have not found any evidence for a growth factor deficiency in scid/scid (C.B-17) mice.

Table 2

Addition to culture	Isotope incorporation (cpm ± sem)	Percent of control
Medium alone	1,518 ± 40	0 %
Rat Con-A spleen supernate	2,862 ± 131	100 %
Control pregnant[a] 1/200 decidual cell supernate	2,053 ± 142	39 %
nu/nu pregnant[b] 1/200 decidual cell supernate	2,014 ± 163	37 %
SCID pregnant[c] 1/200 decidual cell supernate	2,082 ± 103	42 %

a) C3H/HeJ mated to DBA/2 which have a low abortion rate (10 %) represent our standard control supernatant. Supernatants were tested at several dilutions and the values for 1/200 were representative.
b) CD1 nu/nu mice mated to DBA/2 were used.
c) scid/scid (C.B-17) mated to scid/scid were used. In this set of mice, the abortion rate exceeded 30 %.

Suppressor cell activity

The two types of suppressor cells found in the uterus of immunocompetent mice during allopregnancy have already been described. Since the type A cell (pre- and peri-implantation) expresses Lyt 2, we predicted pregnant scid/scid females should be deficient in this activity compared to C.B-17 +/+. As shown in Table 3, this proved correct. Pregnant scid/scid mice had fewer cells and less suppressive activity was recovered per uterus. Interestingly, the reduction was not always as striking as one might have anticipated. No Lyt phenotyping of the residual pre-implantation suppressive activity from scid/scid mice has yet been done. C.B-17 +/+ mice have a higher abortion rate than scid/scid (Table 1) which a deficiency of type A suppressor cells cannot explain. The deficiency might be relevant, however, to a defect in implantation of transferred blastocysts that has been reported in pseudopregnant C.B-17 scid/scid females (Crepeau and Croy 1988).

The type B suppressor activity was assessed by quantitation of soluble suppressive activity in supernatants prepared from decidua. Several observations were made:

(1) Soluble suppressor activity was present in both pregnant C.B-17 scid/scid and pregnant scid/scid.bg/bg consistent with its non T, non NK nature (Slapsys et al. 1986).

(2) The suppressive activity was neutralized by rabbit anti-TGF-beta antibody and by anti-TGF-beta 2 but not by anti-TGF-beta 1 (Clark et al 1989b).
(3) Those mice with high abortion rates generated supernatants with less suppressive activity. Thus, as in other systems, a deficiency of trophoblast-dependent type B suppressor activity correlated with resorption.

Table 3

Induction of pre-implantation suppressor cells by:	Number of experiments	Cell yield per uterus	Suppressive activity per uterus
		scid/scid as % C.B-17 +/+ (range)	scid/scid as % C.B-17 +/+ (range)
Hormonal pseudopregnancy (Brierley and Clark 1987)	3	42 % (8.6-92 %)	33 % (5.8-80.9 %)
Mating	1	12.5 %	38 %

Killer cell activity

Much of the work on suppressor cell activity has employed antigen-specific CTL-type immune responses as the target system for assay in vitro. In view of the potential importance of antigen-non-specific natural effectors in abortion, experiments were done to test for killer cells capable of lysing the Be6 mouse placental cell line that has trophoblastic features (Slapsys et al. 1988). Like freshly isolated trophoblast, this cell line is resistant to antigen-specific CTL and NK cell lysis, but is sensitive to LAKs and to a cell population found in the spleens of immunodeficient nu/nu mice (Clark et al. 1986b; Drake and Head 1988). Spleen cells from scid/scid and scid/scid.bg/bg mice were able to lyse Be6 targets but the scid/scid.bg/bg cells lacked NK activity as assessed using YAC targets (0-5.2% lysis vs 10-54% lysis by C.B-17-scid/scid). This data indicated that SCID mice may possess killer cells capable of damaging trophoblast and that these may not be classical NK cells. Multiplication of % lysis X spleen cell number was used to provide a measure of total cytolytic activity per spleen.

It was important to know however what happened to anti-trophoblast killer cells in pregnant scid/scid mice. A similar analysis was carried out. It was noted that spleen cell numbers increased dramatically (5-10x) with pregnancy. % lytic activity on a per spleen cell basis was decreased slightly in some mice, but with the increase in spleen cell number, total lytic activity per spleen increased dramatically. Although the data were limited, when total lytic activity, suppressive activity, and resorption rate were compared between mice, it was noted that high resorption rate, low suppression, and high total killer cell activity tended to associate. The magnitude of local suppression appeared to be a better correlate of resorption than splenic killer content. The phenotype of the killer cells and their occurrence in the uterus at sites of resorption is currently under study. Preliminary studies suggest the effector is positive for the asialo-GM1 marker and may be present at increased concentration (4-6x) in decidua at resorption sites.

Conclusions

Studies using scid/scid mice have challenged the hypothesis that spontaneous resorption (abortion) is a manifestation of antigen-specific allograft rejection. An obligatory immunotrophic role for maternal T cells in pregnancy has also been disproved. However, experiments using scid/scid mice support a new hypothesis, that abortion is mediated by natural effector cells that are neither T nor B cells but which can attack trophoblast in a setting of low suppressor activity. The scid/scid mouse has been very useful in defining the lineage of unique effector and regulatory cells that may determine the outcome of pregnancy.

References

Athanassakis I, Bleackley RC, Paetkau V, Guilbert L, Barr J, Wegmann TG (1987) The immunostimulatory effect of T cells and T cell lymphokines on murine fetally-derived placental cells. J Immunol 138:37-44

Brierley J, Clark DA (1987) Characterization of hormone-dependent suppressor cells in the uterus of pregnant and pseudopregnant mice. J Reprod Immunol 10:201-218

Carlson GA, Marshall ST (1985) H-2 associated natural resistance against non-syngeneic cells: possible functions of the major histocompatibility complex outside the realm of classical immunology. In: Skamene E (ed) Genetic Control of Host Resistance to Infection and Malignancy. Alan R Liss, New York, p 701

Chaouat G, Kolb JB, Kiger N, Stanislawski M, Wegmann TG (1985) Immunological consequences of vaccination against spontaneous abortion in mice. J Immunol 14:1594-1598

Chaouat G, Clark DA, Wegmann TG (1988) Genetic aspects of the CBA x DBA/2 and B10 x B10.A models of murine pregnancy failure and its prevention by lymphocyte immunisation. In: Beard RW, Sharp F (eds) Early Pregnancy Failure: Mechanisms and Treatment. Peacock Press, Ashton-under-Lyne (UK), p 89

Chaouat G, Menu E, Szekeres-Bartho J, Rebut-Bonneton, Bastany P, Kinsky R, Clark DA, Wegmann TG (1989) Lymphokines, steroids, placental factors and trophoblast intrinsic resistance to immune cell mediated lysis are involved in pregnancy success or immunologically mediated pregnancy failure. In: Wegmann TG, Gill TJ (eds) Molecular Immunology of the Feto-maternal Interface. Oxford Univ Press, New York (in press)

Clark DA (1984) Local suppressor cells and the success or failure of the "fetal allograft". Ann Immunol (Paris) 135D:321-324

Clark DA, Damji N, Chaput A, Daya S, Rosenthal KL, Brierley J (1986a) Decidua-associated suppressor cells and suppressor factors regulating interleukin 2: their role in the survival of the "fetal allograft". In: Cinader B, Miller RG (eds) Progress in Immunology VI. Academic Press, New York, p 1089

Clark DA, Croy BA, Rossant J, Chaouat G (1986b) Immune presensitization and local intrauterine defences as determinants of success or failure of murine interspecies pregnancies. J Reprod Fertil 77:633-643

Clark DA, Croy BA, Chaouat G, Wegmann TG (1987a) Immunological and paraimmunological mechanisms in spontaneous abortion: recent insights and future directions. J Reprod Immunol 12:1-12

Clark DA, Kiger N, Guenet J-L, Chaouat G (1987b) Local active suppression and successful vaccination against spontaneous abortion in CBA/J mice. J Reprod Immunol 10:79-85
Clark DA, Falbo M, Rowley RB, Banwatt D, Stedronska-Clark J (1988) Active suppression of host-versus-graft reaction in pregnant mice. IX. Soluble suppressor activity obtained from allopregnant mouse decidua that blocks the response to interleukin 2 is related to TGF-beta. J Immunol 141;3833-3840
Clark DA, Head JR, Drake B, Fulop G, Brierley J, Manuel J, Banwatt D, Chaouat G (1989a) Role of a factor related to tranforming growth factor beta-2 in successful pregnancy. In: Wegmann TG, Gill TJ (eds) Molecular Immunology of the Feto-maternal Interface. Oxford Univ Press, New York (in press).
Clark DA, Harley C, Book W, Flanders K (1989b) Suppressor factor in murine pregnancy decidua is related to transforming growth factor beta-2. FASEB J (in press)
Crepeau MA, Croy BA (1988) Evidence that specific cellular immunity cannot account for death of Mus caroli embryos transferred to Mus musculus with severe combined immunodeficiency. Transplantation 45:1104-1110
Crepeau MA, Yamashiro S, Croy BA (1988) Anatomical and immunological aspects of fetal death in the Mus musculus/Mus caroli model of pregnancy failure. In: Beard RW, Sharp F (eds) Early Pregnancy Failure: Mechanisms and Treatment. Peacock Press, Ashton-under-Lyne (UK) p 69
Croy BA, Rossant J (1987) Mouse embryonic cells become susceptible to CTL lysis after midgestation. Cell Immunol 140:355-365
Croy BA, Rossant J, Clark DA (1982) Histological and immunological studies of post-implantation death of Mus caroli embryos in the Mus musculus uterus. J Reprod Immunol 4:277-293
Croy BA, Rossant J, Clark DA (1985) Effects of alterations in the immunocompetent status of Mus musculus females on the survival of transferred Mus caroli embryos. J Reprod Fertil 74;479-489
Croy BA, Crepeau MA, Yamashiro S, Clark DA (1987) Further studies on the transfer of Mus caroli embryos to immunodeficient Mus musculus. In: Chaouat G (ed) Reproductive Immunology: materno-fetal relationship. INSERM Colloque, Vol 154, Paris, p 101
Croy BA, Chapeau C, Reed N, Stewart IJ, Peel S (1989) Is there an essential requirement for bone marrow-derived cells at the fetomaternal interface during successful pregnancy? A study of pregnancies in immunodeficient mice. In: Wegmann TG, Gill TJ (eds) Molecular Immunology of the Feto-maternal Interface. Oxford Univ Press, New York (in press)
Drake BL, Head JR (1988) Murine trophoblast cells are susceptible to lymphokine-activated killer (LAK) cell lysis. Amer J Reprod Immunol Microbiol 16:114
Fourgolles AR de, Baines MG (1987) Modulation of the natural killer cell activity in pregnant mice alters the spontaneous abortion rate. J Reprod Immunol 11:147-153
Gendron RL, Baines MG (1988) Immunohistochemical analysis of decidual natural killer cells during spontaneous abortion in mice. Cell Immunol 113:261-267
Lauzon RJ, Siminovitch KA, Fulop GM, Phillips RA, Roder JC (1986) An expanded population of natural killer cells in mice with severe combined immunodeficiency (SCID) lack rearrangement and expression of T cell receptor genes. J Exp med 164:1797-1802
Rolstad B, Benestad HB (1984) The "natural resistance" to bone marrow allografts in normal and athymic nude rats. Rapid cytotoxic reactions both in vivo and in vitro. Eur J Immunol 14:793-799

Slapsys RM, Richards CD, Clark DA (1986) Active suppression of host-versus-graft reaction in pregnant mice. VIII The uterine decidua-associated suppressor cell is distinct from decidual NK cells. Cell Immunol 99:140-149

Slapsys RM, Younglai E, Clark DA (1988) A novel suppressor cell is recruited to decidua by fetal trophoblast-type cells. Regional Immunol 1:182-189

Wegmann TG (1984) Fetal protection against abortion: is it immunosuppression or immunostimulation? Ann Immunol (Paris) 135D:309-312

Wegmann TG (1987) Placental immunotropism maternal T cells enhance placental growth and function. Amer J Reprod Immunol Microbiol 15:67-69

Wegmann TG (1988) Maternal T cells promote placental growth and prevent spontaneous abortion. Immunol Letters 17:297-302

Supported by grants from MRC, Canada.

T Cell-Independent Macrophage Activation in Scid Mice

G.J. Bancroft*, R.D. Schreiber and E.R. Unanue

Department of Pathology, Washington University School of Medicine, St. Louis, MO. 63110, USA.

Macrophages play a critical role in immune regulation and resistance of the host to infection. Of particular importance is the activation of macrophages for microbicidal and tumoricidal activity and expression of cell surface molecules such as class II MHC (Ia) antigens (Unanue 1984). The ability of T cells to mediate these responses has been extensively documented and occurs via the production of specific cytokines, most notably IFN γ (Schreiber and Celada 1985; Unanue 1981). However, the scid mouse now provides an important new model for studying the pathways of macrophage activation (Bosma et al 1983). The scid mutation results in abnormal rearrangement of genes encoding antigen specific lymphocyte receptors and is characterized by the absence of functional T and B lymphocytes (Schuler et al 1986). This defect is specific for the lymphoid lineage since erythroid and myeloid functions are normal. We have therefore used the scid mouse to examine macrophage development in the absence of T cells. Our results demonstrate that macrophage activation can occur in a T cell-independent manner and that this may constitute an important mechanism of natural resistance to infection.

Macrophage function in scid mice:

Our initial experiments examined CB-17 scid (scid) mice for their macrophage content and the phenotypic and functional characteristics of these cells prior to infection. Despite the lack of mature T or B lymphocytes, adherent macrophages with typical morphology were readily obtained from the peritoneal cavity, spleen and liver. Background expression and tissue distribution of macrophages bearing Ia antigens was unaltered by the scid mutation. The frequency of Ia positive cells (determined by immunofluorescence) and the number of Ia molecules expressed (determined by radioimmunoassay) was similar in scid mice and their normal CB-17 coisogenic partner strain.

Analysis of peritoneal cells demonstrated a comparable number of adherent, Fc receptor bearing macrophages in uninfected CB-17 and scid mice ($1x10^6$ vs $0.9x10^6$ respectively). Futhermore, scid peritoneal macrophages responded normally to exogenous activating agents in vitro. Incubation with mitogen induced T cell supernatants or recombinant IFN γ enhanced Ia antigen expression and activated the cells for tumoricidal activity against P815 tumor targets. Macrophages derived from scid mice also expressed membrane bound IL-1 activity following adherence to plastic.

---

* Current address: Department of Clinical Sciences, London School of Hygiene and Tropical Medicine, Keppel St. London, England, WC1E. This study was supported by grants from the National Institutes of Health, Eli Lilly Research Laboratories, Monsanto Company and Genentech Inc.

Current Topics in Microbiology and Immunology, Vol. 152

These data are consistent with reports of normal antigen presentation by scid spleen cells in vitro (Czitrom et al 1985) and demonstrate that scid macrophages respond appropriately to T cell derived stimuli. We have no evidence for any intrinsic abnormality of macrophage function as a result of the scid mutation.

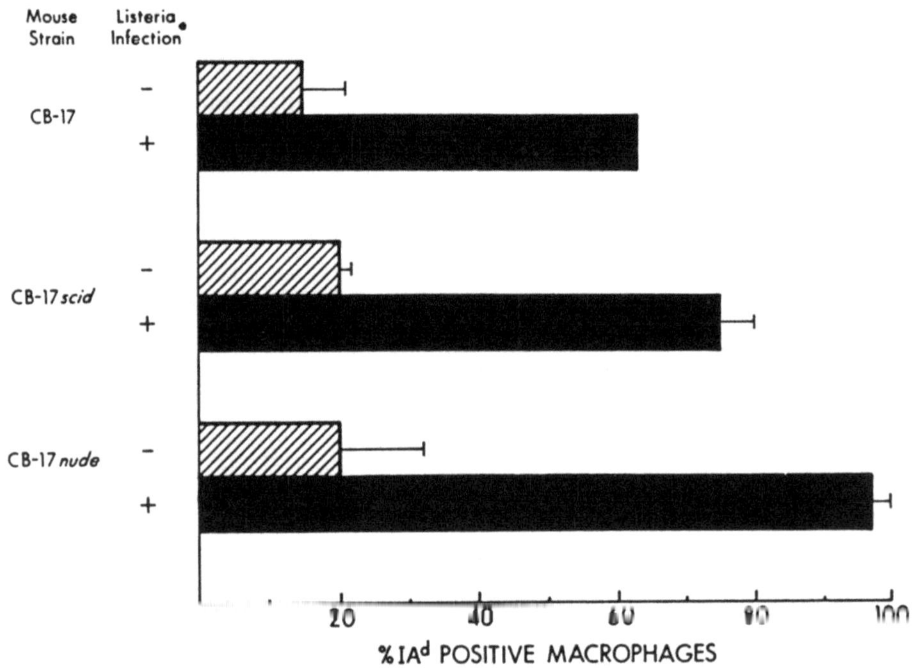

Figure 1: Effect of Listeria infection on macrophage Ia expression in CB-17 vs CB-17 scid vs CB-17 nude mice. Mice were injected i.p. with saline (shaded bars), or infected (closed bars) with $1x10^3$ Listeria (CB-17, CB-17 scid) or $5x10^3$ Listeria (CB-17 nude) and peritoneal macrophage Ia antigen expression assayed 3 days later by double layer immunofluorescence.

Macrophage activation during infection of scid mice:

We next examined the resistance of scid mice to infection with the intracellular bacterium Listeria monocytogenes (Listeria). In immunocompetent mice, Listeria is a classically T cell-dependent macrophage activating agent. Experimental infection is characterised by bacterial replication in the spleen and liver for 72-96 h followed by clearance of the organism from the host. Sterilizing immunity is mediated by activated macrophages and can be transferred to naive recipients by *Listeria* specific T cells (Hahn and Kaufmann 1981). Infection of immunocompetent CB-17 mice with $3x10^3$ *Listeria* i.p. resulted in peak levels of viable bacteria on day 4 ($3.1x10^6$ *Listeria*/spleen) with resolution of the infection by day 12. In contrast *scid* mice showed a progressive increase in bacterial load such that by day 12, spleen *Listeria* counts were over 6000 fold higher than in the CB-17 strain ($6.2x10^5$ vs $<10^2$ *Listeria*/spleen respectively). However, while *scid* mice were unable to eliminate *Listeria* from the spleen and liver, there was partial inhibition of bacterial growth, resulting in a chronic rather than acute infection.

The level of macrophage activation was also assessed in infected *scid* versus CB-17 mice by analysis of Ia antigen expression. Ia molecules are essential for the antigen presenting function of the macrophage and their expression is enhanced during activation of macrophages both *in vitro* and *in vivo*. As expected, infection with *Listeria* significantly increased Ia antigen expression by CB-17 peritoneal macrophages (Fig. 1). Surprisingly, infection of *scid* mice also increased the percentage of Ia bearing macrophages from a basal level of 21% to 76% at day 4 and to greater than 95% on day 12 (data not shown). Identical results were seen in *Listeria* infected CB-17 athymic nude mice which bear a T cell deficiency independent of the *scid* mutation (Fig. 1).

Thus, in the absence of mature T cells, macrophages were activated to express Ia antigens with similar kinetics and magnitude as observed in immunocompetent controls (Bancroft et al 1986). These results were obtained from individually assessed *scid* mice with undetectable levels of serum immunoglobulin. In addition, spleen cells from infected *scid* showed no T or B cell mitogen responses and did not express CD4 or CD8 antigens. Therefore, we have no evidence that infection with *Listeria* resulted in a leaky phenotype and the development of functional T cells.

Activation of *scid* macrophages *in vivo* was not restricted to changes in Ia antigen expression. Peritoneal macrophages from infected mice had an activated morphology and acquired tumoricidal activity. Other microorganisms could also mediate this response. As seen in Fig. 2, administration of killed *Corynebacterium parvum* or BCG enhanced Ia expression to similar levels as immunocompetent BALB/c controls. In contrast, injection of the T cell mitogen Con-A had no effect in *scid* mice but increased the frequency of Ia bearing BALB/c macrophages. Similar experiments using secondary antigenic challenge with the soluble T cell antigen KLH were also unsuccessful in *scid* mice. Clearly, while unable to respond via T cell-dependent mechanisms, macrophage activation could be achieved in *scid* mice via an alternative pathway in response to infection.

Histology of infection in *scid* mice:

Histological analysis performed in collaboration with Dr. K. DeSchryver confirmed our observations of macrophage activation in *scid* mice (DeSchryver et al 1988). Polymorphonuclear cell infiltration was evident in the liver and spleen but mononuclear cell hyperplasia was a conspicuous finding during infection. Mac-1 and ASGM1 bearing cells increased in the nodular and internodular areas of the spleen indicating an expansion of both macrophage and natural killer (NK) cell populations.

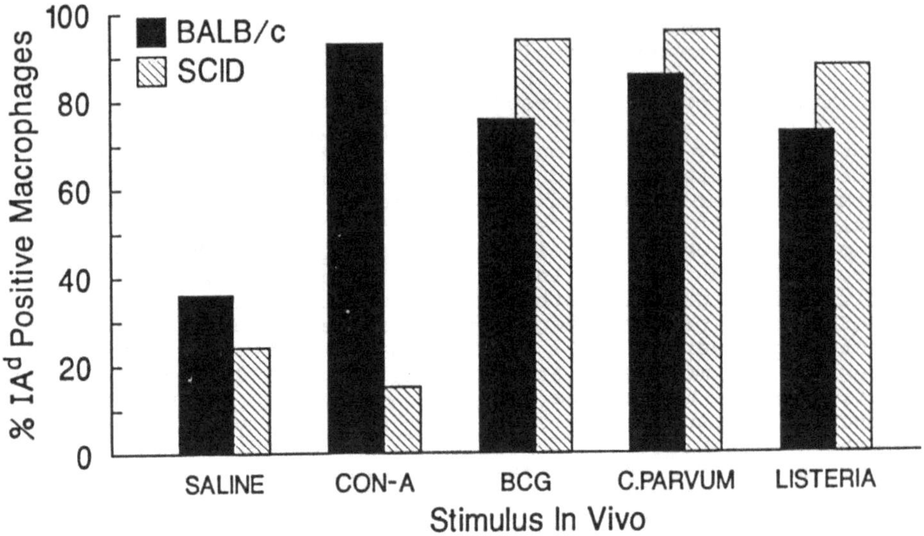

Figure 2: Effect of various activating agents _in vivo_ on macrophage Ia expression in BALB/c vs CB-17 _scid_ mice. BALB/c (closed bars) or CB-17 _scid_ mice (shaded bars) were injected i.p. with saline, Concanavalin-A (100 ug), BCG (10^7 CFU), _Corynebacterium parvum_ (15 ug) or $1x10^3$ _Listeria_ and peritoneal macrophage Ia antigen expression assayed 3 days later.

Widespread infiltration of Ia positive inflammatory macrophages was also observed in the liver, thyroid, kidney, pancreas, salivary glands and gastrointestinal tract. However, unlike immunocompetent controls, _Listeria_ infected _scid_ mice showed neither lymphoid follicles nor germinal centres and were unable to form organized granulomas in the liver or spleen. Our observations parallel those in athymic nude mice which also undergo a chronic infection and suggest that granuloma formation is essential for sterilizing immunity to _Listeria_.

An additional feature was expression of Ia antigens on non-lymphoid cells late in infection of scid mice. The epithelium and endothelium of the hepatic portal tracts (particularly the bile ducts), the pancreatic ducts and the salivary glands which were normally Ia negative now strongly expressed Ia antigens. In contrast, the acute infection observed in immunocompetent CB-17 mice did not result in systemic Ia expression.

The role of IFNγ in T cell-independent macrophage activation:

Studies of T cell-dependent macrophage activation have demonstrated a crucial role for IFNγ both in vivo and in vitro. This cytokine is a potent macrophage activating factor and can enhance microbicidal and tumoricidal activity as well as Ia antigen expression. The importance of IFNγ in the T cell-independent responses of scid mice was examined using two IFNγ specific monoclonal antibodies. The hamster anti-IFNγ antibody H22 binds to an epitope that requires the carboxy terminal region of IFNγ and in vitro neutralizes all its known biological activities (Schreiber et al 1985). In contrast, a control antibody H1.5 binds to an epitope that requires the amino terminus of IFNγ but has no neutralizing activity.

Initial experiments in immunocompetent mice demonstrated that administration of H22 (but not H1.5) antibody prior to Listeria infection suppressed host resistance and blocked macrophage activation (Bancroft et al 1987; Buchmeier and Schreiber 1985). The effectiveness of this strategy allowed us to test whether macrophage activation in scid mice was also IFNγ dependent. After 3 days of infection (Table 1), macrophages from saline treated scid mice showed high levels of Ia expression as previously described. Pretreatment with 100ug of the IFNγ neutralizing antibody H22 completely inhibited this response and enhanced bacterial growth in the spleens of scid mice by approximately 60 fold.

Table I: Effect of monoclonal anti-IFN γ antibodies on Listeria induced Ia expression in CB-17 scid mice in vivo

Pretreatment	Infection	Listeria per spleen log 10	%I-A^d macrophages
Saline	-	-	10%
Saline	+	4.6	76%
H22	+	6.4	14%
H1.5	+	4.9	68%

Mice received saline or 100 ug i.p. of either H22 or H1.5 anti-IFN γ monoclonal antibody 2 days prior to infection with 10^3 Listeria i.p. The number of viable Listeria in the spleen and the percentage of I-A^d positive peritoneal macrophages was determined 3 days after infection. Results are the mean of 3 mice per group.

Moreover, neutralization of IFNγ *in vivo* led to the death of infected *scid* mice within 3-5 days in contrast to the chronic infection normally observed. This was not mediated non-specifically by immune complexes since the non-neutralizing H1.5 antibody had no effect despite binding to IFNγ as strongly as the H22 reagent.

Synthesis of IFNγ was tested directly by culturing *scid* spleen cells *in vitro* and measuring secretion by ELISA assay. Addition of the T cell stimulant Con-A resulted in secretion of IFNγ (155 U/ml) and IL-2 by CB-17 but not *scid* cells (0 U/ml IFNγ) after 48 h . However, *scid* spleen cells could secrete IFNγ in a dose dependent manner when coincubated with heat killed *Listeria* (medium alone: 0 U/ml versus 1508 U/ml at 10^5 *Listeria* per well). Similar responses were also induced by other microbial agents including LPS (454 U/ml at 5 ug/ml) and killed *Corynebacterium parvum* (361 U/ml at 5 ug/ml). Thus, in the absence of functional T cells, *scid* mice produced IFNγ in response to microbial stimuli *in vitro* and utilized this cytokine for macrophage activation *in vivo* (Bancroft et al 1987).

Mechanisms of IFNγ production by *scid* mice:

The experiments presented above demonstrated the existence of a non-T cell source of IFNγ in *scid* mice. Natural killer cells are the only other cell type reported to secrete IFNγ and are enriched in *scid* mice. The involvement of NK cells in the response to *Listeria* was assessed by prior injection of rabbit anti-ASGM1 serum, a potent inhibitor of NK cell activity. At 3 days post infection, *scid* mice pretreated with saline showed an increase in the frequency of Ia bearing peritoneal macrophages (63% vs. 19% for uninfected controls) while treatment with 40ug anti-ASGM1 strongly inhibited this response (26% Ia positive). Anti-ASGM1 did not impair Ia induction *in vivo* following administration of recombinant IFNγ in *scid* mice or the T cell stimulant Con-A in CB-17 mice, excluding a direct toxic effect on host macrophages. Thus, inhibition of NK cell activity in *scid* mice impaired *Listeria* induced Ia expression by macrophages *in vivo*.

IFNγ production by *scid* spleen cells was also abolished by prior removal of adherent macrophages. Adherent cells themselves did not produce IFNγ and NK cell activity was confined to the nonadherent population. This suggested that following interaction with *Listeria*, a macrophage derived product was required for IFNγ synthesis. Analysis of *Listeria* induced supernatants from *scid* spleen cells revealed the rapid appearance of tumor necrosis factor (TNF) which preceded the secretion of IFNγ. Addition of a neutralizing monoclonal antibody specific for murine TNF (TN3-19.12) abolished IFNγ production (Sheehan et al. 1989). Direct addition of rTNFα enhanced the IFNγ response by up to 7 fold in the presence of suboptimal doses of *Listeria* but had no effect alone suggesting that TNF was required but not sufficient for IFNγ production *in vitro*. Finally, injection of the TN3-19.12 antibody prior to infection of *scid* mice inhibited Ia induction on peritoneal macrophages and enhanced bacterial replication in the spleen by 1500 fold. Thus, neutralization of TNF inhibited the synthesis of IFNγ and the activation of *scid* macrophages during *Listeria* infection.

Conclusions:

We have used mice bearing the scid mutation to examine T cell-independent pathways of macrophage activation. Our studies demonstrated that macrophage activation (as measured by Ia expression and tumoricidal activity) and partial control of bacterial replication could occur in the absence of T lymphocytes. This response was dependent upon the production of IFNγ *in vivo* and this could be measured directly by incubation of scid spleen cells with microbial products *in vitro*. Nevertheless, scid mice were unable to respond to T cell mitogens or for specific T cell responses to protein antigens.

These results indicate the presence of two distinct pathways of IFNγ production and macrophage activation *in vivo*. The first, as seen in the scid mutation, is independent of antigen specific immunity and responds rapidly to challenge with microorganisms. This pathway, mediated by NK cells and involving an obligate requirement for TNF, provides the host with a rapid source of IFNγ. We believe that activation of host macrophages via this T cell-independent process is critical for initial resistance to infection. The second pathway of IFNγ synthesis is mediated by antigen activated T cells following clonal expansion *in vivo* and is apparently required for granuloma formation and complete resolution of the infection. In conclusion, the scid mutation provides a unique opportunity to study the mechanisms of natural immunity and the relative roles of these two pathways of macrophage activation in resistance to infection.

References

Bancroft GJ, Bosma MJ, Bosma GC, Unanue ER (1986) Regulation of macrophage Ia expression in mice with severe combined immunodeficiency: induction of Ia expression by a T cell independent mechanism. J Immunol 137:4

Bancroft GJ, Schreiber RD, Bosma GC, Bosma MJ, Unanue ER (1987) A T cell independent mechanism of macrophage activation by interferon gamma. J Immunol 139:1104

Bosma GC, Custer RP, Bosma MJ (1983) A severe combined immunodeficiency mutation in the mouse. Nature 301:527

Buchmeier NA, Schreiber RD (1985) Requirement of endogenous interferon-gamma production for resolution of *Listeria monocytogenes* infection. Proc Natl Acad Sci USA 82:7404

Czitrom AA, Edwards S, Phillips RA, Bosma MJ, Marrack P, Kappler JW (1985) The function of antigen presenting cells in mice with severe combined immunodeficiency. J Immunol 134:2276

Deschryver-Kecskemeti K, Bancroft GJ, Bosma GC, Bosma MJ, Unanue ER (1988) Pathology of *Listeria* infection in murine severe combined immunodeficiency: a study by immunohistochemistry and electron microscopy. Lab Invest 58:698

Hahn H, Kaufmann SHE (1981) The role of cell mediated immunity in bacterial infections. Rev Infect Dis 3:1221

Schreiber RD, Celada A (1985) Molecular characterisation of gamma interferon as a macrophage activating factor. Lymphokines 11:87

Schreiber RD, Hicks LJ, Celada A, Buchmeier NA, Gray PA (1985) Monoclonal antibodies to murine gamma interferon which differentially modulate macrophage activation and antiviral activity. J Immunol 134:1609

Schuler W, Weiler IJ, Schuler A, Phillips RA, Rosenberg N, Mak TW, Kearney JF, Perry RP, Bosma MJ (1986) Rearrangement of antigen receptor genes is defective in mice with severe combined immune deficiency. Cell 46:963
Sheehan KCF, Ruddle L, Schreiber RD (1989) Generation and characterization of hamster monoclonal antibodies that neutralize murine tumor necrosis factor. J Immunol: submitted
Unanue ER (1981) The regulatory role of macrophages in antigenic stimulation II Symbiotic relationship between lymphocytes and macrophages. Adv Immunol 31:1
Unanue ER (1984) Antigen presenting function of the macrophage. Ann Rev Immunol 2:395

Pneumocystis carinii Pneumonia in *scid/scid* Mice

Leonard D. Shultz[1], Peter A. Schweitzer[1], Elaine J. Hall[1], John P. Sundberg[1], Suzanne Taylor[1], and Peter D. Walzer[2]

INTRODUCTION

Mice homozygous for the autosomal recessive mutation severe combined immune deficiency (scid) on Chromosome 16 have proved to be a unique resource in many areas of biomedical research. Since the original report that homozygosity for scid causes a severe deficiency in numbers of functional T and B cells accompanied by a lack of both humoral and cell-mediated immune function (Bosma et al.1983), numerous investigations have focused on the molecular basis of defective immunoglobulin and T cell receptor production and on the selective nature of the lymphocyte depletion in these mice (Shultz and Sidman, 1987). In addition to the value of the scid mutation as a tool to increase our understanding of the molecular events underlying lymphocyte development, recent reports have described the use of scid/scid mice as models to study development of human lymphoid cells following transplantation of human fetal tissues (McCune et al. 1988) and to study interactions occuring during the course of infection of human cells with human immunodeficiency virus (HIV) (Namikawa et al. 1988).

Although scid/scid mice will be increasingly used in many areas of biomedical research, their profound susceptibility to Pneumocystis carinii pneumonia in conventional mouse room environments may limit the scope of proposed experimentation. P. carinii exists widely in nature as an opportunistic pulmonary pathogen of uncertain taxonomy in humans and and in numerous vertebrate species. Recent intensive interest in the growth of this ubiquitous pathogen and in the factors involved in host resistance stems from reports of pneumocystosis in immunosuppressed individuals, in children with primary immunodeficiency diseases and in patients with the acquired immunodeficiency syndrome (AIDS) (Hughes, 1987). The occurrence of pneumocystosis in research colonies of scid/scid mice has provided both a means for the study of P. carinii pneumonia in a controlled experimental system and a hindrance to long-term in vivo experimentation. This paper will describe, by histological, immunocytochemical, and electron microscopic techniques, the characteristics of P. carinii pneumonia in scid/scid mice maintained in a research colony at The Jackson Laboratory. We will also show the successful treatment of this infectious disease with trimethoprim-sulfamethoxazole (TMP-SMZ). Finally, hybridomas producing monoclonal antibodies against P. carinii will be described.

MATERIALS AND METHODS

Immunodeficient mice. C.B-17IcrJ-scid/scid and +/+ mice were housed, two to five animals per cage, in pine bedding in a conventional research animal facility at The Jackson Laboratory. Lexon filter tops were used to minimize airborn infections. They were fed Wayne Lab Blox and chlorinated water ad libitum. Original C.B-17Icr-scid/scid and +/+ breeding pairs were obtained from Dr. Mel Bosma at The Institute for Cancer Research, Fox Chase Cancer Center, Philadelphia, Pa.

[1] The Jackson Laboratory, Bar Harbor, Maine
[2] University of Cincinnati College of Medicine and Cincinnati VA Medical Center, Cincinnati, Ohio

Current Topics in Microbiology and Immunology, Vol. 152

Histopathology. Lungs were fixed in Bouin's solution, embedded in paraffin, and sectioned at 6 μm. Sections from each block were stained with Mayer's hematoxylin and eosin (H+E) or Gomori's methenamine silver.
Electron microscopy. Tissues were fixed in 3% gluturaldehyde, 1% paraformaldehyde in 0.1 M cacodylate buffer, pH 7.4. The tissues were then postfixed in 1% osmium tetroxide, stained en bloc with 0.5% uranyl acetate, dehydrated in ethanol, and embedded in Epon-araldite. Ultrathin sections were stained again with uranyl acetate followed by lead citrate. For transmission electron microscopy (TEM), sections were viewed in a JEOL 100CX-II electron microscope operated at 80kV. For scanning electron microscopy (SEM), the tissues were dehydrated in ethanol and freon. They were then dried from liquid CO_2 with the use of a Balzers Union CPD010 critical point drying apparatus. Samples were coated with gold palladium with the use of a Hummer V Sputter Coater and viewed in a JEOL 35 C electron microscope operated at 7 kV.
Production and characterization of monoclonal antibodies. C.B-17 +/+ mice were immunized with a fine homogenate of lung tissue from C.B-17-scid/scid mice with severe P. carinii pneumonitis. The lung homogenates were emulsified with complete Freund's adjuvant prior to injection. The C.B-17 +/+ mice were challenged with a second injection of lung homogenate emulsified in incomplete Freund's adjuvant 1 week prior to removal of the spleens. Spleen cells from the immunized mice were fused with cells from the nonsecreting mouse myeloma line SP2/0. Supernatants from wells containing aminopterin-resistant hybrids were initially screened for anti-P. carinii antibody by indirect immunofluorescence microscopy. Selected hybridomas were retested by immunoperoxidase staining as described below.
Immunohistochemistry. A modification of the avidin-biotin method was used (Sundberg et al. 1989). Endogenous peroxidase was blocked in deparaffinized 6 μm sections by incubation in 0.3% hydrogen peroxide in methanol for 30 min. Nonspecific staining was blocked by incubation in 10% ovalbumin in PBS for 30 min. Slides were incubated with either media alone or with hybridoma culture supernatants. After 3 washes in PBS, detection was completed using the Vectastain ABC kit (Vector Labs). Diaminobenzidine was used as the chromagen. Light green was used as the counterstain.

RESULTS

Mice with P. carinii pneumonia exhibited progressive weight loss, chronic debilitation, and occasional cyanosis (Fig. 1).

Fig. 1. Female C.B-17 scid/scid mice. The 10-month-old mouse on the left is thin and cyanotic. Histopathologic examination confirmed the presence of pneumonitis caused by P. carinii. The 9-month-old mouse on the right has been treated with TMP-SMZ since weaning.

Retrospective analyses revealed progressive decrease in lifespan of the scid/scid mice born from 1984-1987 (Fig. 2). The scid/scid mice born in 1984 had a mean lifespan of 337 ±13.3 days. The nearly one year lifespan of these mice is in sharp contrast to a mean lifespan of only 97.6 ± 2.0 days for scid/scid mice born in 1987. In order to determine the underlying basis for the shortened lifespan, moribund animals were necropsied and tissues removed for light and electron microscopy.

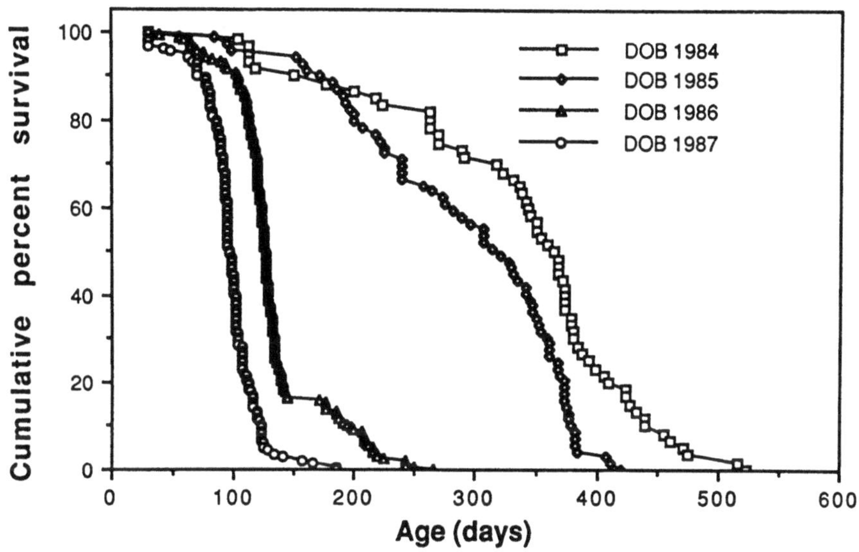

Fig. 2 Cumulative percent survival of C.B-17 scid/scid mice born in a research colony and maintained prior to initiation of TMP-SMZ therapy. Each curve represents survival of 60-164 mice with dates of birth (DOB) from 1984-1987.

Examination of H + E stained lung sections from affected animals revealed interstitial pneumonitis with characteristic amorphous eosinophilic material within alveolar lumens (Fig. 3A). Staining with methenamine silver demonstrated the presence of P. carinii cysts along alveolar walls (Fig. 3B). In contrast to methenamine silver, which stains only the scattered cyst form of P. carinii , we have previously reported that immunoperoxidase staining using polyclonal anti-P. carinii antibody visualizes trophozoites as well as cysts in lungs from affected mice, greatly increasing the sensitivity of staining (Sundberg et al. 1989). In order to develop antibodies for investigation of the antigenic structure of P. carinii antigens, a panel of hybridoma cell lines was produced. Six of these hybridomas secreted antibodies which reacted strongly with P. carinii antigen by both indirect immunofluorescence and immunoperoxidase techniques. Immunoperoxidase staining of lungs from scid/scid mice with these antibodies revealed moderate to intense staining of focal concentrations of trophozoites as well as the cyst form of the microorganism (Fig. 3C).

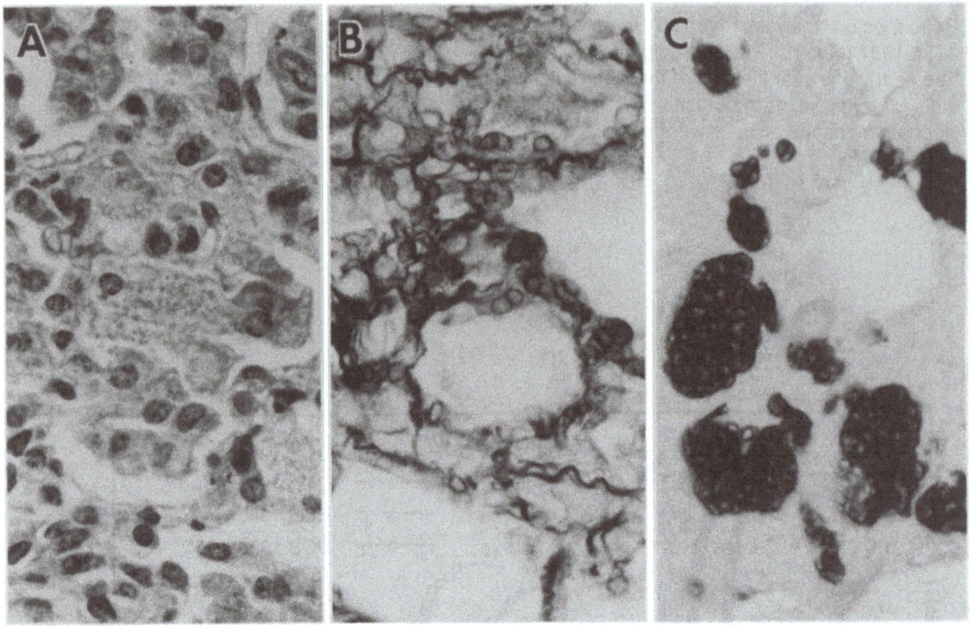

Fig. 3. P. carinii pneumonia in a 12-month-old scid/scid mouse. A. Note thickening of the alveolar walls and the presence of eosinophilic honeycomb-like material in the alveolar lumens (H + E); B. P. carinii cysts stained with Gomori's methenamine silver; C. P. carinii cysts and trophozoites revealed by the avidin-biotin method using monoclonal antibody 2A6 (x 540).

Electron microscopy demonstrated numerous foci of cysts and trophozoites. Fig. 4 shows a group of cysts, two of which contain sporozoites. Scanning electron microscopy of alveoli revealed the presence of clusters of intact and empty cysts together with numerous trophozoites (Fig. 5).

Since it had previously been reported that combination therapy with TMP-SMZ was effective in prevention and treatment of P. carinii pneumonia in humans and in cortisone-treated rats (Hughes et al. 1987; Hughes and Smith 1983), the effect of such therapy on the health of scid/scid mice was investigated. Sulfatrim suspension (Goldline laboratories) was added to the drinking water to supply a daily dose of approximately 500 mg/kg SMZ and 100 mg/kg TMP. The mice were supplied with the sulfatrim suspension ad libitum for 3 consecutive days per week. Treatment with this combination drug has greatly reduced mortality in the scid/scid mouse colony. The overall mortality in mice observed for the first 6 months of treatment has been < 5%. Despite the lifesparing effects of TMP-SMZ treatment, lungs from scid/scid mice killed at intervals after initiation of therapy retained small scattered foci of P. carinii cysts. The failure of such therapy to eliminate P. carinii was not unexpected since TMP-SMZ is not microbicidal for this microorganism (Hughes 1979). Thus, we have been continuing TMP-SMZ therapy of these immunodeficient mice. A separate colony of scid/scid mice maintained under SPF conditions following hysterectomy derivation has remained free of P. carinii.

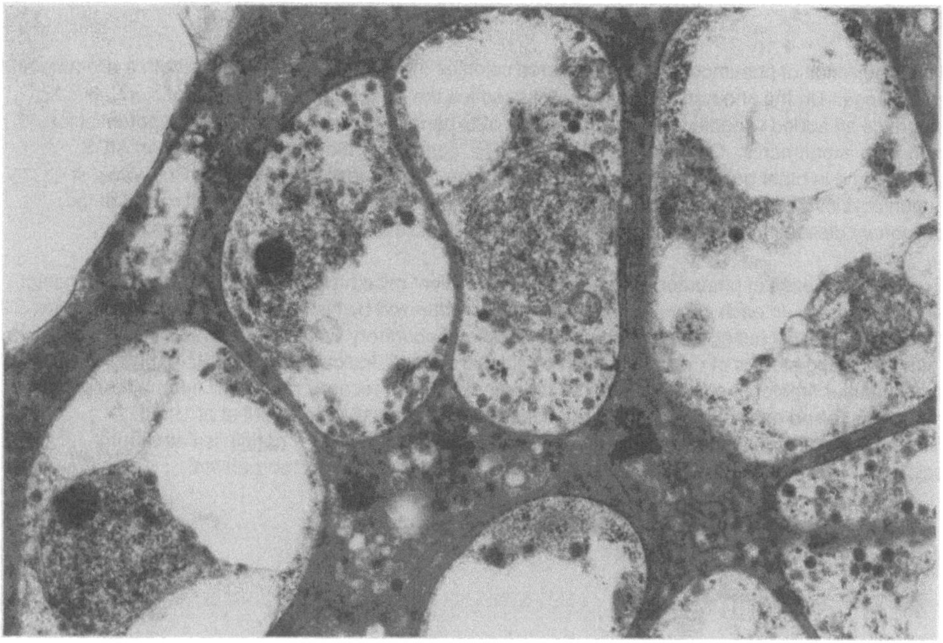

Fig. 4. Transmission electron micrograph of several cysts. Two of the cysts contain intracystic bodies (sporozoites) (x 15,000).

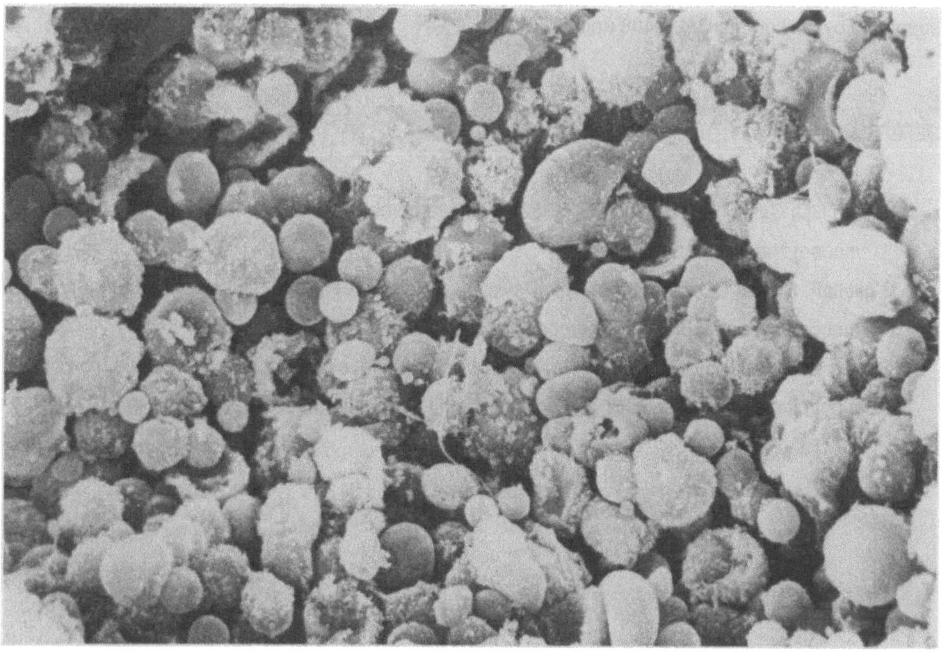

Fig. 5 Scanning electron micrograph showing numerous intact and collapsed cysts along with trophozoites (x 3,600).

DISCUSSION

The occurrence of pneumocystosis in research colonies of scid/scid mice presents both a dilemma and a challenge. On the one hand, such outbreaks reduce the lifespan of the immunodeficient animals, introduce an added variable to the interpretation of experimental results, and limit the potential for long-term experiments. On the other hand, since P. carinii is a major cause of death in AIDS patients and in other genetically or therapeutically immunosuppressed individuals, outbreaks of pneumocystosis provide a manipulatable experimental model to study the immunobiology of the pulmonary disease and to develop more effective diagnostic and therapeutic regimens.

Previous outbreaks of pneumocystosis in immunodeficient mice have occurred in other experimental animal colonies. In each case, the disease was characterized by histopathologic lesions similar to those found in affected scid/scid mice at The Jackson Laboratory (Walzer et al. 1989). Immunoblotting studies of these mice with polyclonal antisera indicated that the P. carinii was of mouse rather than of human origin. Other investigators have recently described murine monoclonal antibodies raised against P. carinii of rat origin (Graves et al. 1986; Gigliotti et al 1986). Current immunoblotting studies with monoclonal antibodies against P. carinii isolated from scid/scid mice are aimed at assessing the antigenic profile of this microorganism.

ACKNOWLEDGEMENTS

We thank Bonnie Hanson, Ann Higgins and Priscilla Jewitt for excellent technical assistance. We thank Robert Evans and Hendrick Bedigian for critical reading of the manuscript. The work was supported by NIH grants CA 20408 and DK 07449.

REFERENCES

Bosma GC, Custer RP, Bosma MJ (1983) A severe combined immunodeficiency mutation in the mouse. Nature 301:527-530

Gigliotti F, Stokes DC, Cheatham, AB, Davis DS, Hughes WT (1986) Development of murine monoclonal antibodies to Pneumocystis carinii. J Infec Dis 154:315-322

Graves DC, McNabb SJN, Ivey MH, Worley MA (1986) Development and characterization of monoclonal antibodies to Pneumocystis carinii . Infec Immun 51:125-133

Hughes WT, Smith BL (1983) Intermittent chemoprophylaxis forPneumocystis carinii pneumonia. Antimicrob Agents Chemother 24:300-301

Hughes WT, Rivera GK, Schell MJ, Thornton D, Lott L (1987) Successful intermittent chemoprophylaxis for Pneumocystis carinii pneumonitis. N Engl J Med 316:1627-1632

Hughes WT (1987) Pneumocystis carinii pneumonitis. CRC Press, Boca Raton

Kovacs, JA, Halpern JL, Swan JC, Moss J, Parillo JE, Masur H (1988) Identification of antigens and antibodies specific for Pneumocystis carinii. J Immunol 140:2023-2031

McCune JM, Namikawa R, Kaneshima H, Shultz LD, Lieberman M, Weissman IL (1988) The SCID-hu mouse: Murine model for the analysis of human hematolympnoid differentiation and function. Science 241:1632-16639

Namikawa R, Kaneshima H, Lieberman M, Weissman IL, McCune JM (1988) Infection of the SCID-hu mouse by HIV-1. Science 242:1684-1686

Sundberg JP, Burnstein T, Shultz LD, Bedigian H (1989) Identification of Pneumocystis carinii in immunodeficient mice. Lab Anim Sci In Press

Shultz LD, Sidman CL (1987) Genetically determined models of immunodeficiency. Ann Rev. Immunol 5:367-403

Walzer PW, Kim CK, Linke MJ, Pogue CL, Huerkamp MJ, Chrisp CE, Lerro AV, Wixson SK, Hall E, Shultz LD (1989) Outbreaks of Pneumocystis carinii pneumonia in colonies of immunodeficient mice. Infec Immun 57:62-70

Immunobiology of Bone Marrow Transplantation: Studies Using Scid Mice

William J. Murphy, V. Kumar, and M. Bennett[1]

Department of Pathology and Graduate Program in Immunology, University of Texas Southwestern Medical Center, Dallas, Tx., USA.

Natural killer (NK) cells are believed to be an effector cell capable of mediating bone marrow allograft rejection in lethally irradiated recipients. However, despite a large body of experimental evidence that suggests that NK cells can mediate specific marrow allograft rejection, the nature of the proposed target antigen present on the marrow stem cell and the mechanism by which NK cells can detect disparities has remained an enigma. NK cells are believed to recognize hematopoietic histocompatability (Hh-1 or Hh) antigens present on incompatible stem cells. Hh antigens are expressed in a noncodominant manner in which homozygosity at the MHC (H-2 in the mouse) is a prerequisite for the expression of Hh antigens (Bennett 1972). In other words, H-2 homozygous marrow grafts are resisted and H-2 heterozygous grafts proliferate. Hh antigens have proven to be elusive to isolate and characterize. However, through the analysis of the rejection patterns using recombinant mouse strains they have been shown to be distinct from class I MHC genes, and are encoded for by genes between the S and D locus of H-2 (Rembecki et al. 1988).

NK cells would, at first glance, seem to be improbable mediators of the specificity seen in marrow graft rejection. Until recently, NK cells have been largely functionally defined as a population of lymphoid cells that spontaneously lyse various tumor targets in an MHC-unrestricted manner. The question arises as to how can NK cells specifically reject marrow allografts? Various hypotheses have been put forth to possibly explain the mechanisms of marrow graft rejection. It has been only with the availability of mice that display severe combined immune deficiency (SCID) that the various theories could be stringently assessed.

SCID mice have no mature B or T cell function due to a defect in the rearrangement of their antigen receptor genes (Schuler et al. 1986). SCID mice will not reject skin allografts and have no detectable immunoglobulin levels (Bosma et al. 1983). However, SCID mice have normal NK cells and their progenitors (Hackett et al. 1986). These mice are thus an ideal model to examine if NK cells alone are sufficient to cause the specific rejection of bone marrow allografts in lethally irradiated recipients. In addition these mice provide an excellent model system to look at possible interactions between NK cells and other immune cells in regulating their function. We decided to use the SCID mouse model and examine their ability to reject bone marrow allografts. In addition, we compared this rejection ability with their +/+ counterparts which have normal T and B cell function to determine if the marrow rejection ability of NK cells is possibly regulated by other cells in the immune system.

[1] Supported by NIH Grants CA-25401, CA-36921, CA-36922, and CA-09082.

The ability of SCID mice to reject bone marrow cells (BMC) was assessed using a short term radioisotope uptake assay that is useful for measuring acute rejection of marrow allografts. C57BL/6J (B6), DBA/2J, BALB/cJ, (C57BL/6 x DBA/2)F1 (B6D2F1), (NZB X C57BL/6)F1 (NZBF1), C.B-17 scid/scid (SCID), and C.B-17 +/+ (+/+) mice were bred in our own colony at the University of Texas Southwestern Medical Center. SCID and +/+ mice were kept in specific pathogen-free conditions until use. The assay for the proliferation of grafted hematopietic cells (Figure 1) has been described previously (Cudkowicz and Bennett 1971). Briefly, recipient mice are lethally irradiated with 450-850 cGy of ^{137}Cs irradiation at a dose rate of 80 cGy/min. The recipient mice are then infused with 1-4 million BMC in the lateral tail vein. The proliferation of donor cells in the spleens of the recipient mice is measured after 5 days by quantitating the splenic incorporation of 5-^{125}I-iodo-2'-deoxyuridine (^{125}IUdR), a specific DNA precursor and thymidine analogue. The data are presented as the geometric means (95% confidence limits) of percentage of injected ^{125}IUdR incorporated in the spleens of groups of 5-8 mice. Irradiated recipients syngeneic with the BMC donor were used to assess the growth potential of the grafted BMC. Both parametric and nonparametric statistical analyses were performed to determine if the geometric values of the various groups differed significantly ($p < 0.05$).

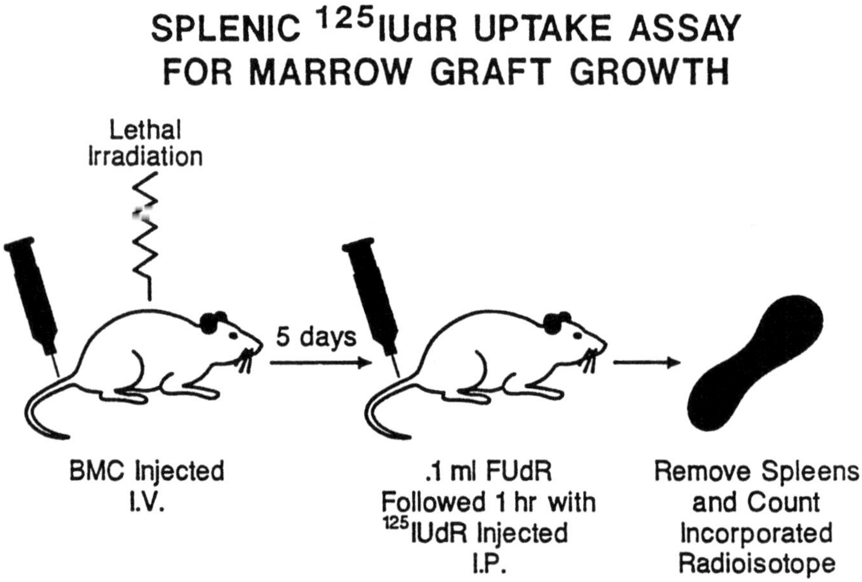

Figure 1. Splenic ^{125}IUdR uptake assay for marrow graft rejection.

When syngeneic and allogeneic BMC were placed into lethally irradiated SCID recipients, the allogeneic marrow was significantly resisted by these mice (Figure 2). Interestingly, even H-2 identical DBA/2 ($H\text{-}2^d$) grafts were resisted by the SCID recipients. DBA/2 has different background genes and has been demonstrated to express an additional minor Hh antigen (HhDBA) which is encoded by genes not linked to H-2 (Cudkowicz and Rossi 1972). Homozygosity at H-2 and this minor locus was required for the marrow grafts to be rejected. Even though both B6 ($H\text{-}2^b/Hh^b$) and DBA/2 ($H\text{-}2^d/Hh^d$,HhDBA) BMC were rejected by SCID mice, BMC from their F1 hybrids (B6D2F1) were not. This indicates that Hh antigen disparities are being recognized by the SCID hosts since the F1 heterozygous grafts are null for both Hh^b, Hh^d, and HhDBA. Since the SCID mice have no functional B and T cells, this experiment demonstrates that these cells are not required for specific marrow rejection to occur and NK cells alone are capable of mediating the specificity seen in marrow graft rejection. Moreover, the lack of serum Ig in SCID mice demonstrates that antibody dependent cellular cytotoxicity (ADCC) is not required for specific marrow rejection by NK cells.

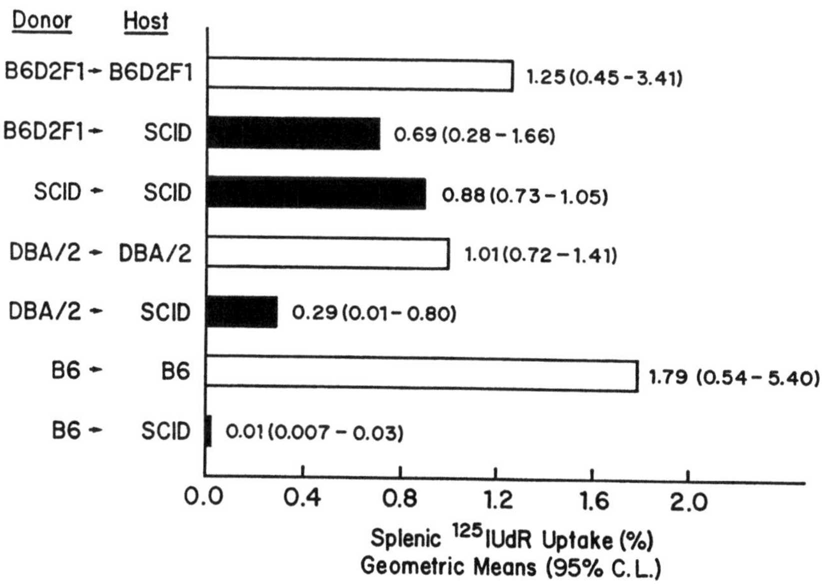

Figure 2. Marrow growth in irradiated C.B-17 SCID recipients. The DBA/2 and B6 BMC grew significantly better ($p < 0.05$) in syngeneic than in SCID hosts, whereas the growth of B6D2F1 BMC in syngeneic recipients was not significantly different than in SCID hosts ($p > 0.05$).

It was of interest that the SCID recipients resisted the H-2 identical DBA/2 marrow grafts since the HhDBA determinant is usually only weakly recognized by BALB/c mice. SCID and +/+ recipients were compared in their ability to reject both allogeneic B6 (Hh^b) and the H-2 identical DBA/2 (HhDBA) marrow grafts. In addition, antisera to asialo GM_1, a glycolipid found on NK cells, was administered to determine if the rejection by the SCID recipients was due to their NK cells. The data in Table 1 demonstrates that both SCID and +/+ recipients can reject allogeneic B6 BMC and that anti-asialo GM_1 can abrogate the ability of SCID mice to reject the graft. This demonstrates that the SCID mouse, although completely deficient in immune function, can specifically reject marrow allografts in an Hh-specific manner and this is due to the functional NK cells present in these mice. When the mice were challenged with DBA/2 marrow grafts, only the SCID recipients were able to significantly resist the graft. This rejection could also be abrogated by the administration of anti-asialo GM_1 serum. Thus SCID mice seem to have an augmented ability to reject marrow allografts as compared to their normal +/+ counterparts. This could due to the lack of immune cells which can possibly regulate NK cell function, in this case marrow rejection. Other investigators, using various model systems, have also reported possible influences of B and T cells on NK cell functions. Brodt et al. (1981) treated mice from birth with immunoglobulin antiserum and noticed an increased NK activity in these mice. Athymic nude mice are also known to have an increased NK cell function. It is possible that various cells of the immune system can regulate NK cell function and that SCID mice lack these regulatory cells. The use of SCID mice can therefore allow for a better understanding of the regulatory mechanisms involved in NK cell mediated marrow graft rejection.

TABLE 1

Differential Ability of SCID and +/+ Mice to Reject Bone Marrow Grafts

Marrow donor		Recipient		Splenic ^{125}IUdR uptake	
				Geometric	95%
Strain	H-2/Hh	Strain	H-2/Hh	mean	Confidence limit
B6	b;Hh^b	B6	b;Hh^b	0.68	(0.15-3.30)
		+/+	d;Hh^d	0.03	(0.01-0.06)*
		SCID	d;Hh^d	0.007	(0.001-0.04)*
		SCID**	d;Hh^d	0.22	(0.06-0.78)#
DBA/2	d;Hh^d,HhDBA	DBA/2	d;Hh^d,HhDBA	1.08	(0.52-2.22)
		+/+	d;Hh^d	1.51	(1.07-2.15)
		SCID	d;Hh^d	0.20	(0.02-1.21)*
		SCID**	d;Hh^d	0.98	(0.63-1.50)#

SCID = C.B-17 <u>scid/scid</u>, +/+ = C.B-17 +/+. All recipients received 1.5 million BMC i.v. after irradiation. After 6 days the isotope assay was performed.

* Geometric mean value significantly ($p < 0.05$) lower than syngeneic controls.

** Mice received asialo-GM1 antiserum on the day of transfer.

Geometric mean value significantly ($p < 0.05$) higher than without treatment with asialo-GM1 antiserum.

To further examine the possible influence of immune cells on NK reactivity, radiation bone marrow chimeras were created using SCID mice as the source of donor BMC. SCID BMC were placed into lethally irradiated (NZB x B6)F1 (NZBF1) recipients. One advantage of using SCID BMC to generate chimeras is that their BMC can give rise to normal numbers of functional NK cells but fail to produce mature B and T cells (Hackett et al.' 1986). Another advantage of using SCID BMC as donor cells is that, by lacking mature T cells, they fail to give rise to graft versus host disease (GVHD) in the recipients. These SCID→NZBF1 chimeras, however, have radioresistant T cells of host origin. Thus it was of interest to determine if the residual host T cells in these chimeras can influence the donor SCID NK cells during repopulation of the animal. The chimeras were created using the method described in Figure 3. When the mice were assayed to determine the extent of chimerism, they demonstrated normal NK function which was demonstrated to be of donor (SCID) origin. When they were assayed for B and T cell function, their spleen cells failed to respond to LPS (a B cell mitogen) and gave a modest Con A response (a T cell mitogen) which was typed and found to be of host origin. Therefore these mice consisted of a SCID hematopoietic system with SCID NK cells in a NZBF1 host which had a small but significant residual T cell population.

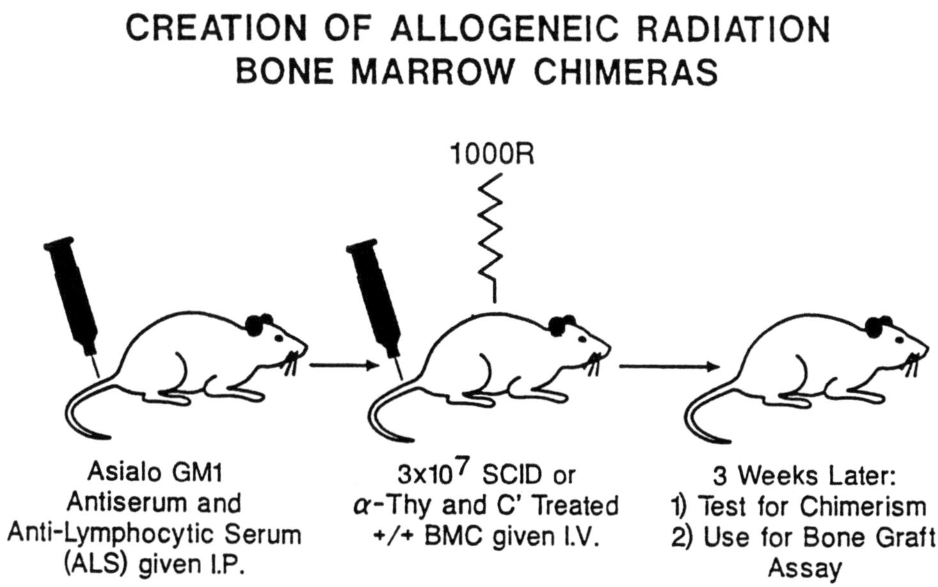

Figure 3. The use of SCID BMC to create radiation BMC chimeras.

These chimeras were then tested to examine the effect of the NZBF1 environment on SCID NK marrow rejection ability.

TABLE 2

Ability of SCID→NZBF1 Radiation BMC Chimeras to Reject Marrow Grafts

Marrow donor		Recipient		Splenic ^{125}IUdR uptake	
Strain	H-2/Hh	Strain	H-2	Geometric mean	95% Confidence limits
C.B-17	d;Hh^{d}	C.B-17	d	1.06	(0.68-1.66)
		SCID	d	1.02	(0.74-1.41)
		NZBF1	b/d	0.22	(0.14-0.36)*
		SCID NZBF1	d	2.24	(1.07-4.03)
B6	b;Hh^{b}	B6	b	0.79	(0.20-3.13)
		SCID	d	0.01	(0.001-0.09)*
		NZBF1	b/d	0.01	(0.004-0.03)*
		SCID NZBF1	d	0.03	(0.01-0.06)*
DBA/2	d;Hh^{d},HhDBA	DBA/2	d	2.71	(1.83-4.00)
		SCID	d	0.26	(0.21-0.34)*
		NZBF1	b/d	0.24	(0.13-0.44)*
		SCID NZBF1	d	2.75	(2.31-3.25)
B6D2F1	b/d;Hh^{null}	B6D2F1	b/d	1.86	(1.34-2.03)
		SCID	d	1.78	(1.17-2.60)
		NZBF1	b/d	3.21	(1.51-6.53)
		SCID NZBF1	d	2.04	(1.10-3.24)

SCID = C.B-17 scid/scid, NZBF1 = (NZB x B6)F1, B6D2F1 = (B6 x DBA/2)F1
All recipients received 1.5 million BMC i.v. after irradiation. After 6 days the radioisotope asay was performed.
* Geometric mean value significantly ($p < 0.05$) lower than syngeneic controls. Underlined value significantly ($p < 0.05$) greater than SCID and NZBF1 groups.

The results in Table 2 demonstrate that the SCID→NZBF1 radiation chimera can still strongly reject B6 marrow allografts. However, even though both SCID and NZBF1 recipients can reject DBA/2 marrow grafts, the radiation chimera fails to do so. It appears that with the influence of the NZBF1 environment (T cells), the SCID NK cells are somehow suppressed enough such that HhDBA disparities are no longer detected. These chimeras now have the rejection pattern simillar to the normal C.B-17 +/+ mouse. Thus the SCID NK cells seem to be able to interact with immune cells present in the host and, in this instance, are downregulated by them. These chimeras will be useful in further characterization of the cell type responsible for this particular suppression of NK activity. In addition, it will be possible to add back specific T cell subsets to SCID mice and determine if this supression can be reproduced.

If one examines the models that have been proposed to explain the rejection of bone marrow allografts, one can see that the fact that SCID mice can specifically reject marrow allografts can rule out several hypotheses. Warner et al. (1985) proposed that natural antibodies were necessary to target the NK cell and cause the rejection of marrow by ADCC. The use of SCID mice proves that natural antibodies are not required for NK recognition of marrow allografts. In addition many reseachers considered the mechanisms of marrow rejection to be identical to those involved in the rejection of solid tissue allografts (i.e. skin). In these systems, T cells are the primary effector cell mediating the rejection of the graft. It is clear, since SCID mice lack functional T cells and fail to reject skin allografts, that the laws of solid tissue transplantation do not necessarily apply with bone marrow transplantation. It is also obvious that a different type of effector cell exists (the NK cell) which must now be considered to play a potential role in the clinical instances of marrow failure.

In conclusion, the use of SCID mice has allowed for the determination that NK cells alone can mediate the specific rejection of bone marrow allografts in the absence of other immune mechanisms. Furthermore, the use of SCID mice and radiation BMC chimeras allows for the study of the regulatory mechanisms in marrow rejection and the interaction of immune cells with NK cells. The chimera data suggests that other immune cells, presumably T cells, can interact with NK cells and regulate their ability to reject Hh-disparate marrow grafts.

Acknowledgements. We wish to thank Silvio and Maria Pena for excellent care of the mouse colony and their friendship. We wish also to thank Dr. Grant Risdon for critically reviewing the manuscript. The authors also are grateful for the helpful and stimulating discussions by the graduate students of the immunology program here at Southwestern.

REFERENCES

Bennett M (1972) Rejection of marrow allografts. Importance of H-2 homozygosity of donor cells. Transplantation 14:289-298

Bosma GC, Custer RP, Bosma MJ (1983) A severe combined immunodeficiency mutation in the mouse. Nature 301:527-529

Brodt P, Kongshavin P, Vargas F, Gordon J (1981) Increased natural resistance mechanisms in B lymphocyte-deprived mice. J Reticuloendothel Soc 30:283-287

Cudkowicz G, Bennett M (1971) Peculiar immunobiology of bone marrow allografts II. Rejection of parental grafts by resistant F1 hybrid mice. J Exp Med 134:1513-1528

Cudkowicz G, Rossi G (1972) Hybrid resistance to parental DBA/2 grafts: independence from the H-2 locus. II. Studies with hematopoietic cells. JNCI 48:997-1001

Hackett JJr, Bosma GC, Bosma MJ, Bennett M, Kumar V (1986) Transplantable progenitors of natural killer cells are distinct from those of T and B lymphocytes. Proc Natl Acad Sci USA 83:3427-3431

Rembecki RM, Kumar V, David CS, Bennett M (1988) Bone marrow cell transplants involving intra-H-2 recombinant inbred mouse strains. Evidence that hemopoietic histocompatibility-1 (Hh-1) genes are distinct from H-2D or H-2L. J Immunol 141:2253-2260

Schuler W, Weiler IJ, Schuler A, Phillips RA, Rosenburg N, Mak TW, Kearney JF, Perry RP, Bosma MJ (1986) Rearrangement of antigen receptor genes is defective in mice with severe combined immune deficiency. Cell 46:963-972

Warner J, Dennert G (1985) Bone marrow graft rejection as a function of antibody-directed natural killer cells. J Exp Med 161:563-576

Growth of Human Tumors in Immune-Deficient *scid* Mice and *nude* Mice

R.A. Phillips, M.A.S. Jewett[1], and B.L. Gallie

Division of Hematology/Oncology, The Hospital for Sick Children, Toronto, Canada M5G 1X8, and Department of Medical Genetics, University of Toronto

INTRODUCTION

The ability to grow human tumors in immune-deficient mice offers offers many research opportunities (Sharkey and Fogh 1984). In addition to simply providing a source of cells for investigation, such animals can provide models for testing therapeutic protocols (Niederkorn 1984; Brunner et al. 1987), for studying metastasis (Dore et al. 1987), and for analyzing changes associated with tumor induction and progression (Sordat and Wang 1984). Although nude mice have been used for many years to grow various human tumors, we expected that scid mice might be better recipients for growth of human tumors because the immune deficiency in scid mice is more severe than that in nude mice (Bosma et al. 1983). To test this hypothesis, we have compared the ability of various different human tumors to grow subcutaneously in nude mice and scid mice.

MICE

Nude mice and scid mice were obtained from the animal colony at the Ontario Cancer Institute. Both strains are maintained in a defined flora, barrier colony (Phillips et al., this volume).

HUMAN TUMORS

Most human tumors were obtained as fresh surgical specimens. These tumors were cut into small, 1 mm^3, fragments and implanted subcutaneously into nude or scid mice. Retinoblastoma tumors were obtained from tissue culture samples. In some experiments, tumors which grew in nude or scid mice were injected into second recipients to determine whether or not the the growth properties of the tumor changed as a result of selection in the first recipient. To conserve mice each mouse was injected in two sites, one on each flank. Since we observed no correlation between growth on one side and growth on the contralateral side, the following tables summarize the results as the number of injected sites tested.

Mice were examined weekly for the presence of tumors. When a tumor had reached a size of 0.5 cm, and maintained this size for at least one month, the graft was considered as positive for growth.

[1] Division of Urology, The Wellesley Hospital, Toronto, Canada, and Department of Surgery, University of Toronto

Current Topics in Microbiology and Immunology, Vol. 152

RESULTS AND DISCUSSION

In total, 28 human tumor samples were tested for their ability to grow in nude and/or scid mice. The data for these tests are shown in Table 1. As expected, none of the breast tumor samples grew in either nude or scid mice. Other investigators have shown that without appropriate steroid manipulation of the recipients, these tumors fail to grow in immune-deficient mice (Brunner et al. 1987; Lippman et al. 1987). Surprisingly, however, none of the ocular melanomas grew in either recipient. Given the ability of cutaneous melanoma to grow in nude mice (Fogh et al. 1977), this result was unexpected. Osteosarcomas grew only in scid mice, but the number of nude mice tested was small. All of the other tumor samples grew in both nude and scid recipients.

Human tumors grow slowly in immune-deficient animals. The average time for the tumors to grow to the threshold size was 4.1 months in nude mice and 3.1 months in scid mice. However, there was marked variability in the time of appearance, and this difference is not statistically significant. There was no apparent increase in the growth rate of tumors, i.e., time for tumor to appear, growing in secondary recipients compared to their growth rate in the primary recipient (data not shown).

As shown in Table 2, tumors grow significantly better in scid mice than in nude mice. However, this overall difference depends entirely on the better growth of retinoblastoma and osteosarcomas in scid mice versus nude mice. All other tumors tested showed no significant preference for growth in either recipient. It is difficult to explain the preference of retinoblastoma and osteosarcomas to grow in scid mice. It is perhaps interesting that these two disparate tumors share an etiological relationship. Both tumors appear to arise when mutations occur in both alleles of the RB1 locus on chromosome 13 (Hansen et al. 1985). Retinoblastoma tumors have previously been reported to grow in the eyes of nude mice (Gallie et al. 1977), but difficulty was previously experienced in growing such tumors subcutaneously. The difference between the previously reported experiments and those described here was the use of pieces of tumor rather than cell suspensions for subcutaneous growth.

The only previous reports on the growth of human tumors in scid mice was the demonstration by Reddy et al. (1987) that human lung tumors grow progressively in scid mice. In subsequent analysis (Bankert et al., this volume), it was shown that tumors which failed to grow in scid mice also seemed to result in engraftment of human lymphoid cells. The possibility that engraftment of patient lymphocytes prevents growth of the tumor or rejects the tumor at an early stage of growth opens many opportunities for investigation. Unfortunately, we did not save serum from the scid mice used in our study and are unable to attempt a similar correlation in our study. Nevertheless, the possible immunological interactions between human tumor cells and autologous leukocytes is interesting and provocative.

In summary, scid mice appear to offer several advantages over nude mice as models for growth of human tumors. First, the mice are better for growth of some human tumors than are nude mice. Second, scid mice offer the possibilities of examining the interaction between human lymphocytes and human tumors. Third, the fertility of scid mice makes them easier to breed than nude mice, and offers the possibility of providing a less expensive animal model for the growth of human tumors.

Table 1. Subcutaneous Growth of Human Tumors in nude and scid Mice

Tumor type	Tumor No.	Source	Growth in nude mice Total	Growth in nude mice No. grow	Growth in scid mice Total	Growth in scid mice No. grow
Retinoblastoma	Y-79 - 6	culture	4	1	4	2
	RB409	culture	10	2	8	3
	"	1st passage	-[a]	-	4	4
	RB383	culture	4	2	6	6
Ocular Melanoma	OM1	fresh tumor	4	0	2	0
	OM2	"	4	0	6	0
	OM3	"	6	0	6	0
	OM4	"	4	0	4	0
Bladder	BLA1	fresh tumor	4	1	4	4
	"	1st passage	4	3	4	0
	BLA2	fresh tumor	2	0	4	0
	BLA3	"	6	0	8	0
	BLA4	"	4	1	6	4
	"	1st passage	4	3	2	0
	BLA5	fresh tumor	4	0	8	0
Prostate	PRO1	fresh tumor	6	2	8	6
	"	1st passage	-	-	4	4
	PRO2	fresh tumor	6	0	8	0
Breast	BRE1&2	fresh tumor	6	0	6	0
Pancreatic Carcinoma	PAN1	fresh tumor	6	6	6	6
	"	1st passage	2	1	4	4
Renal Cell	RTC1	fresh tumor	4	2	6	1
Osteogenic Sarcoma	OS234	fresh tumor	4	0	2	2
	"	1st passage	-	-	4	4
	OS108	fresh tumor	-	-	8	7
	"	1st passage	-	-	6	5
Testicular	TES1	fresh tumor	4	2	4	4
	"	1st passage	4	0	2	2
	TES2	fresh tumor	4	0		
	TES3	"	4	0	4	0
	TES4	"	4	0	4	0
Wilms' Tumor	WIL1	fresh tumor	2	0	2	0
	WIL2	"	4	2	6	2
	"	1st passage	4	2	4	3
	WIL3	fresh tumor	4	0	2	0
	WIL4	"	4	1	4	4

[a] Not tested

Table 2. Comparison of scid and nude Mice for Subcutaneous Growth of Human Tumors

Tumor Type	Number Tumors Tested	Total No. nude Mice	% Grow	Total No. scid Mice	% Grow	'p' value[a]
Retinoblastoma	3	18	28%	23	69%	.01<p<.02
Ocular Melanoma	4	18	0%	18	0%	N.S.
Bladder	5	28	29%	36	22%	N.S.
Prostate	2	12	17%	20	50%	N.S.
Breast	2	6	0%	6	0%	N.S.
Pancreatic	1	6	100%	6	100%	N.S.
Renal Cell Carcinoma	1	4	50%	6	17%	N.S.
Osteogenic Sarcoma	2	4	0%	20	90%	.001<p<.01
Testicular	4	20	10%	14	43%	N.S.
Wilms' Tumor	4	18	28%	18	50%	N.S.
Totals:		134	22%	167	44%	<.001

[a] Chi-square test - nude vs. scid

ACKNOWLEDGEMENTS

This research was supported by grants from the Medical Research Council and by a Terry Fox Program Project Grant from the National Cancer Institute of Canada. B.L.G. is a Scholar of the Ontario Cancer Treatment and Research Foundation. Excellent technical assistance was provided by Mrs. J. Chin.

REFERENCES

Bosma GC, Custer RP, Bosma MJ (1983) A severe combined immunodeficiency mutation in the mouse. Nature 301: 527530

Brunner N, Osborne CK, Spang TM (1987) Endocrine therapy of human breast cancer grown in nude mice. Breast Cancer Res Treat 10: 22942

Dore JF, Bailly M, Bertrand S (1987) Metastases of human tumors in experimental animals. Anticancer Res 7: 9971003

Fogh J, Fogh JM, Orfeo T (1977) One hundred and twenty seven cultured human tumor cell lines producing tumors in nude mice. J Natl Cancer Inst 59: 2216
Gallie BL, Albert DM, Wong JJ, Buyukmihci N, Pullafito CA (1977) Heterotransplantation of retinoblastoma into the athymic "nude" mouse. Invest Ophthalmol Vis Sci 16: 256259
Hansen MF, Koufos A, Gallie BL, Phillips RA, Fodstad O, Brgger A, Gedde DT, Cavenee WK (1985) Osteosarcoma and retinoblastoma: a shared chromosomal mechanism revealing recessive predisposition. Proc Natl Acad Sci USA 82: 62166220
Lippman ME, Dickson RB, Gelmann EP, Rosen N, Knabbe C, Bates S, Bronzert D, Huff K, Kasid A (1987) Growth regulation of human breast carcinoma occurs through regulated growth factor secretion. J Cell Biochem 35: 116
Niederkorn JY (1984) Enucleation induced metastasis of intraocular melanomas in mice. Ophthalmology 91: 692700
Reddy S, Piccione D, Takita H, Bankert RB (1987) Human lung tumor growth established in the lung and subcutaneous tissue of mice with severe combined immunodeficiency. Cancer Res 47: 24562460
Sharkey FE, Fogh J (1984) Considerations in the use of nude mice for cancer research. Cancer Metastasis Rev 3: 34160
Sordat B, Wang WR (1984) Human colorectal tumor xenografts in nude mice: expression of malignancy. Behring Inst Mitt 74: 291300

MIX
Papier aus verantwortungsvollen Quellen
Paper from responsible sources
FSC® C105338

If you have any concerns about our products,
you can contact us on
ProductSafety@springernature.com

In case Publisher is established outside the EU,
the EU authorized representative is:
Springer Nature Customer Service Center GmbH
Europaplatz 3, 69115 Heidelberg, Germany

Printed by Libri Plureos GmbH
in Hamburg, Germany